Jane Hume Clapperton

Scientific Meliorism and the Evolution of Happiness

Jane Hume Clapperton

Scientific Meliorism and the Evolution of Happiness

ISBN/EAN: 9783337417185

Printed in Europe, USA, Canada, Australia, Japan

Cover: Foto ©berggeist007 / pixelio.de

More available books at **www.hansebooks.com**

SCIENTIFIC MELIORISM

AND THE

EVOLUTION OF HAPPINESS

BY

JANE HUME CLAPPERTON

"Not as adventitious will the wise man regard the faith which is in him. The highest truth he sees he will fearlessly utter; knowing that, let what may come of it, he is thus playing his right part in the world; knowing that, if he can effect the change he aims at—well; if not—well also: though not *so* well."—HERBERT SPENCER

> "The good want power, but to weep barren tears.
> The powerful goodness want, worse need for them.
> The wise want love; and those who love want wisdom;
> And all best things are thus confused to ill.
> Many are strong and rich, and would be just,
> But live among their suffering fellow-men
> As if none felt: they know not what to do."
> SHELLEY'S *Prometheus Unbound*

LONDON
KEGAN PAUL, TRENCH & CO., 1, PATERNOSTER SQUARE
1885

(The rights of translation and of reproduction are reserved.)

I Dedicate this Book

TO THE MEMORY OF MY EARLY TEACHERS,

GEORGE ELIOT AND JAMES CRANBROOK,

WHO HAVE JOINED

"The choir invisible
Of those immortal dead that live again
In minds made better by their presence : live
   .   .   .   .   .
In deeds of daring rectitude, in scorn
For miserable aims that end in self,
In thoughts sublime that pierce the night like stars,
And with their mild persistence urge man's search
To vaster issues."

And I Inscribe it to

GEORGE ARTHUR GASKELL,

THE LIVING FRIEND

BY WHOSE AID MY AIMS HAVE SHAPED THEMSELVES TO

SCIENTIFIC MELIORISM.

PREFACE.

IN bringing before the public this book, which contains my view of Scientific Meliorism, I venture to deprecate classification with Lentulus, in whom "the various aspects of 'motive' and 'cause' flitted about among the motley crowd of ideas which he regarded as original, and pregnant with reformative efficacy. . . . The respectable man had got into his illusory maze of discoveries by letting go that clue of conformity in his thinking which he had kept fast hold of in his tailoring and manners. He regarded heterodoxy as a power in itself."

The synthetic arrangement of ideas that are not original, but the common property of the age in which we live, is the purpose I have had before me in writing this book; and if in it I do not prove to my reader that my thinking is conformable with the facts of life and the facts of thought around me, I shall have failed in the major part of my undertaking.

I shall deem it a success if, on the other hand, my reader is struck by the familiarity, not the novelty, of my views, and especially if perchance this book should bring to him, as its only revelation, the recognition of himself as Scientific Meliorist.

We are told of George Eliot that "to a friend who once playfully called her optimist she responded, 'I will not answer to the name of optimist, but if you like to invent Meliorist, I will not say you call me out of my name.'"* This appears (so far as I can learn) to have been the first use of the term "meliorist."

In George Eliot's "Life," vol. iii. p. 301, I find that, on June 19, 1877, she wrote in a letter to James Sully, "I don't know that I ever heard anybody use the word 'meliorist' except myself. But I begin to think that there is no good invention or discovery that has not been made by more than one person."

In the *National Reformer* for September 21, 1884, the able writer who signs himself "D." says in a note to one of his articles, "The meliorist, as such, neither asserts that life is good or bad, but contends that it can be made worth living."

In the article entitled "The Great Political Superstition," published in the *Contemporary Review* of July, 1884, Mr. Herbert Spencer has this passage: "But if we adopt either the optimist view or the meliorist view— if we say that life on the whole brings more pleasure than pain, or that it is on the way to become such that it will yield more pleasure than pain—then these actions by which life is maintained are justified, and there results a warrant for the freedom to perform them."

And in the second volume of Mr. Ward's "Dynamic Sociology," at p. 468, I find the writer saying, "From humanitarianism it is but one more step in the same direction to meliorism, which may be defined as humanitarianism *minus* all sentiment. Now, meliorism, instead of an ethical, is a dynamic principle. It implies the improvement of the social condition through cold calculation, through the adoption of indirect means. It is not content merely to alleviate present suffering; it

---

* Miss Edith Simcox's article on "George Eliot" in the *Nineteenth Century* for May, 1881.

aims to create conditions under which no suffering can exist." To the word " meliorism " this note is appended : " The language seems to be indebted to George Eliot for this much-needed word, and if it is employed here with a slightly different shade of meaning from that which she originally assigned to it, it is at least one which is not only not supplied by any other word, but one which is in harmony with its etymology."

The dedication of my book was written some months before the publication of the " Life of George Eliot." The blessing of her personal intercourse was never mine ; but had she not lived, my mind must, under the action of other formative influences, have shown a different result ; and if my work has vitality, and takes its place as a true bud of the progressive civilization to which humanity tends, the roots from which it sprang, the germinal forces that gave it birth, ought, I conceive, to be made known.

Nor am I linked to George Eliot subjectively only ; objectively also there are links.

Into the delightful social circle with which the first volume of her life makes the public acquainted I was generously and cordially welcomed. The dear friends of her youth are my valued friends ; and that Mr. Bray did not live to see the publication of a work the prosecution of which he steadfastly encouraged, is a grave personal disappointment that overshadows its completion.

As compared with the life of George Eliot, that of the Rev. James Cranbrook was unhistoric. Yet to him her words apply : " The growing good of the world is dependent on " *such*. " The effect of " his " being on those around was incalculably diffusive." " That things are not so ill with you and me as they might have been, is half owing to the number who lived faithfully a hidden life, and rest in unvisited tombs." To Mr. Cranbrook I owe stimulus to self-culture. He gave me that inspiration to effort that comes from intimate intercourse with a pure and elevated nature, and that sympathetic instruction which makes study easy.*

Our friendship renders a tribute of thanks to George Arthur Gaskell inappropriate, but if my "Scientific Meliorism" is ultimately valued by any minds, they are entitled to know that physical debility and its consequent mental inertia would long ago have triumphed, to the relinquishment of an enterprise that entailed much labour, but for Mr. Gaskell's patient and persistent encouragement and aid. His thorough mastery of evolution philosophy has afforded me confidence in the working out of the details of my synthesis. Each chapter as it was written has been submitted to him, and every point of difficulty discussed with him. In dealing with the land question, and elaborating the retrospect and prospect of Socialism, his assistance has been direct and wholly invaluable. The important central generalization of the chapter on "The Land," viz. the causal relation between landlordism and civilization —a generalization which is, I think, calculated to throw fresh light on the land question—was first brought forward by Mr. Gaskell, in lectures to working men delivered in Leeds, Manchester, and other places in the years 1882-3, and published in the *Brighouse News* of October 25, 1884. The two generalizations, "sympathetic selection" and "social selection," incorporated in chapter xviii., will be found by the reader formulated as laws of race in a correspondence with Mr. Darwin given as appendix to the same chapter. Also the contribution to Sociology, named "the law of the elimination of evil" (p. 396),

---

* The Rev. James Cranbrook was author of "Credibilia" and "The Founders of Christianity."

belongs to Mr. Gaskell. It was arrived at some years ago, and is here published for the first time.

I need only say further, that, looking upon society as a living, growing organism of vast complexity and incalculable capacity, I have tried to distinguish between the forces which are antagonistic or destructive to the true health of that organism, and those forces, on the other hand, which directly tend to the creation of conditions under which "no suffering can exist." The possibility of evolving superior social conditions is, to my mind, a scientific certainty, dependent upon psychic effort; and since I am one of those to whom it is "a source of constant mental distortion . . . to go on pretending things are better than they are;" since I am compelled to admit that "life, though a good to men on the whole, is a doubtful good to many, and to some not a good at all," it seemed my simple duty to stimulate (so far as personal ability permitted) to this necessary psychic effort.

"And here I make an end. And if I have done well, and as is fitting the story, it is that which I desired: but if slenderly and meanly, it is that which I could attain unto" (2 Maccabees xv. 37, 38).

JANE HUME CLAPPERTON.

October, 1885.

CONTENTS.

CHAPTER		PAGE
I.	HAPPINESS	1
II.	THE SURVIVAL OF BRITISH MILITANCY	13
III.	THE TRANSITIONAL NATURE OF SOCIAL CONDITIONS	27
IV.	DEVELOPMENT IN MORALS	39
V.	THE POOR	50
VI.	PATRONAGE OF THE POOR	65
VII.	POVERTY IN RELATION TO GENERAL WELL-BEING	81
VIII.	SOCIAL PRESSURE IN THE MIDDLE CLASSES	103
IX.	THE EVOLUTION OF MODERN SENTIMENTS	120
X.	THE PERIOD OF YOUTH	141
XI.	THE SUBJECTIVE REQUIREMENTS OF SOCIAL LIFE	166
XII.	INDIVIDUAL RIGHTS	194
XIII.	OUR UNORGANIZED SOCIAL LIFE	219
XIV.	LEGISLATIVE VERSUS VOLUNTARY METHODS OF REFORM	244
XV.	HOME, SWEET HOME	261
XVI.	THE EXPANSION OF OUR DOMESTIC SYSTEM	277
XVII.	MARRIAGE	297
XVIII.	HEREDITY	323
XIX.	EDUCATION VERSUS CULTURE	343
XX.	THE TREATMENT OF EVIL-DOERS	364
XXI.	THE LAND	378
XXII.	SOCIALISM VERSUS INDIVIDUALISM	389
XXIII.	RELIGION	409
XXIV.	SCIENTIFIC MELIORISM	425

SCIENTIFIC MELIORISM

AND THE

EVOLUTION OF HAPPINESS.

CHAPTER I.

HAPPINESS.

"Add to the power of discovering truth, the desire of using it for the promotion of human happiness, and you have the great end and object of our existence."—HERBERT SPENCER.

To establish in the general mind a correct theory of life, is to create an essential condition of human progress.

Without correct thinking, right acting is little likely to take place; and in an age characterized by a chaos of ideas regarding the principles and conduct of life, it is in vain to look for harmonious and consistent action tending to universal happiness.

Amongst the numerous theories of the present day, there exists a proposition or generalization based upon experience and firmly held by a few individuals, whilst the attitude of most minds towards it is one of indifference, if not of prejudiced dissent.

The proposition I mean, is, that happiness for all at all times, is the primary object of life.

Now, people accustomed to analyze their own feelings will, perhaps, readily admit that what they are hoping for and inwardly craving is happiness in some form or other; but they will hesitate to make the admission that all mankind is in precisely the same condition, unless they have reasoned out the subject, and are convinced of the perfect propriety of this mental and emotional state. Few individuals, however, are accustomed to self-analysis. The generality of men receive from teachers some theory or other, which lays down what they *ought* to desire, and *ought* to make the prime object of life, and they never test the theory by their own individual experience and practice.

Unfortunately, popular teachers hold conflicting and diverse opinions, and the mass of the unthinking coming under their influence are swayed hither and thither, and, unable to judge for themselves which teacher is right and which is wrong, the general mind becomes thoroughly confused, and evils which are enormous in extent and disastrous in consequences, result alike to individuals and to society at large.

Christianity has had for centuries a clear-cut compact body of dogma, a set of propositions concerning what we are to believe and what we are to do; and these have been tested by time and tried by practice; so that one would naturally expect that Christian teachers at least would give no uncertain sound, but be united and harmonious in their teaching, strong in their mental position, able to enlighten the inquiring, and to still all wavering and doubting minds. This is not so, however. Within the Church, as well as without, there is a vast diversity of opinion. Into these differences I do not propose to enter. It is sufficient for my purpose to point out what is the general result of Christian teaching upon individuals who are docile, earnest, devout, and unreasoning. This general result is to drive them into an ideal life, to keep their minds fixed on abstractions, and their hearts quivering with emotions, far removed from the interests of this commonplace, practical world, in which they are not to seek happiness, but only power of endurance to bear all manner of trials, temptations, sorrows, and sufferings till life be done. Then, and not till then, will happiness be reached; for happiness is the ultimate end, the goal to which Christian teachers point in the distant future, though not in the present. The religious theory is, not that happiness for all at all times is the primary object of life, but that happiness in the future is to be the aim of all, and will certainly be achieved by some.

From religious let us turn to literary teachers, and first I speak of Thomas Carlyle. He drew no ideal picture of a happy future life, either for the race or for the individual, but turned attention from ideals to the real, from abstractions to the actual world around, the present social state, pointing out its miseries and its shams, and powerfully inveighing against its corruptions. The burden of his message was—awake from dreams, bestir thyself, and do what thy hand findeth to do with all thy might; *this* in contrast to one class of religious teachers who had said and sung, " Lay thy dreadful doing down." " Doing ends in death." His teaching and example inspired manliness, sincerity, and self-reliance, and many noble natures were kindled into sympathy and action. " Be indifferent," he said, " alike to pleasure and to pain; care only to do work, honest, successful work (no futilities), in this hurly-burly world." Outwardly an objective existence of active usefulness, inwardly a subjective state of quietism and stoicism—*this* was Carlyle's philosophy of life. But mark, while there was to be no direct seeking for reward, he believed the result of such a course would be blessedness, and *that* he counted something purer, nobler, more desirable than happiness. If we take his own individual history, set forth in the " Reminiscences," as typical of this result, however, we perceive the theory breaks down. *He* toiled and plodded on, doing work, successful work, to what appeared the end of a noble and victorious career; but the blessedness never came, or if it did, it was not nobler and purer than happiness; it was a state bankrupt of hope and love and charity, and full of gloomy anger and querulous unsatisfied egotism. A grand old hero was Carlyle, and yet the lesson of his life seems this— mechanical efficiency is not enough to make the vessel hold Promethean fire undimmed and safe throughout a long career.

Another teacher who has passed away, George Eliot, gave no distinct utterance regarding her theory of life, and formulated no rule of life, in all her many works of genius. But the action of her influence has been to develop social and sympathetic feeling, to make individuals tolerant and tender towards their fellows, judging none without due regard to his or her surroundings. She has accustomed her thoughtful readers to the scientific aspect of human nature and of social life, to watch the manifold relations between the two, the action and interaction of forces without and within, and to see the continuity of causation along with the reforming effect of ceaseless changes. The evolution theory underlies all her work. Her pictures of life are realistic: no false colouring and vain delusions, no perfection in individual character, but aspiration, effort, and broad humanity; no perfect happiness attained, but indications that she thought the social state wanted altering, and that some readjustments would conduce to nobler life and greater happiness. The Carlyle hero-worship was foreign to her pages; his admiration for despotism and rough-and-ready methods of reform finds no sympathy with her; she hopes for progress through the slow and gradual process of natural change in outward social system, and in inward human nature. " What I look to," she once said, in conversation with a friend, " is a time when the impulse to help our fellows shall be as immediate and irresistible as that which I feel to grasp " (and as she spoke she grasped the mantelpiece) "something firm if I am falling." Although she formulated no theory, George Eliot, I conceive, held the belief that happiness for all at all times *is* the object of life, partly to be achieved by further development of the altruistic or sympathetic side of human nature.

Some writers teach that *culture* is the thing to be desired. The rapid growth of wealth in this country, and consequent reign of capital, has forced upwards in the social scale a class of people destitute of culture and refinement. These dominate society, and take the lead in fashion. Luxury and ostentation are everywhere prominent, extravagant modes of living prevail without the comfort of the former simpler and more genial social modes; and all this is side by side with poverty and destitution that do not decrease. Patronage, with its demoralizing influence on both classes, appears the most prominent bond between the wealthy and the poor, and vulgarity of mind characterizes the age. There is little to surprise us in the fact that gentle and refined natures turn with disgust from both. In the effervescence of youthful enthusiasm they had plunged into manifold schemes of philanthropy, to be thrown back upon themselves by the futility of the work, and the vulgarity of the workers. Slowly and gradually they withdrew into a narrower sphere—not necessarily a selfish one, but a sphere bounded and circumscribed by their own personal tastes and temperaments. They found solace and relief in intellectual pursuits. Art and culture yielded them a pure elevated enjoyment; and, conscious of the breach between themselves and the great mass of human beings around, they adopt the theory that what is wanted is true culture. Sweetness and light are held up as the panacea for the ills of life, and to elevate the masses, as the only path of progress ! Sitting serene in their high altitude of intellectual light, they think they solve the problem of life for millions of toiling workers struggling for existence, when they indicate that what *all* want is happiness, and happiness means a culture which will take whole ages to secure.

But other teachers, thinking less of intelligence than of moral sentiment, point, not to *culture*, but to *perfection*, as the aim of life. They recognize the individual differences in human nature. Some intellects are slow and dull, incapable of being kindled into fervour or brightened into swift reflection, and culture for such is hopeless; but surely in God's sight all men are equal. Birds without song have brilliant plumage, or if not *that*, some other gift to make life precious; and so it is with man. "The law of *compensation*," they say, "holds throughout humanity." The business of *all* is to call forth in *each* what is noble, lovely, and of good report; and for the most part hearts are deep and tender even when heads are dull. The finest works of literature and art may fail to give one pleasurable sensation where there is no special faculty to apprehend their beauty; but "kindness makes the whole world kin;" and when the noble and the generous, the tender and the sympathetic, part of human nature is appealed to, the response is ready and complete. Happiness, then, seems with this class of teachers to be the aim of life; but happiness means individual excellence of character, moral goodness, and not the intellectual elevation of the masses.

Pessimists, again, are clear and definite in *their* solution of the enigma of life. Happiness for all at all times *is* the object and the aim, nay, the only justification, of existence; but with the knowledge of this fact has come, they think, the further knowledge that it is unattainable. Evil is not overbalanced by good in the lives of average individuals, and life is to be condemned, because it results, and must inevitably result, in more of pain than pleasure.

On the other hand, Positivists are optimistic in *their* theory of life. They have bright hopes for the future of the human race, though ages may elapse before the culmination of its destiny. Meanwhile, the state of misery and wretchedness that prevails is deeply felt, graphically described, and loudly deplored. The earnest exponents of this creed are eloquent, cultured, and refined, and the teaching of the school has strong attractions for the best class of minds. It says, we want a new religion; without *that* no progress can be made. At present the public mind is quite at sea, floating in a chaos of all fixed beliefs. To come to some settled convictions and formulate a creed is the necessity of our times, and *we* are doing *this*. Religion is a scheme of thought and life, whereby the whole nature of individual men and societies of men is concentrated in common and reciprocal activity, with reference to a Superior Power which men and societies alike may serve. The Superior Power lies in Humanity itself and man's true function; his religion, in short, consists in understanding this, and in seeking to perfect our common humanity.

Such are some of the diverse opinions and conflicting theories that prevail upon this momentous subject of human life, and in *two* points at least there is accordance. All show the conviction that there is misery around us, and that happiness in some form or other is what the world so desperately needs. Meanwhile, apart from theories, close observation will reveal that *practically* all of us, from the lowest to the highest, the youngest to the oldest, are engaged in the pursuit of happiness, *i.e.* in avoiding pain and securing pleasure. A child shrinks from lessons and seeks play, because the one causes painful efforts, the other gives pleasurable sensations, unless there be the beginnings of an intellectual sense and the child is what we call studious, when the sense of effort is overcome by the pleasure of learning, and there is no unwillingness. Or if the representative faculty is strong, the thought of a parent's or teacher's approval may be so clear in the young mind, as to make the future happiness counterbalance the present effort. But it is always pleasure at the moment, or pleasure in anticipation, or fear of punishment, *i.e.* avoidance of pain, that gives the stimulus to work. The human nature of a tender mother is much the same. She hates to hear her offspring cry; she loves to see them smile. She *seems* to sacrifice herself to them, but in reality it is not so; for her greatest pains and pleasures reach her through them. Her personal desires, her dearest hopes, are centred in her children. She is proud of their acquirements, ambitious for their future, happy in their success. If she strives earnestly to check and discipline them, it is because she dreads, for them and for herself, some baneful consequence, should she refrain. She does not act for a selfish end. Her nature is more complex, far wider and deeper, than the child's; but still her action is essentially the same. She is avoiding painful and seeking pleasurable sensations, present and future, for herself and for her children.

Nor with the poor man is the case widely different. The pain of hunger or the dread of hunger, for himself or for those beings whom he loves, stimulates to a life of continuous and wearing toil. If he submit to present pain, it is that he may avoid remote pain, and secure the satisfaction of his most pressing appetites.

The leisured classes, again, are differently situated. With conditions that remove them from the struggle for mere existence, they seek enjoyment according to their individual characters and tastes. What interests their minds and stirs their emotions most is sure to be pursued. We may speak of each career as guided and controlled by genius, ambition, noble sentiment, pure benevolence; but in every case these qualities of mind have brought about the choice of life which gives gratification to the individual and brings its own reward.

People differ in their thoughts of *what is* happiness, and differ in their methods of attaining it, but all pursue it in some form or other. For God (that is to please Him), for Humanity, for the lower animals, for friends, for self, according as the nature is religious, philanthropic, sympathetic, or selfish, for the present or the future, for here or hereafter—simple happiness is universally the aim of man. Just as when plants are placed within a darkened room they turn towards the light, they grow towards it, all their bloom and beauty tend that way, so similarly human beings seek sunshine of the heart, the light and warmth of happiness. This is a law of life and nature. Are we to reverence and call it instinct in the lower organisms, whilst we condemn it in the higher? Surely not. The *real* question, and the only one on which we may express a moral judgment, is concerning the *methods* we adopt for the attainment of that happiness we are entitled to seek.

If we say goodness is the aim of life, we must allow that goodness means, to aid and bring about the happiness of all. Religions signifying less than this are unworthy of the name. It is emphatically the good who suffer in the present evil state, for, as it has been truly said, "the contrast between the *ideal* and the *actual* of Humanity lies as a heavy weight upon all tender and reflective minds." There are, without doubt, natures too superficial to perceive the widespread miseries around us, and natures which, because not personally involved, are too egoistic to feel or care about them. But the *good* both see and feel. They know that their own goodness has depended largely on the conditions of their lives, whilst thousands of their fellow-creatures, born in degradation, brought up amidst vile surroundings which they did not choose, have had no chance to be good at all! To natures of this kind it is no comfort to point to a future happy state, to immortality for *themselves*, perhaps annihilation for those who are calling out their infinite compassion. Such consolations are an insult and a pain. The *actual* is what concerns them; their feelings get no rest, their intellects surge with perpetual efforts to conceive some means of *radical* reform, some method to secure more goodness and more happiness for *all*, *i.e.* for every woman, man, and child alive in the present day.

Not only in the lower classes is there misery, poverty, crime, monotonous lives of ceaseless toil, indifferent, often bad work and drunkenness (which, after all, seems the very natural outcome of an uncertain future and a present state of general hopelessness); but even where wealth abounds there is little real satisfaction. The customs and habits of life are artificial, the pleasure-seeking languid. There is an absence of simple joyousness, and the presence of much repining and discontent. There is an absence of manly and womanly public spirit and the noble quality of life which gives to society true dignity; there is the presence of mental and moral debility of tone.

Surely there must exist some deeply rooted *cause* or *causes*, obstructive of general happiness. The evils that surround us compel attention, and persistently suggest that somewhere there lies hidden a canker that is destroying our social welfare.

If that be so, then patient endurance is not a virtue suited to the present age. Nor is mere culture the thing to be desired;/ unless, indeed, culture means enlightenment,/ to point out what the evil is, and aid in the discovery of an efficient cure, a cure that will bring happiness in some sort to *all*, and not only to the *few* who are intellectual or æsthetic.

James Russell Lowell has pertinently remarked, that people cannot be argued into a pleasurable sensation; and to the masses who have neither time nor taste for culture, is it not cruel irony to tell them that they want " perception," and they require " to know the best that has been thought and said in the world," when they know their real want is happiness, and happiness with them means more of the necessaries of life, and more of life's sweet humanities, some freedom from anxiety and care, some leisure from constant and too monotonous toil. It is to be deplored that men of leisure, intellect, and earnest thought, appear to get entangled in the meshes of their own high culture, and to fail of their true mission to teach and guide the ignorant. In the present condition of our social life to the practical mind, the theories of such men seem castles in the air, and to the hungry, what they offer is stones for bread.

If teachers unanimously believed in, and set before the public, the simple proposition that the greatest possible happiness for all at all times is the primary object of life, the *force* contained in earnest minds throughout society would be brought to focus, and intellectual light would stream in and play upon the central question, viz. *the causes* of our present evil social state.

As it is, the waste of force in thought and action is enormous; for derivative and secondary questions dissipate the energy required to dive down to the foundation of things, and side channels divert out of its true course the current of thought and feeling which should move in the direct line of progress, and by its massive strength overcome every obstacle that lies in the path.

To make my meaning plain. Many minds at present are occupied with working out the problem, how children may acquire most rapidly a smattering of every branch of general education; many minds with the problem, how to keep alive the workhouse breed of pauper children, so weakly and diseased; many, again, with the problem, how to give an hour or two of innocent amusement to the labouring class (almost too weary to enjoy it); and many with the problem, how to bring the law to bear upon young criminals without that awful risk of adding to the evil and spreading the taint of crime. Other problems (too numerous to mention) engage public attention; but observe, all of them relate to *particular evils* without enquiry into *general causes*, and every thinker knows that criminals and overworked labourers and pauper weaklings show no tendency to diminish in their numbers, and that, deal with the present sufferers as you may, there comes up from the hidden spring or source an ever-fresh perennial stream.

Now, what I ask is this—ought we to rest contented with mere ameliorations of our evil social condition? Are we certain that the causes of misery cannot be reached? that no methods exist, or can be devised, by which we shall secure that in this wealthy England we may see a slow, but steady diminution of poverty, crime, ignorance, and hereditary disease, until gradually they are thinned out and die away, and there exists for us an entirely different social state, viz. a state of moderate independence for all the industrious of the lowest class, a state of comfort such as will remove temptation to brutality and crime, and, as a rule, sound constitutions and strong vitality for every normal British child; a social system giving a fair field and no favour, but an equal chance to all, *i.e.* to the clever and the stupid, to the energetic or ambitious, and the plodding or phlegmatic, to the muscular, materialistic natures, and the sensitive or spiritual—to *all*, I say, an equal chance for the achieving of a successful life to the full measure of personal capacity, and for the securing of happiness according to the taste of individuals without hurt or injury to the happiness of others? Is *this*, as a picture of what might be, impracticable? If thinkers say it is so, then we must face the fact, and relinquish all dreams of *curing* social evils. We must be satisfied with mere ameliorations; but at least we may ameliorate under better conditions of co-operation and *unanimity*.

In the present day, widespread as are ameliorations in the form of philanthropic schemes that intersect the entire country, few *logical* minds are able to take real interest in them. The word "philanthropist," has almost come to signify an individual stronger in the heart than in the head, one who is good, generous, tender-hearted, but not hard-thinking or far-seeing; and we cannot but perceive the reason why. It is because philanthropy never fully accomplishes the purpose it has in view. Its measures are not lasting and effectual, and to the thoughtful *failure* is stamped upon its efforts.

The method of popular philanthropy is to break up society into small sections, in order to minister to its wants. It has innumerable hospitals for special diseases, for general diseases, and for incurable diseases. It has societies for the sick, for the destitute, for indigent old men, for decayed gentlewomen, for itinerant minstrels; girls' friendly societies, and societies in aid of distressed foreigners. It has homes for the blind, for the deaf and dumb, for the insane, for servants, for governesses, sailors' orphans' homes, and colleges for orphans of the clergy, and so on, in endless numbers. Now, under a systematic teaching of the principle that the thing to be aimed at is universal happiness, the public mind will cease to be satisfied with this fractional method of dealing with society; it will view *society as a whole*, and what causes the *widest* and most *general* misery will demand the first consideration; whilst the remedy desired will be such as tends to *uproot* the evil, and not merely mitigate by lopping off its branches. I am not insinuating that any human suffering is trifling, or unworthy of our highest thought and endeavours to remove; *but* we must see things in their due proportions if we are to overcome social evils in all their magnitude.

And this is not all. Thinkers of the present time have yet another task before them. They have first to ascertain the true conditions of healthful and happy social life, and then, by study of these conditions in all their bearings upon humanity, they have to reason out right principles of social action, and to lay down maxims of conduct adequate to guide the masses into a happier and better course.

Medicine, in former days, had no pretensions to be regarded as a science. Remedies were tried upon disease almost at haphazard, and observations made of the supposed results, without any knowledge of the true action of the drug, or of the nature of the organs it was meant to act upon. Rules were laid down and habits formed as to medical treatment in all cases of illness; unless, indeed, the symptoms were so peculiar as to force upon the doctor the suspicion that his patient's case could not be classified, and treated by ordinary routine. The whole proceeding was superficial, and no real progress was made until a new and better method had been adopted, viz. the studying of human nature in its *healthy* conditions, instead of merely experimenting upon it when diseased. Anatomy, Physiology, Histology, were found to be *necessary* foundations of medical science. And *now*, in dealing with a sufferer, the doctor has within his mind an approximately true picture of what ought to be the normal state. He knows the nature of the whole organism when its functions are orderly and healthful; and, as a rule, he can detect in each patient's case the cause or centre of disturbance.

In social life we are no further advanced than medicine was in those early days, when all was tentative and purely experimental.

Of Dr. Lydgate, George Eliot says, "he brought a much more testing vision of details and relations into this pathological study than he had ever thought it necessary to apply to the complexities of love and marriage, these being subjects on which he felt himself informed by literature, and that traditional wisdom which is handed down in the genial conversation of men."

This irony might be extended to all mankind. Surely there is folly in the fact that men so little care to make a study of the complexities of social life! What other interests can at all compare with these?

Until the laws of individual life, with the conditions of happy industrial and social life, are thoroughly investigated, no maxims, or practical rules of right conduct, can be established, although our present social condition is in every respect unsatisfactory; for, besides the general unhappiness, another evil stares us in the face. The *moral* nature of some classes in our midst is tending to deteriorate.

Men who began their business or professional careers with noble impulses and lofty hopes, will frankly make admission that the struggle for existence, the pressure of keen competition all around, has made them lower their high moral standard, and live habitually and consciously below the level of what they know to be true and honest and just. Now, in these cases, there must be loss of self-respect, and that *alone* brings injury to sensitive human nature. Again, the gentle and benevolent amongst us do not all possess a patience that will make them work on hopelessly for ever. Many get hardened in the process. Their best efforts are frustrated, their sympathies exhausted, and their moral tone debased. Nor is this all. Our children are growing up without distinct and well-directed moral training; for we are in the midst of the decay and death of a regulative system no longer fitting for the age; and thinkers and teachers have as yet done almost nothing to aid us in the dilemma.

It is more than time that *all* should put their shoulders to the wheel, and, grappling with the problems of our complex social life, find true solutions that will reveal what social conduct ought to be, and give a scientific basis for a wide-reaching modern moral code, subserving general happiness.

CHAPTER II.

THE SURVIVAL OF BRITISH MILITANCY.

"There is no creature whose inward being is so strong, that it is not greatly determined by what lies outside it."—GEORGE ELIOT.

IN various parts of India and in the Malay peninsula, there are to be found certain native tribes which are wholly unwarlike. The occupations and habits of these tribes are peaceful, and travellers and missionaries, who have come into contact with them, describe what an admirable spirit of brotherly love and concord exists within the groups, the individuals showing mutual kindness, gentleness, and confidence.

Of the Jakuns it is said, they are "inclined to gratitude and to beneficence;" their tendency being not to ask favours, but to confer them.

Of the Arafuras: "They have a very excusable ambition to gain the name of rich men, by paying the debts of their poorer fellow-villagers." And one gentle Arafura, who had hoped to be chosen chief of his village, and was not, met his disappointment with the spirit of a philosopher and philanthropist, saying, "What reason have I to grieve? I still have it in my power to assist my fellow-villagers."

The domestic as well as social relations of these peaceful tribes are pure and simple. Not polygamy, as we should be apt to suppose, but monogamy is the order of their married life. Of the Lepchas, Hooker says, "the marriage tie is strictly kept." Of the Bodo and the Dhimals, Hodgson tells that adultery and concubinage are not tolerated; but chastity is prized in man and woman, married and unmarried. They treat their wives and daughters with confidence, and the latter are free from all out-door work whatever. The

Santál also "treats the female members of his family with respect." *

Now, with the not unnatural bias of civilized man, the pleasing general impression that manners and morals hang upon civilization and a certain amount of high culture, *we* should feel no surprise if we had to learn concerning these tribes, that when brought into relations with men of different type—I mean men hardy, fierce, and truculent—they readily showed the white feather.

Gentleness in savages we associate with timidity, and the Bodo and Dhimal, the Santáls, Lepchas, Arafuras, and Jakuns, might prove cringing and fawning, base, treacherous, and morally weak, without exciting any surprise in us. The very opposite of this, however, proves to be the case. They show themselves independent and courageous, although not warlike. The amiable Dhimal resists injunctions that are urged injudiciously, "with dogged obstinacy." The simple-minded Santál has a "strong natural sense of justice, and should any attempt be made to coerce him he flies the country." The Jakun is said to be "extremely proud," but the pride shows itself in refusals to be domesticated and made useful to human beings of a different race, and therefore alien to himself. They have perfect respect for the claims of property. Hooker says of the Lepchas: "In all my dealings with these people they proved scrupulously honest"; and of the Hos (which belong to the same group as the Santáls) Dalton writes: "A reflection on a man's honesty or veracity, may be sufficient to send him to self-destruction." The Santál is courteous and hospitable, whilst at the same time he is firm, and free from cringing. The Lepchas "cheer on the traveller by their unostentatious zeal in his service," and when a present is given to them it is divided "equally among many, without a syllable of discontent, or a grudging look or word." In short, the virtues which we have been accustomed to associate with civilization are present with these tribes, whilst the vices we attribute to barbarism, viz. bloodthirstiness, cruelty, dishonesty, treachery, and selfishness, are glaringly absent.

In the desultory reading of books of travels we are interested in the accounts given of these gentle uncultured savages, and we have momentarily in our mind's eye a picture of Arcadian life in all its simplicity and charm, but we are apt to miss

* "The Industrial Type of Society." By Herbert Spencer. *The Contemporary Review*, October, 1881.

entirely the *great significance of the facts*. Happily for us, Mr. Herbert Spencer has pointed us to the lesson.* Human nature is not dependent upon culture, nor even upon religion, to humanize and make it lovable. There is *that* in its essence, the very groundwork of its nature, which makes it capable of developing under favourable conditions, into what is admirable, pure, and gracious. I use these terms advisedly, for the traits given us of these peoples show virtue, truth, generosity, moral courage, and justice; and no nobler, more elevated sentiments have as yet been found in, or thought of by, the civilized man.

But what are the favourable conditions which have made these peoples what they are? An entire absence of warlike surroundings, warlike proclivities, and warlike training, with the presence of the opposite conditions; for "they have remained unmolested for generation after generation, they have inflicted no injuries on others," and all their unselfish feelings have been fostered and nourished by the sympathetic intercourse of a peaceful life.

Now let us turn to our own social state. We have passed through the purely *military* stage, and are now *industrial* in our type of national life. The dominant genius of the British is not warlike. We are lovers of peace; and (although many recent events appear to conflict with this statement) we have no *national* desire to extend our territory, or to conquer and rule over those races which are established on other portions of the earth's surface, no matter how vastly superior to our own may be their climate and their quarters.

It is true that we annexed the Transvaal, contrary to the wishes of the Boers; we tried to take a slice of Afghanistan, and call it our scientific frontier; we have assumed protective power in Central Asia, with a spirit arbitrary and self-interested; and these, and other political actions, belong to the old *régime;* for although the national type is industrial, we are full of survivals of the past. The military spirit lingers, it is diffused throughout society; and in every crisis or dilemma, which presents new difficulties to surmount, the tendency of nations, as of individuals, is to relapse, to slide backwards to the order and the method of a former time.

In spite of these relapses, however, we are an unwarlike people. Should it so happen, that our social misery bring revolution, anarchy, and chaos (as some philosophers antici-

* *Contemporary Review*, October, 1881.

pate), and in the pitch and toss of circumstance a man of hard despotic nature take the lead, and, carrying on the work of reconstruction, rule the nation for a time, establishing himself as despot—can we suppose, if such an one developed personal ambition, and, like the first Napoleon, wished to play the *rôle* of conqueror of Europe, bribing John Bull with military glory, whilst seeking for himself a safely seated dynasty—can we suppose, I say, that John Bull would consent? No! for to do so would be *far more* than a relapse; it would be to change the hue and whole complexion of the national mind. The scheme could only bring revulsion, and the despot's overthrow. A sanguinary blood-stained path of so-called glory, has *of itself* no fascination for the British race, and the hypothesis proposed, we may reject as an evil not within the region of the possible.

The purely military epoch is past for *us*. The spirit of that age is dead, although its dispersing vapours, or I might say its grave-clothes, hang about and hamper us in our progress.

The fact is, in our social state, as well as in our theories of life, we are transitional, confused, and incoherent. Hence our political actions are confused, and form an incoherent, incongruous whole. But we are moving, although with a limping gait, forward, always forward; and that means towards a time when swords will be beaten into ploughshares, and spears into pruning-hooks; when the modern huge engines of destruction, that men are now so proud of, will seem as useless and contemptible as Don Quixote's military accoutrements, so bravely donned to fight the evils of *his* day, but which only served to hide his folly, and afford to us a gently cynical amusement.

Meanwhile, in our transition state, a work of sifting has to be done. Throughout the whole relations of our national life, relations political, commercial, social, domestic, educational, and religious, forces, or rather tendencies exist, belonging to the two *régimes*—the military or warlike, and the industrial or peaceful. We ought to find out which is which, and whilst repressing, checking, and controlling the tendencies which are survivals of the past, we ought to nourish, foster, and encourage those which belong to the era we are approaching, and which will help to carry on transition, till we emerge at last a peaceful nation in the highest sense, with every shred of warlike proclivity shaken off, and showing a brave front, yet speaking a coherent political language before all Europe, in a foreign policy consistent at all times with itself.

Of course, whilst other European nations remain armed to the teeth, we must keep up our army, and be prepared for efficient self-defence; but necessity alone rules in this matter. and no *aggressive* action can be justified. We have had proof of this in the failure of the Jingoism of a few years ago, and in excuses that were offered for attacks upon the Zulu king—that he would not permit his soldiers to marry, and such like. They were ludicrously paltry, and would never have been put forth, but that the leaders (in the policy), themselves aggressive, felt that they had behind them a nation non-aggressive, to which they would require to justify the action in some other sort than by plainly saying, "Let us annex our neighbours' territory." They trusted to the confusion of the general British mind, and played upon its weakest side, viz. that kindly, self-deluding notion, which John Bull lays as flattering unction to his soul, that *he* is the world's grand missionary, equipped by Heaven to overthrow false deities, and bring about a universal rule of Christian creed, and a Christian code of morals. Turning from the subject of what surviving militancy exposes us to, in *politics*, let us see if we detect it in our commercial and social life.

In a purely military state the sympathetic feelings are necessarily repressed, whilst the bold and keen, the hard and cruel side of human nature is developed. To hate all enemies, and avenge all injuries, are highest virtues. The predatory instincts of mankind are fostered and encouraged, whilst treachery and cunning are not discredited. It is the man who can take a mean advantage of his enemy, and successfully intrigue against him, who is a hero. The man who is rapacious in appropriating plunder, who never falters in his keen pursuit of prey, and in the hour of victory is cruel and relentless—it is *he* who is thought godlike and heroic. No thought of individual happiness or misery affects him; military glory is his aim, and when for a short time war is over, his iron nature finds its sphere in ruling with tyrannic power, while slavish followers bow down and cringe before him. This is a very meagre picture of militancy and its appropriate sentiments, but sufficient for my purpose.

Now, let me take three characteristics of the military stage—rapacity, intrigue, antagonism—and ask, Have we not striking proof of the existence of these tendencies in our commercial world? Lately we have been hearing much of what is called a "cotton corner." This is not the place for a minute description of this modern phenomenon; but here is what a recent writer on the subject says. "Dealings in futures require little or no capital for their conduct, and it is open to any individual, or to a 'ring' by the extent of their operations, or by fictitious proceedings, to raise and depress prices at their pleasure, and establish situations of a most disorganizing character—*and this is done.* This fatal facility of trading in paper contracts, lying over many months without the dread necessity of having to provide one single penny to pay for the large value involved, has developed an extent of speculation, to which the term gambling may be righteously applied. *This* is bad enough; for the irresponsibility of the operators, unknown to the public, is continually setting home and foreign markets in courses for which no intelligible reason can be found by legitimate traders, and creating universal embarrassment and loss; but when to this is added operations of a gigantic character, in which one man or a 'ring' deliberately sets to work to manipulate markets, effects are produced, the disastrous character of which cannot be measured, for the influences are so widespread, and touch so many various interests throughout the world. The electric telegraph flashes the fictitious values, produced by these operations, north, south, east, and west, raising false hopes, and producing needless despair—for what? The interest of one man, or a small group of men, to whom the operation may bring a profit, the value of which, as compared with the cost of the disturbance to the entire commercial world, is as a drop to a bucket of water!"*

Now let us here recall to mind two pictures: first, that of these simple unlettered savages, of whom it is said—they display unusual respect for the claims of others; the pure Santáls with whom "crime and criminal officers are unknown," and who "never think of making money by a stranger;" and second, the picture given of a warlike age with all its fierce characteristics, and its savage type of human nature, and say which do *we*, in our industrial stage, resemble most?

It cannot be unjust to take as typical of our commercial class, the London stockbroker, and the Liverpool merchant. They are in the very front and at the top of our industrial system. The tendencies which they display permeate and filter down through every stratum of that social section, and their example makes conditions which *must* foster and increase

* "On Commercial Corners," William Halhead. *Nineteenth Century*, October, 1881.

similar tendencies in traders of the lower grades. The fact is this. Amongst us the strong predatory instincts of mankind survive—rapacity and greed are not discredited. A perfectly unscrupulous forgetfulness of human brotherhood (in trade) is felt as no dishonour. And if success attends upon a man in his commercial warfare; if his intrigues are only wide enough to give him plunder, on a vast scale, he passes for a merchant prince, the rightfulness of whose transactions is little questioned, and men poorer, but of nobler sentiment, extend to him the hand of fellowship and call him gentleman! The British race has got beyond the point of being blinded by empty military glory, to all the horrors of a blood-stained field; but it is *fascinated by success in the pursuit of wealth ;* and when, as often happens, success goes to the mean, the grasping, the hard-bargaining, the utterly selfish—it is *fascinated* still! and overlooks the broken hearts, and broken lives which may lie at the foundation of a gigantic fortune.

The Bodo and Dhimal are morally above a devilish rapacity, but so is not as yet John Bull! *His* god is wealth, and he is capable of sucking the blood and trampling out the life of human happiness in keen pursuit of gold. Even whilst within his Christian temples he bows himself and calls the love of that secret god of his "the very root of all evil." Of course, when men of wealth and station show a vile example, it would be too absurd to look for uncorrupted morals in the lower social classes. I do not say benevolence and generous kindness are not widespread and real. The forces that belong to both *régimes* are present; but the prevailing dominant force in our commercial world is military, and it is the hard and keen and cruel side of human nature that is developed in the strife of competition. In the transactions of mercantile rings coercion is as much the order and rule as in a battle-field. The trader who detests the system, and would choose to buy and sell on reasonable profits and at little risk, is *forced* to do as others do, or leave the market. Yet these merchants boast of the liberty of the British subject, and would bristle with indignation were a Russian to say, "Your trade and commerce are as despotically ruled as is our nation. British individual freedom in industrial life is only your delusion." This would not be to overstate the fact. The men who "twirl the markets round their fingers" are tyrants and despots in the field of commerce, and secretly they trample down, demoralize the nature, and crush the liberty of every honest trader. In the mental atmosphere of an exciting mercantile game honesty is not thought of as good policy, and any policy that brings success is looked upon as honest; the claims of others to a fair share of coming trade is unscrupulously disregarded. Such social justice is a thing unknown. Antagonism, not fellowship, is the spirit of commercial life, and brutal egoism the vital spring of action.

It is a strange picture we present. Outwardly a peaceful nation, averse to carry on destructive activities against external foes, yet within ourselves full of these destructive activities; for in the habits of our industry, the customs of our business life, we plainly show the predatory instincts of uncivilized man; and all our culture and education are of no avail to counteract the tendencies.

But let us now turn to the field of education, and inquire what are the influences there.

In the series of observations and experiments made by M. Pasteur, which have led up to his discovery of the means of saving sheep and horses from the fatal disease called charbon, one step in the process has a special interest. Certain localities were haunted by the disease, although the pastures were excellent and apparently free from any unhealthy conditions. M. Pasteur ascertained, however, that some ten years previously sheep had died of the disease, and been buried there at a depth of from ten to twelve feet below the surface. He was aware that the germs of the disease are all but indestructible, and that therefore the dead carcases could infect the living sheep if brought into contact with them. But how could that occur? To his acute, divining mind the tiny earth-worms presented a possible means of conveyance. The fields were covered with their little heaps, and soon the microscope revealed that he had hit upon the actual fact. These small creatures had been unconsciously employed in deadly work, conveying poison germs from ten feet deep up to the surface of the ground, to find fresh victims to the fatal charbon. Here was a new condition, a danger never dreamed of, whereby the living might be hurt and injured by survivals of the past.

In reading this account it occurred to me that we ourselves are somewhat like these earth-worms, busy sowing poison germs from a past age in the unsophisticated minds of children, whilst ignorant of any harm, and full of good intentions. To any one who has never reflected upon the absence of method in training the young, this notion will seem absurd. But let me explain. Most people are agreed that education means something more than the three R's, or than cramming the mind and loading the memory with facts of history, geography, and grammar. In the Prussian national system it is spoken of as the harmonious and equable evolution of the human powers; and in Stein's longer definition these words occur: "The impulses on which the strength and worth of men rest are to be carefully attended to."

Whether we call it education or not, the character of every individual largely depends upon the training in ideas and in habits which he receives during the period between infancy and adult life. This training comes through direct teaching, but also through example and through all the various influences which surround the child, whether in the home, the school, or the playground. The environment or conditions of his life make up the raw material from which at every moment he assimilates a portion, and, according to his native bias, builds up his individual character. Now, these conditions are not all *outward*. The vivid pictures in a child's *imagination* play an important part in building up his character, and it is *here*, I think, we plant poison germs.

No sooner has a boy of six or seven mastered the difficulties of reading little words, than his mother puts into his hands a story-book with pictures, and is delighted when she sees him leave his hobby-horse to read his book. This is natural; she knows that it is chiefly by his own efforts, and by practice, that he will acquire facility in reading; and naturally also she chooses a story highly coloured and sensational, for otherwise the hobby-horse would still be preferred. The book, then, contains Blue Beard, Jack the Giant Killer, Jack and the Beanstalk, and Ali Baba and the Forty Thieves.

The chamber of his infant mind is furnished now with pictures, which make him live within a world of wonders far removed from what he sees of actual life around him. Giants and fairies, bold warriors and lovely ladies, castles and forests, mysterious dark closets, hens that lay golden eggs, and harps that cry out "master!" all mix and mingle with the facts of everyday occurrence, his lessons and play; and out of this queer medley the little fellow draws his happiness, and shapes his conduct. Whether it is wise at all, to *force* a child's imagination, and not simply leave it to natural development, and let it find its own food in common things that interest a child (of which there are always plenty) is a question I shall not here discuss. The mother's object was merely to help her child to acquire facility in reading, and no doubt she gained that end. I think myself that six or ten months more of hobby-horse, and of distaste to books, would be no loss to general culture, when the boy became a man; but this is not my point. I want to look at what these story-books accomplish in the way of giving a sound training and direction to "the impulses on which must rest the strength and worth of men." Our children are not like the offspring of the Bodo and the Dhimal, who for generation after generation have known no fighting and committed no injuries on others. They are inheritors of a past history wholly different; and we ought to recognize the fact, that in them there are sure to lie latent instincts of cruelty, hatred, revenge, treachery, and cunning; qualities that were required when their progenitors were fighting their way to freedom and peace. What we have then to guard against is, the stirring or exciting of these latent anti-social instincts, whilst we fill the child's life with what will rouse and play upon his social non-malignant feelings. In purely physical phenomena, the rule is this: exercise an organ, and it grows strong and supple; leave it alone, and it dwindles into powerlessness. This method in the training of a child has never been adopted.

After reading Ali Baba, a boy's mind is full of shifting scenes that call into play his predatory instincts, his greed, his tendencies to cunning and revenge, and that exercise his latent power of bearing thoughts of bloodshed and pain without flinching, and so tend to make him cruel. He thinks of the forty robbers and that delightful cave of theirs, filled with their plunder; he glories in the thought of Ali Baba helping himself so lavishly, and wishes *he* were there to do the same; and then, the little scene at home when robbery is called "good luck," and Ali Baba's wife gloats over the plunder and counts the money piece by piece! If we had seriously wished to train our boy to play an unscrupulous part in the great commercial rings, we scarcely could have hit upon a more effectual course, than feeding his bright innocent imagination with such a tale as this.

Again, Cassim's mal-adventure causes the boy to strive to picture to himself a man killed and quartered! a task to which he is further stimulated by such details as: "when Ali Baba reached the cave, he was surprised to see blood spilt outside the door." Ali Baba's horror when he found his brother's quarters within, however, did not prevent him seizing further plunder; he loads again two asses with bags of gold, and the little reader thinks this natural and quite commendable! Then follows an offensive picture; and we, who would preserve that child from witnessing the sad accompaniments of real death and burial, yet let his young imagination play about the sewing together of a quartered body, getting ready the coffin, etc. After that comes a double intrigue in which cunning fights against cunning, and revenge is the feeling excited in the boyish reader. He thinks, how clever and delightful of the robber chief, to trump up the story of a benighted traveller, and so to win from Ali Baba a hot supper and good bed! But more delightful still, of clever plotting Morgiana to outwit him! The cruelty of her deed, the lies she tells, escape all notice, because the *instinct of revenge* is satisfied, and I maintain the satisfaction, though purely ideal, will make *that instinct grow*. To us the notion that out of one kettle burning oil could come to kill thirty-seven men is so absurd that we do not perceive its action on the boy's mind. There is no absurdity to him, the whole affair is real. He cannot but admire the murderess of thirty-seven men, who crowns her brave exploits with a sublime finale. The scene is fascinating; bright lights, festive music, a gaily dressed, lovely, dancing girl, who in a moment draws from her girdle a sharp poniard and plunges it into her enemy's heart. Then comes her reward—gratitude for saving life (not a word of the lives she has taken) and marriage! This is her reward! To the child, therefore, marriage presents itself as a pleasant sugar-plum that Morgiana has earned by fearlessness and cunning. The whole story from beginning to end is barbarous. Its sensational pictures, its intellectual ideas, and, above all, its moral tone, are barbarous. They belong to the barbaric period of our history; and we are like the earth-worms poisoning living minds by bringing to them germs out of the decaying carcases of the dead minds of a former age! Could any action be more suicidal? more obstructive to moral progress? And mark, the direct teaching that the boy's mother gives conflicts at every point with these ideas. She bids him love his enemies, and never seek revenge. She tells him Ananias and Sapphira were struck dead for telling lies, and that the man who robs and plunders must be put in prison, and so on. Imagine the confusion of the infant mind! No wonder that what is right and what is wrong take no clear shape within him; and since all his malignant feelings are excited and made active, what wonder if in the moment of temptation he acts at once from these. Jack of the Bean-stalk disobeys his mother. He pleases *himself* on all occasions, and when she prays to him with tears, he turns away and leaves her broken-hearted. Nothing but good, however, results from this behaviour. He leads a charming life of free adventure, plundering and killing giants, and getting petted by a fairy; until at last he settles down, a worthy, a noble man! The youthful reader thinks he may do likewise. His mother's words sound tame to him and foolish; what matters it if he deceive her now and then, and go his own way; perhaps he too may meet with strange adventures. I do not say that consciously he reasons thus, but that unconsciously he does; and even when without the smallest motive for deception, a clever child is questioned on some simple matter, he is very apt to let his vivid fancy colour and embellish all the facts, and think there is no harm in being thus untruthful.

Now, should these pages fall into a mother's hands she may object that I, whilst pointing out the evils to avoid, propose no substitute for barbarous tales, and, naturally, she shrinks from reducing her child's happiness, or limiting his pleasures. But my conviction is, that if we made a holocaust of all these survivals of a savage past, and started on a new and rational mode of dealing with the young, one of our earliest steps would be to give them greater freedom, and so vastly to increase their happiness. For, whilst we make mistakes in spurring them to artificial pleasures, and furnishing their minds with hideous pictures of an unreal world, we just as constantly err on the other side by checking and repressing them, when, if only left alone, they would quite innocently grope their way to pleasures that are simple, natural, and beneficial. This is, however, a wide subject, and not to be entered upon at present. I only touch upon a single point.

A niece of mine, whose age is ten, announced nine months ago that dolls had lost their charm for her. She never, never meant to play with dolls again; her brothers had laughed, and called it a childish amusement! She put them all away and steadily kept her resolve for some few months. Then came a fit of doll attraction, which she dares not now indulge in public. She finds a hard sofa-pillow suitable in shape, and with it draped in a crib-blanket, she steals into a room where boys must not intrude, and wanders up and down, the clumsy baby in her arms, crooning to it, and singing, as happy as a bird. Here is 'an innocent, harmless play', of which the child is made ashamed; a source of pure enjoyment which she foregoes, or only snatches at in secret; yet this play usefully calls out the budding instincts of her womanhood, and strengthens these through exercise. Why are our boys not taught to reverence sex distinctions? I see no reason for not telling them distinctly that a sister's love of dolls arises from the difference of sex; it shows her tender, gentle nature, and points to future fitness to fulfil the sacred duties of a mother. If elders took a tone like this boys would soon cease to molest a little sister, and a foundation would be laid in character for right relations between men and women. As it is, what better dare we expect from lads whose ideal heroines are of the Morgiana class, than that they should feel contempt for play that is not wild and savage in its character.

But even our best guides in the great work of education make grave mistakes in this important matter. One of the most learned of these guides says : " Predatory pursuit excites us from the earliest years, and any incidents embodying it will waken up the feelings, and exercise the imagination in a bloodthirsty chase; thus enlivening the dull and dreary exercise of learning to read and spell." * Therefore, this high authority approves of making a child's first reading-lesson sensational—about a cat and rat; or he suggests that a mouse would be better, as a cat's torturing play with a mouse before eating it comes within the range of the child's own experience, and is to him "one of the rarest treats." † Is this not atrocious? The child has an emotional as well as an intellectual nature. Are we deliberately to sacrifice the former to the latter? Are we to teach him to read at the expense of fostering and strengthening his anti-social, predatory, savage instincts? If all we aim at is to make our children keen and clever in commercial competition, this method may be wise ; but if our object is to *carefully develop all the impulses on which must rest the strength and worth of men*, then, certainly, the method is altogether wrong.

It is our ignorance of the supreme importance of social conditions in the building up of character that so misleads us. Favourable conditions are an all-powerful factor in the making of a noble human being. We want our children to be something more than Bodos, Santáls, or than gentle Arafuras ; and we possess the means to make them highly cultured ; but before high culture we must put, as of *still* greater value, all the virtues

* " Education as a Science," by Alexander Bain, LL.D., p. 244.
† *Ibid.*, p. 246.

that the Arafuras show; and from the history of these uncultured savages, and the fact that they have risen to a moral state that puts our boasted civilization to shame, we ought to learn this lesson :—The surroundings of our race in childhood and youth are full of adverse influences, and character is daily suffering injury that direct moral teaching will never rectify.

We have to search for and discern what all these adverse influences are, and to control them, or cast them out.

CHAPTER III.

THE TRANSITIONAL NATURE OF SOCIAL CONDITIONS.

"The Ascidian throws away its tail and its eye, and sinks into a quiescent state of inferiority. . . . We are as a race more fortunate than our ruined cousins—the degenerate Ascidians. For to us it is possible to ascertain what will conduce to our higher development, what will favour our degeneration."—Professor E. RAY LANKESTER, F.R.S.

THE doctrine of evolution is calculated to produce a twofold effect upon the human mind. It is capable of giving intellectual clearness and emotional beneficence. To do this, however, it must be thoroughly understood and logically applied to our present social state. It must be regarded as explanatory of things as they are, and prophetic of things as they will be, and should be.

When the doctrine is so regarded and applied, the evolutionist no longer feels despairing in face of a society, whose perplexing miseries and many-sided evils had hitherto overwhelmed him. A flood of light enters the field of human action. He perceives the social forces that are at work amidst apparent confusion. He apprehends their approximate causes or antecedents, and he sees that in the enlightened intellect of man there lies latent power, not to overcome, but to direct and control these forces.

This knowledge alters his emotional state. Anger and disgust are undermined. They die away, and into their place is born sympathy, beneficence; for, however great may be the wrong-doing of humanity, individuals are not to blame. An individual is good or bad according to the conditions of his birth, the method of his education, the circumstances of his life; and by alteration of these objective phenomena (under the ruling of an enlightened and persistent reason) the subjective phenomena of individual human nature and of social relations will be redeemed, and raised to nobler ends and purer morality.

At the present stage of social evolution the militant qualities of human nature are everywhere prominent, and disfigure our civilization. We have lively predatory instincts, keen pursuit of gain, treachery, cunning, gross injustice, brutal selfishness, rampant in our midst.

Individuals of the highest type, those who are by nature good and gentle, are inwardly grieved and outwardly buffeted, by the perpetual outcome of vicious, uncivilized tendencies. If ignorant of the great doctrine of evolution, they are certain to feel despairing of the immediate future, and regretful of the past. They sigh for the " good old times," when, as they fancy, work was conscientiously done, not scamped as it is now; when the necessaries of life were pure and unadulterated; when servants were less exacting and more honest; when wages were not perpetually rising; when labourers were not discontented; when children were not rebellious; when doctors were less required; when lawyers did not charge exorbitantly; and when comfort and ease were more abundant, and far more liberally enjoyed.

Of these "good old times" we shall presently speak, but what I would here point out to my reader is, that whilst with the good and gentle this state of mind only begets despondency and a measure of whining, thousands of human beings of a different type are thrown by it into anger, jealousy, or some other turbulent emotion. In society at every point one comes across either passive discontent, or active indignation, accompanied by the intellectual conviction that some individual or individuals *must* be to blame for the unsatisfactory state of our industrial and social life.

A few days ago a grocer's assistant charged a lady customer a price for candles above the market value. The lady remonstrated, remarking, "at the co-operative store the price is so and so." The grocer immediately stepped forward, and explained that he had lowered the whole of his prices to the level of the store's, but that as yet his shop lads were apt to make mistakes. "Probably," said the lady, "had the prices been raised, not lowered, the lads would better apprehend the change." Her irony gave no offence. The man at once became confidential and communicative. "These abominable stores," he said, "threaten to drive honest tradesmen like me out of the field. It is all the fault of clergymen and lawyers." "Indeed !

how so?" the lady asked. "These men," he replied, "are clever and idle. They do uncommonly little work themselves, and charge enormously for the little they do. There is no other class so eager to buy cheap, and they have set their wits to work and invented these horrible stores, where *they* buy cheaply to the ruin of such men as me. The lawyers," he went on, "are the worse men of the two. Their own work is perfectly useless. If two men quarrel about a bit of money, the lawyer may settle the dispute; but he pockets the whole of the money and calls it his fees!" "Well, well," said the lady, "if no lawyer existed, at least you must allow that the men would fight about the money, and very probably homicide would be the result. Surely bloodshed is a worse evil than the loss of money."

The grocer's condition of mind is extremely common in the present day. Men ignorant of evolution, and pressed upon by social changes that make the struggle for existence severe, and too often overwhelming, have a vague sense of injustice done them. They reason quite illogically, each one from his own centre of difficulty, and are alike in cherishing a spirit of vindictiveness, a readiness to strike out in any direction that prejudice or bias may prompt.

This grocer is certain to make church disestablishment a vital point in his political creed. His dislike to the clergy, and his hatred of co-operative stores, aided by faulty logic, will become *dynamic force* in his exercise of the franchise.

I am of course aware of the general opinion that education is the panacea for ignorant vindictiveness, and every other social evil. With State education made compulsory, progress, it is argued, must necessarily be rapid. It may be so—but progress in what? Will it be progress in all the benign influences of a happy civilization?

I venture to assert that the *first* effect of education upon the masses is naturally to increase the general sensibility to the many disabilities of their lot.

Hitherto the great working-class of this country has borne a strong resemblance to the patient, toiling Issachar—crouching down as a strong ass between two burdens, and bowing his shoulder to bear; but, as education throws into activity the massive brain, Issachar will no longer crouch. He shakes off his submissive lethargy, his passive contentment with a meagre, squalid existence. He demands, with a growing consciousness of strength, and a growing irritation of feeling, why, to the poor man, the conditions of life are wholly different from what they are to the rich man. "The conditions of *your* life," replies public opinion (amongst the wealthy classes), "are dependent upon *wages;* and wages have been steadily rising for many years. This boon you owe to capitalists, that class of men whose abstinence has created the power of payment of your labour. What you owe society is gratitude, not rebellious discontent; and the science of Political Economy, carefully studied, will prove to you the perfect justice of our relative positions."

Now, in reference to this public opinion, it is a significant fact that John Stuart Mill, a master in the science of Political Economy, declared, in the year 1869, that the great questions involved in Socialism were destined by the inevitable tendencies of modern society to become more and more prominent, until fully discussed, and finally settled.

In Socialism the fundamental question turns upon the justice of the long-established arrangements of society in reference to property, and to the production and distribution of wealth.

In 1884 we are forced to recognize the correctness of Mr. Mill's prevision.

It is not Issachar alone who complains of injustice. We have not now to deal only with unorganized masses whose sense of grievance smoulders under the weight of dull wits, unable to formulate a clear and definite claim.

We have a body of Socialists, small in number perhaps yet, but sharp-witted and keen-sighted, who will no longer permit these all-important questions to be tyrannically disposed of, or coolly ignored. "The present arrangements of society," say these Socialists, "are not justified by the fact that they have been long established. They are wholly inadequate to general requirements or to promote the welfare and good of *all.* They act injuriously upon the happiness of every social class, and hurt the moral nature of humanity at large. By these arrangements success in life depends upon the accident of birth, and afterwards on accident of opportunity; and no degree whatever of personal good conduct can be trusted to, to raise one in the world, without the aid of fortunate accident."

Neither must we think that the race is to the swift or the battle to the strong, and that, viewed from *afar*, and called "survival of the fittest," we may hail the issues as a triumph to redeem our present sense of unfitness, shame, and sorrow. It is not so. Vices, not virtues, help one on, servility and sycophancy, the lies and tricks of trade, gambling and downright knaveries; and unless our consciousness of right and wrong can be reversed, society must enter on some different path, ere we can say that all is going well to bring about in human nature moral, intellectual, and physical "survival of the fittest."

Let me just revert here to our good friend the grocer. His little trick was palpable, and the allusion made to it told on him as a soft impeachment, calling for no sense of shame. It was quite evident to the customer that the man's own feeling justified it on the ground of self-preservation.

To put it plainly, the social pressure is so great in trade, that cheating is justifiable; because every man willing to work has the right to get his living somehow, by honest means if possible; if not—then by dishonest means! Competition rules the destiny of every individual in trade, and morals must succumb to profits, *if* the daily bread depends on profits. In former times it paid a man to keep his customers by strictly upright dealing. His field of trade was limited, and if he lost his character for honesty by shuffling and petty fraud, he soon would lose his customers. Thus the forces brought to bear upon him as a tradesman, or small distributor, tended to keep his moral nature pure. But now conditions are altered, and the forces tend the other way. The slow and constant trade of regular customers pays less well than periodic bursts of active trade, dependent on *chance* customers whose poverty or cupidity keeps them ever on the strain to nibble at and rise to the bait of *cheapness*. Annual sales, and constant advertising and puffing are the order of the day, and tricks of trade, adulteration, petty fraud, are natural, because profitable. If you detect an adept in this art, he knows he is no worse, and rather boasts he is no better than his neighbours in the trade, and should you treat him with harsh condemnation his sense of justice is aggrieved. The conscience then in trade admits of cheating up to the point of doing as others do, when necessary to maintain the business niche and keep one's own profits on a level with those of competitors.

But meanwhile *human nature suffers*. It is not possible to have two moral standards for one's daily life, to cheat in business, and be above cheating in all private matters; therefore at the present time strict honesty and downright truthfulness are rare in all transactions into which money enters.

A gentleman was travelling some little time ago from Scarborough to York in a third-class railway carriage where every seat was occupied. A young couple present were mutually engrossed at first in talk of Scarborough, the aquarium, the music at the Spa, and so forth, when suddenly the young gentleman began to inform the general company how he had got the better of, and taught "a valuable lesson" to, the money-taker at the gate of the Spa. He had presented himself for entrance, put a shilling down, and waited for the change. Sixpence was handed to him, but the shilling was not swept into the till. The shrewd young gentleman kept his eye upon it, waited until the money-taker retired for a moment, and then returned it to his pocket; and so was able with evident pride and satisfaction, in presence of his young lady friend, to relate the circumstance as though he felt it was greatly to his credit, and had every reason to anticipate that others would feel the same! Nor was he much mistaken, for, with the one exception, all his fellow-travellers received his statement with a smile of sympathy, and some with hearty words of gay approval. These people were not of a low class. Gentleness and some refinement of appearance showed that fact distinctly. The gangrene of a low morality, this tendency to trickery and petty fraud, has made its way in all directions, and shows itself with open face where one would least expect it. Some young ladies of the upper ten thousand, when skating-rinks were fashionable, were known to take advantage of their social status and their sex, and, entering with a crowd, to push their way into one of these rinks not once, but many times, without payment! These girls must have received what is considered education of the highest kind, both moral and religious; they were outwardly refined and cultured. How comes it that the sense of simple honesty and truth was not within them? Had some poor urchin in the crowd snatched a purse, they would have known it was a crime and readily denounced it; but thieving in the *lower* class is tangible enough to blunted minds!

Police and all our paraphernalia of protection point us to transgressions of the poor, whilst in the middle and upper classes similar transgressions pass unnoticed.

Upon this point of public morals the indictment of Socialism against the present system is too true. With a useless, and often unhappy, though wealthy upper class, a middle class eagerly strained and overworked, and a lower class reckless because so suffering—the moral nature of mankind has little chance to prosper, the nobler elements of humanity are crushed and trampled on. If Socialists are correct in their impression that we are travelling onwards towards a new feudality, viz. that of the great capitalists, we may anticipate that all the fierce, militant tendencies of human nature will long survive and flourish. But that we *should* continue long to travel in that direction is to *me* inconceivable.

Let us glance for a moment at the history of the past. For two centuries after the Conquest intestine war and feudal oppressions embittered the life of the British labourer. At any moment he might be called from the plough to take up arms in his master's quarrel; and if he did sow seed and see his fields ripen, the harvest of his hopes might still be cut down by the sword of the forager, or trodden by the hoof of the war-horse. He was bondsman and slave, defenceless in the hands of the lords of the soil, who, at the best, protected him in the barest necessaries of a scanty livelihood—a hut without a chimney, its furniture a great brass pot, and a bed valued at a few shillings.*

In Edward the Third's reign a change for the better came, for the plague of 1348 and perpetual war with France made men more valuable by diminishing their number.

Edward freed the bondsmen to recruit his armies, and the forced services of villeinry were gradually exchanged for service paid by wages—the wages, however, often fixed by statute. By the middle of the eighteenth century, wages, after many ups and downs, stood at the ratio of about a bushel and a half of wheat for one week's labour. By the middle of this century they had fallen to what could only purchase one bushel of wheat.† Now, why was this? Because, meanwhile, two men had lived and died whose great discoveries had given an impetus to industry beyond all previous experience, and had made Great Britain rich—I mean Arkwright and Watt. The barber had used his brains to purpose, and mechanism took the place of human fingers. Were human fingers, then, thrown out? Not so. Cheap supply created more and more demand. Machinery and human labour, side by side, were under strain and stress to meet the call of new desires. Cotton and wool and flax were woven into fabrics, and poured out of Great Britain to every quarter of the globe; capital was coined, and wealthy capitalists bid against each other for more labour still;

---

* Wade's "History of the Working Classes."
† Robert Dale Owen, "Threading my Way," p. 220.

agriculturists flocked into towns, and huge factories sprung up in all directions; population rapidly increased, and children were sucked into the vortex; health and happiness were never thought of when work was freely offered. Outwardly the British world had vastly altered; internal wars were over, and no longer was the war-horse visible in harvest-fields.

What the scene presents now resembles rather a huge hive of bees secreting honey and amassing stores for future use. Beyond her shores Great Britain has assumed a foremost place among civilized nations. Everywhere her power is felt, for the resources of her wealth seem boundless. She can thin the ranks of her adult population, and swell her army to put down the tyrant Emperor Napoleon, who is carrying war and bloodshed across the plains of Europe, and still keep at work the huge Leviathan of her own trade and commerce by the deft fingers of her little children. Summer and winter find her tiny bees—infants of seven and eight—at labour in the factories from 6 a.m. to noon. One hour for dinner is allowed, and then they toil again till eight at night! Now, were these the "good old times" of which we talk? These were, at all events, the times when England's greatness was built up, and unexampled fortunes arose, founded, alas! upon the industry of young children, sweating in factories for thirteen hours a day. Let critics of the present, who are staunch admirers of the past, read of *that* period for themselves and judge more wisely.

In the year 1815 the two Robert Owens, father and son, made a tour of inspection of all the chief factories in Great Britain. "Not in exceptional cases," writes the son, "but as a rule we found children of ten years old worked regularly fourteen hours a day, with but half an hour's interval for the mid-day meal, eaten in the factory." In the fine yarn cotton mills the "temperature usually exceeded 75 degrees;" and in all the factories the atmosphere was more or less injurious to the lungs. In some cases "greed of gain had impelled the mill-owners to still greater extremes of inhumanity, utterly disgraceful to a civilized nation." Their mills were run fifteen, and sometimes sixteen hours a day; and children were employed even under the age of eight! "In some large factories from one-fourth to one-fifth of the children were cripples or otherwise deformed. . . . Most of the overseers openly carried stout leather thongs, and frequently we saw even the youngest children severely beaten."*

* "Threading my Way," p. 102

What do we say to this picture of life now? Whom do we blame for it as we look back upon the past? Were the parents of these infants to blame, or their employers? The two reformers who inspected and exposed the evil, cast no blame on either. Robert Owen never indulged in personal abuse. He wrote to the Earl of Liverpool, "It would be clearly unjust to blame manufacturers for practices with which they have been familiar from childhood, or to suppose that they have less humanity than any other class of men." Instead of blaming, he strained every nerve to bring about an alteration in the system. He wrote and spoke and agitated for protection of children by the law, and for compulsory education; and he pointed out and demonstrated all the evils that arise from *competition* left unshackled by the law, and absolutely free to regulate itself at any cost of human life and health and happiness.

In four years' time the first point he had aimed at was secured; public opinion reached his standpoint, and the oppression of infants was put down by statute. His second point was gained in 1870.* And ever since his noble and unselfish life was lived, minds have been here and there awakening to his third point, viz. the grave and sad drawbacks attached to free competition. Robert Owen proved that if all the branches of the cotton, woollen, flax, and silk manufactures were included, the machine-saved labour in producing English textile fabrics exceeded, in the year 1816, the work which two hundred million of operatives could have turned out previous to the year 1760.† The world was richer by all this enormous producing power— a power surely sent down from Heaven, he thought, to assist man in his arduous toil, and set him free from the old curse, that in the sweat of brow alone should he eat bread. But what had actually resulted? No leisure from the toil—no freedom from the curse! Throughout the old world the new, senseless, dull machinery *competed* with the living sons of toil! A contest, Robert Owen says, goes on "between wood and iron on the one hand, and human thews and sinews on the other; a dreadful contest at which humanity shudders, and reason turns, astonished, away."‡ *His* reason grappled with the problem, Why does this rapid growth of wealth enrich the few, and leave the many in their misery? *Nay*, more than that, it presses down to deeper depths of poverty and degradation.

* The Government Bill passed for National Education.
† "Threading my Way," p. 215. ‡ *Ibid.*, p. 218.

Are there no means, he asked, by which mankind can work together for the benefit of all, and as the world grows richer, every son of toil be elevated and made happy? He turned to Political Economy to aid him in the search for a true remedy. *Its* panacea was to lower the taxes, and his common sense rejected that as futile. Himself a manufacturer, he had every opportunity of studying the question. He was in sympathy alike with masters and their men; and, after years of keen inquiry and ardent devotion to the cause, the method he accepted and believed in as the *only* remedy was a form of Socialism. Free from all personal greed of wealth, and filled with noble wishes for the happiness of man, he tried to give the world a brave example of his method, and embarked his fortune in a bold experiment, which proved a failure. It is not my purpose to enter into any description of the simple communistic settlement of "New Harmony," or to try to discover all the causes of its failure. With Socialism as a remedy I have nothing to do at present. My object is to show how vast and real are the grievances, evils, and miseries on which the *argument* for Socialism is based, and to reiterate the statement made by Mr. Mill in 1869, that the fundamental questions involved in Socialism, viz. what relates to property and to the best methods of production and distribution of wealth, have sooner or later to be thoroughly investigated.

As regards Socialism itself, Mr. Mill's opinion was, that in some distant future time communistic production might possibly be found a method well adapted to the wants and nature of mankind, but that it would require a high standard of moral and intellectual education; and the passage to that moral and intellectual state must necessarily be *slow*. Meanwhile, the sorrows of our labouring class that vexed the righteous soul of Robert Owen remain just as before. The problem is unsolved on which he spent his life and wrecked his fortune. We have *all* the evils that he deplored amongst us still. We have an overwhelming mass of fellow-creatures toiling to the utmost of their strength, tied hand and foot by poverty, and often weighted by a sense of heavy dull despair, in spite of an intervening half-century that has been full of national prosperity and energy of life; in spite of all that science does to lighten manual work; in spite of intellectual power which pushes education to the front; in spite of wide benevolence and boundless human sympathy, ready, if only it were possible, to embrace mankind at large, and give enduring happiness to all. Recourse to poor-laws has resulted in a failure quite as great as Robert Owen's remedy of Socialism, and individual efforts to ameliorate the condition of the poor have accomplished almost nothing.

In Robert Owen's day the evil was confined to wealthy Britain; now it has extended to that new world, which, with its vast stretch of rich untrodden soil, formed the sheet-anchor of our old world's hopes.

Mr. Henry George's picture of poverty dogging the footsteps of progress in America I need only refer to. Its truth has been frankly acknowledged; and Professor Goldwin Smith speaks in a similar strain. "It is a melancholy fact," he says, "that everywhere in America we are looking forward to the necessity of a public provision for the poor." Nor does *he* think that general education will mend the matter. On the contrary, he warns the public that "there will in time be an educated proletariat of a very miserable and perhaps dangerous kind;" for "nothing can be more wretched and explosive than destitution, with the social humiliation which attends it, in men whose sensibilities have been quickened, and whose ambition has been aroused."

This great problem of poverty in the midst of wealth—and observe, it is a poverty which mars the happiness of the rich, as well as the poor—demands solution. It is forcing itself upon public attention, both in the old world and the new. There is no escape from it; sooner or later this problem must be thoroughly examined by educated reason, grappled with, and solved by the exercise of a cold calculation.

Our civilization moves rapidly; but in what direction it will continuously move is by no means certain. We have reached a stage of civilization in which are to be found "elaborate arts, abstruse knowledge, complex institutions; and these are results of gradual development from an earlier, simpler, and ruder state of life," but the stage is not one of satisfactory life or general happiness.

Nevertheless, to bemoan the past is to betray an illogical mind, or an uncomprehensive state of sentiment or feeling. You or I, my reader, may be moving downwards in the social scale through press of circumstance. It may be true for us to say, that our grandparents had such good old times of ease and comfort, as make our days of mortifying life, of toil and worry and anxiety most painful by the contrast; but these days were full of misery for many. Every age has had its favoured few, surrounded by its struggling, suffering many. No age can be worthy of the name of good, until the happiness of all human beings becoming the aim and object of each, through combination in well-sustained and well-directed effort, the goal of a state of *universal well-being*, is steadily approached.

In reference to this future golden age, one vital and certain point is this—a high moral standard of adult conduct in industrial and social life, with a high standard of education for the young, form its all-essential, absolutely necessary conditions. Therefore, the smallest advance made by us in the elevation of general and individual morals, is a sure step upon the path that leads to that distant goal.

CHAPTER IV.

DEVELOPMENT IN MORALS.

"There is no private life which has not been determined by a wider public life, from the time when the primeval milkmaid had to wander with the wanderings of her clan, because the cow she milked was one of a herd which had made the pastures bare."—GEORGE ELIOT.

I MUST remind my reader that what we all desire and vaguely seek is happiness, although many of us are unconscious of the fact; and to others who at least suspect it in themselves, it appears meanly selfish; they therefore try to disguise and cover up the truth. But such is human nature; it craves the satisfaction of its whole being, and if the whole being is noble and dignified, its satisfaction cannot be other than an adequate aim.

Were conditions necessarily such that each individual had to live an isolated solitary life, and satisfaction meant the free and spontaneous exercise of all the lower faculties of his animal nature only—if to sleep, to eat, to laugh, comprised man's whole happiness—then we who stand upon a wholly different platform, would judge by heart and mind that the game was not worth the candle.

But the actual fact is far otherwise. Man is a social being, demanding for his happiness an infinite variety of tender human bonds. He is linked to his fellow-creatures round and round, not by outward iron chains forged on the anvil of hard necessity alone, but by silken cords of inward sympathy and feeling. If in his keen desire for happiness he overlooks, or selfishly forgets these cords, what happens? Inevitably *this*— jarring and inward discord arise; the man has done violence to his own nature, and has missed the path that leads to satisfaction. He is torn, lacerated, harassed, and in the indulgence of his ignorant selfishness, he finds himself most miserable.

We need not fear, then, to face the simple truth, or feel ashamed concerning it. The fundamental facts of life are not degrading; if we think them so, it is by some misapprehension of their nature.

To seek for happiness is a legitimate object of humanity, and if we understand humanity aright we shall perceive that the pursuit, the *conscious* effort to achieve that end, entails a discipline, a self-control, an ordering and a regulating of our life which in itself ennobles. It is the universal, blind, unconscious groping after happiness that we have cause to fear, or what is worse—an aimless, purposeless, existence, which in its blank indifference leads to unscrupulous action on the part of the individual, who thus encroaches upon the rights, and mars the happiness of others.

Given, then, a vast concourse of social units, that is, of individuals who cannot be happy if they live to themselves alone, one great problem to be solved is, What are the outward or objective favourable conditions? In other words, what is the form of integral life that will best conduce to happiness?

But now there is another point made clear to human reason, not by inference alone, but by evidence. It is this. Man's goodness as well as happiness is conditioned by his life. That is, the outward form or social system of human life, not only in a great degree controls the destiny of each individual, but it acts upon and moulds his nature. The existence of the gentle uncultured tribes of savages which have been described, presents to the world an objective proof of this fact, whilst subjectively we all may find some proof within the range of individual experience.

These tribes have quite outrun the British race, and reached a goal of goodness and happiness beyond our own achievement. I do not say *that* goal would suit the British race; for we are not simple—we are complex. We are not uncultured—we have a grand inheritance of intellectual tastes, and intellectual work accomplished by our forefathers; and it is not by returning to the simple savage nature (as Rousseau wildly dreamed), but by pursuing our *own* path, burdened by, or at least carrying with us, all that is worthy of respect and reverence from an almost hoary civilization, that we must fight our way onwards to a loftier goal of goodness and happiness.

There cannot be drawn a full and fair comparison between

ourselves and Bodo, Dhimal, Santál; for outward life with them has lain within narrow limits, and the adjustment of their conduct and nature to these limits was as the opening of a simple flower to meet the sunshine which could not fail to give it healthful, pleasurable existence. Our conditions from the first were very different. With a much wider range, obstructions to a peaceful life were continually encountered, and in the effort to overcome obstruction and the strife engendered, there came to be developed in our race the fierce and truculent side of human nature, a warlike tone and spirit foreign to the Bodo, Dhimal, and Santál.

Our island in the sea had little of the wild luxuriant productiveness of nature's gardens, where sunshine pours to make it always afternoon, and allures to dreamy days of half-inert existence; and just as little was the development of our branch of the great human tree like the opening of a simple flower. All along the line of our past history, hard and continuous struggles have been the very terms of life; and up to the beginning of the present century, the type that came through these struggles and was able most triumphantly to perpetuate its kind, was the hardy, the brave, the bold, the aggressive, and *not* the type in which gentleness and goodness were the prominent qualities. The latter type was worsted in the fight, and, as a rule, succumbed; never dying out entirely, but kept down in numbers, whilst the other preponderated in number, and was dominant in rule.

Meanwhile, the action of climate upon the vigorous active type must not be overlooked. It had a twofold effect in shaping the conditions that surround us now. Remembering our fundamental fact of human nature—that every individual without exception is groping after personal happiness, it cannot surprise us that dull gray skies and frequent outward gloom turned inwards the joy-seeking tendency. *Inwards*, both as regards the family group and as regards the individual. Home brightness made amends for outward gloom, and nowhere in the world has home life been made more rich and full, or has human nature developed greater faithfulness and depth of tenderness in all domestic ties.

In the individual the inward tendency has led to meditation and reflection. Intellectual joys have been pursued when sensuous enjoyments could not be obtained; and steadily has great activity of brain and clear intelligence increased, until the British have ceased to be creatures of pure animal propensities and savage instincts only; and, although these instincts and propensities remain, the race has won, besides, an aptitude for sympathetic happiness and intellectual pleasures that has altered—in a word—the *quality* of life.

In possession, then, of this wider nature, it has passed from the barbaric, warlike stage of civilization into the industrial, more peaceful stage; and here its bright intelligence has found a new sphere of active enterprise. The effort to make life more easy and comfortable, has led to the study of the phenomena of nature. Forces hitherto unknown, have been found capable of working in man's service, and science has taken its place in human history, and proved a powerful factor in the development of the new industrial life. Nevertheless (as I have already pointed out), progress has been thwarted, and happiness and goodness marred by survival of the qualities of warlike life in man, qualities no longer absolutely necessary for national preservation, and which have led to inconsistent public action and frequent temporary relapse.

In foreign politics the fierce aggressive spirit of the race has shown itself at intervals. We have trampled on the rights of other nations, have interfered in questions not affecting our own country's interests, have made ourselves the terror of races less advanced than we are, and often seized their land, to find it no advantage, but a burden and embarrassment.

In trade the warlike spirit, and savage predatory instincts have been even more conspicuous; but *here* a new condition intervenes, a condition which, arising naturally and found in many respects beneficial, is now clung to as a necessary feature of our social system. I mean *free competition*. Under pressure of a rapidly increasing population made possible by growth of capital, competition has fostered cruelty, rapacity, and greed. The subjection of outward foes has not caused the struggle for existence within the nation to diminish. On the contrary, it was never more intense, nor in many individual cases more hopeless, than at the present day.

National existence has been fought for and won, and during that struggle the two great classes of the nation clung together, and formed a solid though not homogeneous mass. The rich protected, guided, ruled the poor. The poor laboured for and supported the rich. The advantage was mutual, and the aims were alike. But now the nation is composed of heterogeneous elements. Not two classes but many live side by side, and rival one another in the struggle for effective life. The peaceful bond of wide protection on the one side, and helpless slavish dependence on the other, almost nowhere in the present day exists. Parental love, as we have seen, gave way under the strain of competition, and even infants were used, and frightfully abused, in purely industrial pursuits. It appears almost a retribution, that now, in factory districts, children in their teens, barely touching manhood and womanhood, no sooner become self-supporting, than they throw off parental control, and frequently establish themselves outside the home of infancy, slighting and discarding its protection.

The sexes, we might think, would always retain the peaceful bond of strong protection and of weak dependence; for surely *nature* has decreed that man shall be the bread-winner, the supporter, whilst woman simply clings to his strong arm, and does her part in sweetening and embellishing his life.

But theories go down before stern facts. Bread-winners of both sexes are equally numerous. In every field of remunerative labour into which woman has been able to force her way she is struggling for her own existence and competing with man; whilst the old theory of man's protection, which lingers as a mere survival of the past, adds to her difficulties in many ways, and is as a stumbling-block laid across her arduous path.

The two great classes of producers and capitalists are held together by a link of mutual dependence, *absolute* in its nature. Without capital no production can take place; without labour no capital can be employed; yet that absolute link does not suffice to maintain peace in these relations. The history of British manufacture is a history of perpetual conflict. No doubt amongst the ranks of employers of labour many noble and disinterested men have acted justly and uprightly, and intelligent labourers have here and there shown their ability to recognize the fact. But as a rule these two classes, compelled to aid each other, feel mutual antagonism, a mutual suspicion, jealousy, and hatred. Employers have aimed at keeping wages low, whilst amassing enormous fortunes for themselves—nothing but compulsion has made masters' profits fall and workmen's wages rise.

Workmen have had to fight for every inch of vantage ground, to show their growing strength in Chartist riots, to combine in trades unions, to turn out on strike, and suffer frightful hardships in the cause of forcing wages upwards; and meanwhile the belligerent attitude of the class has told enormously upon the sensitive human nature of the race. The children of that class instinctively rebel against and suspect the motives of their masters. Resistance to authority is with the rising generation a valid *esprit de corps*, and dominated by it alone, they will resist all efforts at reconciliation, and stubbornly hold out beyond the bounds of reason and common sense. Throughout the struggle neither capitalists nor labourers have taken a wide and rational view of the interests of the community (in which their *own* are necessarily involved), or been able so to adjust their mutual claims, as that both might equally expand and diminish with trade fluctuations, and capitalist and labourer share alike in mercantile prosperity and mercantile reverses. We may impute this inability, to deficiency of intelligence or to deficiency in moral sentiment—the sentiment of justice and equality of rights. But there the fact remains, and *now* the thoughtful and unbiassed onlooker for the most part has lost faith in reformation on the old lines. He looks for change of system and believes that in some form of co-operation alone shall we probably find the effectual way of reconciling the conflicting claims of capital and labour, and also the antidote to our great moral backslidings which to our shame and loss are fast becoming national.

Another great section of our social community where change of system is now imminent, and indeed has begun, is that composed of owners of, and labourers on the land.

Free competition among the wealthy, who desire land for other purposes than production, combined with rapid increase of our population, have given an undue value to landed property, and forced up rents beyond the point which admits of a fair remuneration to the farmer.

In Ireland the mass of the people have for centuries laboured on the soil, for no other industrial employment in their own country, offered them the means of subsistence. Landlords as a rule have screwed out of them a rent that left no margin of profit to the labourer, to be spent in making life enjoyable. When bad years occurred landlords have not remitted payment of rents, but held their tenants bound for past arrears; and these unfavourable conditions have so acted upon character, as to undermine the motive to personal exertion, and extinguish the spirit of independence.

The Irish are a pleasure-loving race. What wonder is it, that during long centuries of hopeless toil, they, shut up to seeking happiness in animal propensities, have imprudently, nay recklessly, populated at all times up to starvation point, until the climax has been reached, and, as a consequence of pressure from above and below, a period of rebellion is now in progress which is bound to initiate decisive change.

A social revolution is taking place, begun in Ireland, but that will not end there—a revolution of profound, far-reaching kind, and in its course hostility and keen antagonism show themselves in agricultural as in all mercantile relations.

Again, amongst distributors as amongst manufacturers not peaceful association, but fierce competition reigns. This field of active life is specially over-crowded, and the general interests imperatively require that the numbers in it should be reduced.

It must always be for the benefit of a community that the price of commodities should be as low as possible—that such articles as are universally desired should reach the purchaser or consumer as cheaply as accords with a fair remuneration to producers and capitalist. The price of most articles, however, embraces much more than this; it includes various profits to distributors, wholesale and retail. One can clearly see how that part of the price may be saved to the consumer. Facilities of movement and of communication throughout the country are constantly increasing, and these facilities bring nearer together the consumer and producer, and make many of the links in the chain between them no longer absolutely necessary. Already in the country districts, ladies, by means of telegraph or letter, purchase wearing apparel from the large shops in London or from co-operative stores in the nearest towns; and small shopkeepers of their own immediate neighbourhood have lost their custom. This is a movement certain to grow, and although entailing suffering on many individuals, it is desirable that it should grow; for not only do the general interests require that commodities should be cheapened, but another tendency exists and must be met. I mean the tendency of wages to rise.

The actual producers in a flourishing community must be adequately recompensed for labour. It is only common justice that the growing wealth of a country should reach the workers in the form of increased value of labour; and whilst in this country it is possible enough that interest on capital has reached a maximum, it is quite certain that wages have not done so.

In order, then, to cheapen commodities, distribution must be simplified, and all profits to unnecessary intermediaries be cut off. This implies the crushing out of the means of livelihood of thousands of human beings, who, in the struggle for existence, are already feeling the action of this social force, viz. the new adjustment which slowly but surely will have its course.

Men in the position of the grocer, of whom I spoke in my last chapter, are incapable of judging calmly and justly of our social system. They see the wheel of Juggernaut approaching to annihilate them, and it is no wonder that the very love of wife and children, the longing for domestic happiness and comfort, acts like an acid on their relations to the outer world, corroding and consuming all brotherly love in trade association, all healthy public spirit, and turning gentleness and goodness into hostility and spite. That grocer feels hatred towards lawyers and clergymen, a hatred founded upon a grievance purely imaginary; but the real pain and bitterness of spirit that the man endures give pathos to this error of his judgment. It was referred to because it illustrates the antagonisms rankling in the bosoms of men of his class, which are caused by social pressure, and aggravated by unlimited competition.

Another class in which, with infinitely less apparent excuse, the moral nature is degenerating, is that of stockbrokers and speculators. Cotton-corners and commercial rings are developments of our civilization that give us cause to blush. Many speculators are men of wealth and social position. The pressure of a hard necessity, the fear of wanting daily bread is *not their* stimulus to keen pursuit of plunder and selfish public action. A love of sheer excitement, a gambling spirit, possesses them, and is so strong that it will carry them to all extremes of reckless speculation, that may result in ruin to thousands, whilst it necessarily presses upon and imperils all traders who, preferring a safe and strictly honourable course, refrain from running risk of ruin by building upon pure expectation.

I have said a gambling spirit is the cause of these commercial rings, but that does not explain the whole of the phenomena; for why should men of some humanity, some social feeling, some refinement and culture, be led by pure excitement into action that is brutal, cruel, unjust, utterly regardless of the interests of others? There must be something lacking in the lives of men when such a state of things occurs; when, notwithstanding all the civilized attainments of man, the latent savage instincts come to the front and cause relapse into barbarous conduct. If human nature were satisfied with healthful, pure enjoyment, and time occupied in ennobling pursuits, this passion for wild venture in the marts of British commerce would be recognized for what it truly *is*—no better than the brutal instincts of a wholly barbarous tribe, or the zest that Spaniards feel in witnessing a bull-fight.

Once more, then, we are brought to see that want of happiness is sapping the foundations of our national morality. The British are not happy; therefore they are not good. And as an approximate means for moral regeneration, we, as a nation, must exert ourselves to ascertain the *causes* of our social misery and discontent. So long as these causes remain hidden, we may anticipate that all the barbarous qualities of man will survive and flourish, while the gentler human virtues will languish and decline.

But national consciousness is everywhere growing. *Therein* lies our ground of hope in view of the future of our race. No single class throughout the social system is so sunk in degradation as to be ignorant of its own misery, or to be slavish in the endurance of it. A slave will bend before the scourge and turn and fawn upon his master; not so the British subject. On all sides murmurings and loud complaints are heard, and nowhere is there dumb and abject subjection. Each class is critical of the class above its own, and intellectual life in *some* sort is stirring everywhere.

Hatred of rulers, and an ignorant belief that *no-rule* would make the misery less—jealousy of the rich, and childish prejudice against some special class—dangerous attempts to right what is conceived to be wrong by arbitrary measures—these are all symptoms of a wide awakening to consciousness of misery; and however inconvenient in themselves, they are nevertheless healthful, hopeful signs, in view of what our future may bring forth.

Should my reader here ask: "Are you weak enough to dream that such a world as ours, a world of social life so wholly out of joint, can be speedily remodelled?" I reply, that I have no such dream. I am convinced of the truth of the words spoken by the wise John Stuart Mill—"there is *no one* abuse or injustice now prevailing, by merely abolishing which the human race would pass out of suffering into happiness." No! there is none. The world is not a stage on which a transformation scene can be enacted. The work must be a slow, gradual, measured process, dependent upon the growth of knowledge which will guide to rational and radical methods of reform.

The public consciousness is everywhere awake to misery.

It will in time awake to the causes of that misery and find that many of these causes can be avoided, if not overcome. Individual members of society, learning the true nobility and dignity of seeking happiness, will consciously pursue that path, and in *this alone* will lie a safeguard from many errors of the present. Personal conduct will adapt itself to bring about the general welfare, when this knowledge is widespread, that individual happiness greatly *depends upon* general welfare.

Lastly, and let this be noted, small alterations in individual conduct, which, taken separately, appear insignificant, bear with them great results when widened into universal action.

In my second chapter I took a particular fact in the training of children and enlarged upon it, in order to illustrate this point. The evolution view of human nature and social life is modern, and novel to the majority of minds. It is not yet established in the general consciousness. When it is, the utmost pains will assuredly be taken to avoid the direct fostering of such instincts as are adverse to the nobler life we aim at and desire.

At present, parents in their ignorance reverse this principle. They thrust into children's hands the crude and barbarous survivals of our savage epoch, and so unconsciously give to the young, false views of life and a distorted preparation for life's duties. It may appear a trifling error to give a child "Jack and the Bean-stalk," or "Ali Baba;" but if a tender mother will only make herself familiar with the action of such tales upon the infant mind, and see how all that *she* holds worthless is exalted there, I feel convinced she will pronounce the consequences not likely to be trifling. An idle, disobedient boy is certain to encounter charming adventures! Lords and ladies are the people to admire, and all the gaudy glitter of an Eastern court, the empty pomps and vanities of life, are thought of as delightful! To plunder, plot, and scheme, bear no disgrace, if by such action wealth can be secured; and even murder may lead up to rich reward in brillant marriage!

I need not recapitulate, but let me call upon all parents to be critical of what their children read. This is a grave matter, for on it depends the bent of the young nature. Mental pictures in an infant's mind form one of the great conditions that day by day and hour by hour build up his character, and lay the sure foundations of his adult life; and I must repeat, that when throughout society, individuals everywhere are acting on this knowledge, the small and apparently insignificant alterations in conduct that it must entail, will tell enormously on human nature in the aggregate.

The nineteenth century stands, I believe, at the threshold of a new form of social life, and on the eve of a fresh departure.

Hitherto the race has stumbled forward, fighting blindly, struggling manfully for life. Now the epoch before us is one of consciousness—the open-eyed, dignified manhood of our race. Power, possession, both are ours; we only pause for knowledge, which will enable us to apply them to the good and happiness of all. The struggle that goes on within the nation is unworthy of its manhood. I do not mean that *work* will ever cease—far from it. Consciousness of what true happiness consists in, in the case of vigorous and social human beings, must bring to light this fact: that "the finest pleasures of life are to be found in the world of action," and voluntary work will therefore never lose its charm. But compulsion in our work—the bond of mastership and servitude—the fierce inhuman struggle to obtain a share of work, when aptitude and willingness exist—these will certainly subside, and classes now at strife will find new social joys in peaceful and industrial combination.

To make forecast of what we may anticipate, however, is not my present purpose. My aim is purely practical. By the study of evolution I think it possible to guide the thoughtful and earnest in our midst to personal conduct which will tend to bring about a happier social state.

Our greatest miseries are Poverty, Disease, War, Drunkenness, and Premature Death. If I can point to any causes of these miseries which may be modified and altered by our individual action and example, and if my readers are convinced by reason and inclined by moral sentiment to act upon the evolution lines, my object will have been attained.

Especially on *Education* is there much to say; for, whilst we seek to work out steadily the welfare of our race, and bring about a different social state, we have no less to train our children into fitness for the happier and purer life, and free them from all gross survivals of our barbarous ancestry.

CHAPTER V.

THE POOR.

"What do we live for, if it is not to make life less difficult to each other?"—GEORGE ELIOT.

"Becky Sharp's acute remark that it is not difficult to be virtuous on ten thousand a year has its application to nations; and it is futile to expect a hungry and squalid population to be anything but violent and gross."—HUXLEY.

IN the city of Glasgow, where cotton-factories, iron-works, and other industries abound, it sometimes happens that masters give an annual excursion party to their operatives, and pay for their transit *en masse* to Rothesay, Innellan, Dunoon, or some other of the numerous watering-places on the noble river Clyde.

Tourists on their way on a summer morning to feast their eyes on Nature in all her solitary grandeur in the deep Highland glens, will travel down the river in a steamboat crowded with human beings, and gaze with interest on the scene. A huge excursion party crammed into the steerage, like sheep packed closely in a pen—the men respectable, the women for the most part bright with vulgar finery, the children not over-clean, but always burdened with the full complement of articles of wearing apparel customary in genteel society—not a neckerchief, or glove, or feather in the hat awanting—and all expectant in a high degree of some hours of great enjoyment.

The return to Glasgow in the evening is a very different affair. The scene is often hideous and degrading even to look upon, and this, perhaps, when cloudless weather has afforded no palpable excuse for drunkenness, or any artificial aids to stimulate enjoyment.

Some years ago, two girls of the middle class were spending

the late part of the summer at a small watering-place on the Clyde. The season was wet, and not propitious for excursions of any kind. yet week after week these factory excursionists were landed at the place in droves of hundreds at a time, to spend the day, as best they could, on muddy roads and under dropping skies.

It made the girls miserable. They could not see wisdom or kindness in the action of the masters, and they plied all business men of their acquaintance with the questions, "Why do not masters wait till settled weather gives some prospect of a really pleasurable holiday?" "Why are operatives not released in groups of six or ten at a time, rather than all at once? for surely in that case all could not be disappointed of good weather on their holiday, and those that were unfortunate enough to have wet weather, would not be too numerous to be offered shelter in the houses of natives of the coast towns."

The replies to these questions seemed to them, in their youthful ardour of benevolence and ignorance of all the difficulties of manufacture, miserably inadequate.

It was said that weeks before the holiday the master had to charter the steerage of the steamer, and that to change the day would entail great loss on him; that the labourers would not care so much for a mere day of leisure spent with wife and children—they preferred these huge excursions; that the custom had grown up naturally; it was the cheapest way for employers, the happiest for employed, and so on.

At last, one morning when the barometer was low, the ground already soaked, the clouds very threatening, there landed at that place two large excursion parties—hundreds of men and women, with young children, even babes in mothers' arms. They spread themselves in all directions, climbed the wet hill, formed crowds upon the sea-road, laughing, joking, trying to be merry. It reminded our young friends of Dickens's Mark Tapley in his brave efforts to appear jolly amidst the horrible surroundings of the malarious swamp at Eden.

It was eleven a.m., and until six p.m. they knew they must remain just where they were, without a chance of shelter if the clouds broke overhead. In half an hour the flood-gates, alas! were opened. The rain came down in torrents, and so continued pitilessly all day long.

Now, in that place, not nearly so large a town as Rothesay or Dunoon, there was no public hall or building of any description capable of sheltering these people except the churches, and these, unlike the churches in most foreign lands, stood with closed doors and gates securely barred. The only hospitable open door was that of a large hotel, and there the drinking-bar was quickly crammed with men, who, beginning with a moderate glass to justify their entrance, soon ended in excess—a low and despicable debauch.

The little row of shops afforded here and there a chair to some poor mother burdened with a babe in her arms, perhaps a child of three or four upon her lap as well. In one shop alcohol was sold, and here young men were treating (as they called it) young women to this poison, to keep out the wet, but thereby implanting, it may be, seeds that in their future homes would develop into fruits of misery and wretchedness.

The sight was, to the eyes of our young friends, so odious and distressing that it stung them into action. They went to ministers and elders of the churches, and entreated that the doors of these should be thrown open to give shelter to the miserable excursionists, who by that time resembled half-drowned rats. A hundred objections were made—it was Saturday, and there would not be time enough to clean again and make the buildings fit for Christian congregations on the Sabbath; Bibles and hymn-books might be stolen; and, besides, a step so grave and unusual could not be taken without the full consent of the kirk session. These reasons for rejecting the proposal proved decisive in two cases, but at last the third church was opened to receive the dripping, houseless wanderers and give them shelter for two hours.

The sad experience of that day made an impression on the girls never to be forgotten. For the first time they had come into close contact with the poor, had felt some of the difficulties and sorrows of their lot, and the bitter disappointments they are too often exposed to.

As they watched them embark at six o'clock—reeling husbands, screaming children, angry mothers, excited, half-tipsy girls in draggled, tawdry finery—the sense of deep compassion overcame disgust, and as they turned away they gently said to one another, "Poor things! poor things! they are not nice, they are not good; but then, it is *because* they are not happy."

This was a youthful judgment, and it is its converse that mature minds, for *the most part*, hold; nor is this fact surprising, for disappointment is so frequently the *one* result of earnest effort to ameliorate the condition of the poor that patience gets exhausted, and robust philanthropy, pronouncing judgment, "They are not happy *because* they are not good," retires, and leaves an open field to those religious workers who put morality upon the Christian basis and are quite content with small ameliorations, since they think it a Divine decree that the poor should be with us always.

It seems to me that, just as in a stereoscope we place two pictures of one object in order that we may obtain a perfect picture standing out in full relief, so we should act in judging of the poor. We must consider the two aspects of our subject, and form no judgment till in imagination we can combine the two and see the object in its true proportions. We must come near to them, get into sympathy with them, understand aright their nature and their true desires; then from a distance study their surroundings, and see how, within the lines of social justice and the general well-being, adjustments may be made that will conduce alike to elevate and make them happier.

In reply to observations that the poor should have more pleasure and less work, that toil with them is too monotonous, life much too unvaried, and so on, I have often heard it argued that if changes in that direction are to be carried out, we must prepare ourselves for greater degradation, drunkenness, and vice amongst them; it is only constant occupation that keeps them out of mischief; they are not fit for leisure. *This* and much more is said, and then excursion holidays are pointed to as proof that idleness invariably causes self-abuse, and that for *their own* sakes, as well as for that of the community at large, things should remain just as they are, and men are cautioned to beware of interference.

Now, the nature of this argument is illogical and the proof irrelevant. I freely grant that periodic holidays are often wholly misspent, and prove injurious to the physical and moral well-being of the pleasure-seekers. I think they foster drunkenness, extravagance, and reckless licence, and *no* huge excursion such as I have described can possibly take place without some positive evils resulting. But then, we treat our operatives very much as cattle. We turn them out to grass in crowds, and at wide intervals; we send them here and there in ill-selected droves; and so long as this is done I cannot see that we have any right to judge them by *human* standards or to expect from them the virtues of a civilized race. Besides, have I not shown how, on the day in question, every influence was adverse to individual comfort and contentment, to kindly social feeling and a high moral state? By the law of parsimony, therefore, we are logically bound to recognize these antecedents as quite enough to explain their ill-conduct, at least on *that* occasion, which leaves us without basis for the assumption of an ineradicable and innate human depravity.

The truth is this: capacity for recreation must be cultivated, like all other human powers, and for its cultivation frequent leisure is required; but after months of monotonous toil, when workers are set free for twenty-four hours, joy will not come at sudden call, and in their efforts to obtain it, drinking and strong excitement, ending in collapse, are all they know of pleasure. It is a *natural* result—a strong rebound from tension of the string that held the human being upright at his wheel too long. And when we view the matter calmly we perceive that other methods than those we use at present are required to accomplish the great end of giving, in that section of our social body, pleasures that will not injure and corrupt.

But here we must take up a question hitherto untouched. We must define happiness, and see what its *essential* conditions are. An individual, to be happy, must, in the first place, be free from pain and positive discomfort; in the second place, he must have free play to all his instincts, and free exercise of all his powers, in such degree as will not injure him as a whole. Let us apply this to ourselves, whilst bearing in mind the history of our race which I have tried to sketch, the phases of its past development, and the impress these phases have left upon its nature.

We have seen that whilst we are social beings on the whole, and social life must be maintained, anti-social qualities abound. No single individual is exempt from these; therefore repression is absolutely necessary, and every one must bear the yoke and feel the curb, in just so far as his anti-social nature tends to the injury of others. In short, the perfect man nowhere exists, and perfect happiness is impossible. But the practical problem for us is how to give to *every* individual, imperfect as he is, *all* the happiness *possible*, although we must repress and counteract in him the vicious tendencies that cause unhappiness around.

In those excursion parties bodily discomfort largely mingles, and the anti-social qualities assert themselves; the conditions necessary to happiness are not bestowed, and instincts that are low and barbarous get exercised. The system is an error in the present state of human nature, and with an uncertain climate such as ours.

Let us turn from holidays and look at daily life amongst our working classes. Bodily discomfort and frequent mental anxiety are in a great measure the lot of the married women. A few days ago I heard a health lecture delivered to some hundreds of these women by a well-known doctor. On the subject of alcohol, he said—it was the grave opinion of the best authorities, that drinking among women is greatly on the increase; that amongst men the causes of drinking are various—exposure to inclement weather, conviviality, social feeling—and although he did not tender these as any excuse for men, he wished to point out that the case of women is entirely different. They fall victims to the vice from two causes mainly; either they inherit the predisposition, and the number of that class is small compared to the other, or they begin to drink in order to relieve pain. To counteract the feeling of weakness they resort to it again and again, until it becomes with them a fatal habit. He exhorted them to pay far more attention to their general habits, to be more *selfish*, as they would think it, attending to their own food and comforts, and not always considering first their husbands and children. He told them to take rest, and said, he firmly believed that poor women often endure excessive pain when patiently pursuing their daily toil; he advised them to take food instead of alcohol, and to seek variety in their food.

All this was admirable, but I thought he might have said much more. In innumerable cases these women fall into bad health from their children being born too rapidly, and their number being too great. It is simply impossible for poor mothers to rest sufficiently long after each confinement. Internal discomfort arises, and by the time they have three or four children, the daily life has become a drudgery, a constant strain upon the nerve force. They strive to keep up to the mark by dram-drinking, or tea-drinking. The one induces loss of self-control and self-respect, and ends in utter degradation; the other induces horrible dyspepsia. Mental depression sets in, and the constitution gradually gives way.

Now, mark the consequences. Unhappiness of wife and mother means unhappiness all round; for she is the very pivot on which turns the wheel of life in the home circle. Her husband becomes alienated, the children are crossed and thwarted, if not neglected, the anti-social instincts are not repressed, the gentle and loving qualities are not fostered and strengthened, the evil spreads out widely, and in the end society at large is sure to suffer by the ruin of some member or members of that family. The prison or workhouse will receive them at the last.

This is no exaggerated picture. I speak from personal knowledge and from the knowledge of trustworthy friends. Some of these tell me that in a factory district in the centre of England the usual number of a workman's family is eight, in some cases ten, twelve, or fifteen, and in a few even twenty; and mothers have been heard to say: "I've 'ad my fifteen or my twenty on 'em, but thank 'eaven the churchyard 'as stood my friend!" This was not spoken by hardened women, incapable of mother-love, but by gentle, patient drudges, who had felt the burden of their number incompatible with their social position, and with simplicity and frankness spoke out the honest truth. It shows us that a *large* family, in that class at least, not only strains the physical endurance of the mother, but also it unduly strains her emotional nature, her power of tenderness, and of parental love.

This is *important* in all its bearings upon human nature. Knowing as we do the benign influence of sympathy and affection upon children, we must perceive how desirable it is for social well-being, that philoprogenitiveness or parental desire should be the prime motive to continue from generation to generation the human race. But in those cases where children are born long after parental desire is exhausted, the *conditions* are unfavourable to the child, both before birth and during childhood, when its existence is felt to be a burden and not a blessing.

Mr. Jenkins, in his touching little satire, "Ginx's Baby," has put a father's state of mind broadly before us when a thirteenth child is in question; but he contrasts with it the inexhaustible (as *he* thinks) mother-love! Ginx "frankly gave her" (his wife) "notice that, as his utmost efforts could scarcely maintain their existing family, if she ventured to present him with any more, either single or twins, . . . he would most assuredly drown him." This is a delicious touch of irony, for it is so wholly a man's view of the position—hers the blame! his the burden! Later, when the arrival of number thirteen is imminent, the poor mother being unable longer to hide the impending event, Ginx "fixed his determination by much thought and a little extra drinking. He argued thus: 'He wouldn't go on the parish. He couldn't keep another youngster to save his life. He had never taken charity and

never would. There was nothink to do with it but drown it.'"
Another effect of an over-crowded home upon the father is also well told. Ginx "acted honestly, within the limits of his knowledge and means, for the good of his family. How narrow were those limits ! Every week he threw into the lap of Mrs. Ginx the eighteen or twenty shillings which his strength and temperance enabled him continuously to earn, less sixpence, reserved for the public-house, whither he retreated on Sundays after the family dinner. A dozen children overrunning the space in his rooms was then a strain beyond the endurance of Ginx. So he turned out himself to talk with the humble spirits of the 'Dragon,' or listen sleepily while alehouse demagogues prescribed remedies for State abuses."

We have no need, however, to take a fancy sketch of life amongst the poor. If we do not visit them ourselves, we are all at least familiar with Miss Octavia Hill's labours, and her simple, truthful account of *facts*. "Our lives in London," she says, "are over-crowded and over-strained; we all need space —we all want quiet;" and again, "a friend at the East End has said, 'the winter does not try us half so much as the summer, for in the summer the people drink more; they live more in public, and there is more vice.'" They are driven, poor creatures, out of their hot, close, crowded rooms, into lanes, courts, or narrow streets, where good and bad must herd together, and where the very children cannot be protected from sights that must necessarily corrupt the moral nature. In short, the street affords an admirable school for developing inherited brutality and every anti-social instinct. Let us realize this fact. The summer, which poets glorify and celebrate—

"Child of the Sun
From brightening fields of ether fair disclosed,
Refulgent Summer;"

which brings refreshment in some shape to all other classes, which is the very hey-day of enjoyment to the children of the upper class, that season is for the poor of London a time of dread, of fresh discomfort and misery, worse in some respects than the cruel biting winter—a period of great temptations and of sinking lower in the moral scale. In this temperate zone, where none of *us* at least have over-much of summer, how infinitely *sad* to know of human beings who feel that, however keen may blow the wintry wind, it is in their lot not so unkind as lovely summer! And how are we able to anticipate any great *moral regeneration,* so long as homes are over-crowded and home-life fails to give bodily comfort and enjoyment?

Another evil of the over-crowded homes is the quick withdrawal of youths and maidens from the domestic circle. At sixteen, or even earlier, many lads and girls go into lodgings, and throw off both the restraint of home life and its useful responsibilities. Fathers and mothers can scarcely claim to be aggrieved by such action, for the young people are self-supporting, and if during childhood they have seen that parents thought them a burden, if their young emotional natures were not fed to the full by mother-love, they are not likely to be held to home by any strong tender tie, and their sense of justice does not impel them to carry babies, scrub floors, or stand at the wash-tub for mother in their leisure hours, when pleasure lures them to the theatre or some enticing place of recreation. To themselves, however, and especially to girls, the loss is great. A bread-winning occupation as a rule gives *no* training for the duties of a wife and mother, and many a girl marries, as ignorant as a babe of everything she ought to know, and on the knowing of which her own and her husband's happiness largely depend. An honest decent workman once told me, he had married two girls out of a gutta-percha factory, and that the only thing these girls could do (to be called domestic work) was make a crochet collar! They could not sew, they could not cook, they could not light a fire! But said he, "before my second wife came to me, I took good care to engage a tidy, active, neighbour housewife, to teach her everything." Too frequently, however, it is not with respectable husbands that these poor girls who early leave their homes find refuge. They fall easy victims to those who traffic in youth and innocence, and are dragged into a life which their forsaken mother grieves over as a deep degradation, and which fills to the brim the cup of unhappiness of her married life.

I do not say that crowded homes is the only cause of these evils. I have already pointed out that the young of this generation have an instinctive, sturdy spirit of independence, and are by nature inclined to rebel against all authority. A crushed, over-burdened mother is certain to find it difficult to control her offspring. But what I argue is, that the counteracting force of tender love and cherishing, the influence of a bright and happy home, are doubly necessary in the present day, and that it is *impossible* to secure these blessings, so long as families are not limited to a number that the father can support, and the mother can give birth to without injury to her constitution, and can attend to without ruin to her home comfort.

I have traced in the history of industrial life, from the beginning of this century, the various forces that have led up to our present condition, the impetus given to manufacture by the introduction of machinery, the removal of men to the field of war and their replacement by children, the impetus consequently given to rapid populating, the introduction of the mercantile spirit into family life, the tendency to regard children as so much capital in hand, the custom of looking upon large families as in all ways desirable, the break-down of parental love and parental authority, and the absence of any great improvement in the condition of the labouring classes. If we try to gather up the *general* results, these are an enormous increase of national wealth, but extreme inequality in the distribution of it, and in regard to the *comforts* it gives, a much greater inequality than in the earlier feudal times. Again, the nation has *solidified* and taken foremost rank amongst European powers; but within it classes have become strongly antagonistic, and the family has lost its solidarity. The patriarchal group, with its privileges and its responsibilities, is disintegrated, and there is no firm and well-assured Patriæ Potestas to control and mould the young,* who for the most part receive no training whatever in domestic and social virtues. The habits and customs of the present day in many cases resemble those of birds. There is a hurried pairing, before rapid breeding, followed by the scattering of the brood when the young are barely fledged; and later on, these young ones of the human race, broken into harness at the wheel of our industrial life, are treated, as I said, like cattle, and have no outward surroundings calculated to ennoble or exalt them in the scale of being, making it, therefore, all the more important that humanizing influences should be around them within their homes.

Now, let us see if other obstacles to happiness lie in the path of our working-classes.

We are often told that wages are extremely high in this country, that workmen have it in their power to live in perfect comfort, if they choose to expend their money frugally, to look before them a little, and by saving and economy prepare to

* The principle of Patriæ Potestas is explained in the fifth chapter of Sir H. S. Maine's "Ancient Law."

meet inevitable contingencies, such as sickness, or being thrown out of work for a time. Lately, indeed, Mr. Mallock has spoken at some length upon the want of motive power amongst our labouring classes. He says: "If they are poor, squalid, and dependent, it is because they have no efficient desire to be anything else." Then he draws a contrast between an imaginary cluster of poor curates' dwellings, and a rude colliery or pit village. "In the former," he says, "we should find every modest improvement that was possible. We should find such riches as there were made the most of. In the latter we should find dirt and disorder. We should find every symptom of outward penury; and yet all the while, as it were, there would be money spilling itself in the gutter. The reason is that the riches of the colliers would be in excess of their desires. They would have the power to get many luxuries, but they would be unable to conceive what luxuries to get. They would throw part of their money away on objects that gave them no pleasure; another part they would squander in excessive eating and drinking. They would have what in their own language is called a 'spend-out.' In this way they would again reduce themselves to that degree of want, which would once more set them to work."* This is in many cases no doubt true. Extravagance and waste are common enough in the lower, as they are also in the middle and upper classes. In the manufacturing district of the Midlands, before spoken of, my friends met with many wives and mothers who were in the daily habit of frying sausages in fresh butter! These poor people believed that such was the custom in the houses of the rich! Emulation of those above them in social station, and ignorance of wherein true happiness consists, combined with the absence of any sort of rational training to domestic life, bring ruin in their train; and were a sudden and universal rise of wages to take place, I do not doubt that dirt, disorder, and every symptom of squalid misery would remain much as before. The desire for neatness, clean and bright surroundings, and solid comfort, must be instilled in early years, or rather it will of itself grow up within the young if they are firmly planted in homes affording them these blessings; but whilst an opposite state to this prevails, and young people are set adrift at fifteen or sixteen, spare money will certainly be spent in dress, in gaudy fripperies or in follies that debase and vitiate the taste. There is *no*

* *Contemporary Review* for February, 1882. "The Functions of Wealth," by W. H. Mallock.

immediate and sudden cure for the state of things in Mr. Mallock's pit village. We can only look forward to a gradual alteration in the homes; for these must be the schools for *motive power* if future colliers ever rise to a moral elevation equal to that of the model curates.

Meantime, instead of harping on faults and follies of the poor, which, after all, are but the natural results of our industrial system, with the customs it has introduced, and the habits it has engendered, our practical work is *this*—to reason out our social problems, and *initiate* changes in outward conditions that will act on human nature for its good. Even Mr. Mallock would allow, that amongst the poor a certain number, at least, do possess motive power, and in effect show a strong desire to lift themselves out of their low condition. The impulse to thrift is not by any means wholly absent. Widespread and extensive operations of Friendly Societies, Burial Societies, Industrial Assurance Companies, etc., clearly prove the contrary. One of these, named the Prudential Insurance Company, had grown so remarkably within a very few years, that the commission appointed to investigate and report upon Friendly Societies called attention to it. In the year 1867 it had 358,000 members, and in 1872 it had no fewer than 1,013,041 members, showing that in five years it had nearly trebled the number. Others have made almost equal progress. Now, conceive how largely the interests of the labouring classes are bound up in the honest and economical management of these institutions. I mean intellectually as well as materially, for the impulse to thrift has to be stimulated, encouraged, and established. We must never forget a fact that evolution teaches, viz. human nature is like a plant that constantly throws out new shoots and tendrils significant of healthy life; but if the atmosphere is all ungenial, if the soil gives no deep root, if nowhere tendrils can lay hold on strong supports, the healthy life is blasted, nipped in the bud, frustrated, and progress is arrested. Now mark, the management of these societies has been frequently bad, and always extravagant. In the Prudential Insurance Company, out of a premium income of £1,608,849, no less a sum than £753,455 was spent in salaries, office expenses, and other business charges. Judge from this fact of the expensive nature of this kind of society. I must not enter into details; these are given in the *Fortnightly Review* for December, 1881, in an article called "Thriftless Thrift," and to it I refer my reader. The writer, however, there shows that the society's collectors go from door to door and find contributors amongst the very poorest and most ignorant, who are often unable to read the rules presented to them. Then by neglect of these rules they lose their membership in a few years, or months, perhaps, and derive no benefit at all from the society which, moreover, has taken money from them. It is, unfortunately, to the interest of the society to gain new members, and then let them drop off; consequently, poor contributors (too ignorant to understand the merits of the case) can only feel defrauded of their savings, and the impulse to *thrift* in them is checked, and terribly frustrated.

But let us see what the nation does in the way of fostering and cherishing this precious tendency to thrift. It offers to insure the poor man's life, but not for any sum lower than £20. This excludes a number of the labouring class, men who have no wish to provide a small inheritance for their children—possibly they may think them better without inheritance—but who do desire to provide for their own burial, and not put their children to expense on that account. We can trace down from patriarchal times the ambition to be decently if not ostentatiously buried. A pauper burial is felt by the poor to be a degradation and disgrace, and surely this is a sentiment we ought to guard, and not permit to be stamped out. As early as 1868 it was perceived that Government arrangements for life insurance did not meet this natural desire, and Lord Litchfield brought in a bill to enable the working classes to insure their lives for the sum of £5. But the bill was opposed by insurance companies, and those societies whose interests were at stake, and this important measure for the benefit of the poor remains in abeyance to this day. Burial societies, then, are still resorted to, and a calculation has been made by which it appears that if a man keeps up his penny a week subscription to one of these for forty years, he has actually paid £8 13*s.* 4*d.*, which, with interest, amounts to £16 7*s.*, for the sum of £6, which his family receives at his death. "Therefore," says Mr. Tremenheere, "it is fearfully manifest that a vast number among the humble, the most confiding of the wage-earning classes, are blindly led to spend in the pursuit of an object that does them honour, a sum enormously beyond what would procure for them that object, could the desired facilities be put in their way by the Government." There is no doubt that the weekly savings of our industrial classes fall a prey to expensive societies and clubs that very frequently become insolvent; and bitter disappointment is added to the anxiety of toilers who have striven to forestall their future, and provide for what they feel to be a serious evil of their lot. It is true *we* might judge differently, and think more space and comfort in homes of far greater consequence than decent burials; but it is *not* our business to judge for the poor. We have to observe rather what are their *own* wishes and feelings, and when these are in no way adverse to the general well-being, our simple duty is to aid them in their purpose, assured that if we add to the innocent happiness of mankind, we are taking the right course to help in moral regeneration.

Philanthropy commits grave error when it tries to patronize and force upon the poor ameliorations and reforms, or even "a cultus" they do not feel the need of or desire. With classes, as with individuals, the right attitude of mind is that which Dorothea found in Ladislaw. "He was a creature who entered into every one's feelings, and could take the pressure of their thought, instead of urging his own with iron resistance." We have too often urged with iron resistance our own views of reform upon the poor, and when we find them not at all plastic in our hands, we lose all faith in human nature, *their* side of it, at least, though not our own.

To return to Mr. Mallock; he makes a statement which, as a general truth, I look upon as quite misleading. He says: "The condition of the labouring classes is, within certain limits, proportionate to their faculty of desire." Of course, to what extent he means the words, "within certain limits" to qualify and quantify his proposition I cannot say, but, unless they nullify it altogether, the statement is untrue. Happy contentment does not prevail amongst the labouring classes, and a majority of individuals have the faculty of desire extremely active. Thousands are rising steadily in the social scale, carrying with them a sense of grievance against those classes that have wielded the legislative power. From the records of the Probate Court, and by comparison with the Registrar-General's death-roll, Mr. Mulhall clearly proves that the middle class has been largely reinforced from the ranks of the lower, and that since the year 1840 the progress, especially in Scotland, has been quite remarkable. Nevertheless, he concludes his interesting paper with this strong caution to the public: "We *must* elevate the masses, morally and socially, not for philanthropy or Quixotic sentiment, but for the same motive that we carry out our sanitary improvements—the instinct of self-preservation."

Now, I maintain that patronage is not the means by which we shall ever elevate the masses, or materially improve them morally and socially.

The things we must look to are the following:—A gradual rise of labourers' wages; a steady lowering of the price of necessary commodities until they almost reach the cost of production; a gradual change from the competitive and capitalist system to co-operative methods of production; the narrowing of hours of labour until time is left for daily recreation; the growth and spread of rational opinions concerning population, parental health, and parental duty, with the strengthening of all family ties; the opening of facilities for life insurance, and for aiding personal efforts in accumulating wealth, proportional to desire, on national security; and lastly, education based upon the principle that the habits, dispositions, and sentiments of children have to be formed *for* them; and that all education must bear upon, and lead up to, the right conduct of life.

In my next chapter I will show how Patronage is at present demoralizing the poor, and thereby revealing itself as a force adverse to the well-being of society.

CHAPTER VI.

PATRONAGE OF THE POOR.

Professor Huxley said, at the opening of Sir Josiah Mason's College, Birmingham: "If Priestley could be amongst us to-day . . . the kindly heart would be moved, the high sense of social duty would be satisfied, by the spectacle of well-earned wealth, neither squandered in tawdry luxury and vainglorious show, nor scattered with the careless charity which blesses neither him that gives nor him that takes, but expended in the execution of a well-considered plan for the aid of present and future generations of those who are willing to help themselves."

WHEN mention is made of our great middle class, there immediately arises in some minds visions of lordly mansions where wealthy merchants dwell, or magnificent, often untasteful houses erected in the suburbs of manufacturing towns, where stiff and stylish dinner-parties are given, that for luxury and show surpass anything found amongst the higher social class, but where the "feast of reason and the flow of soul" seldom, if ever, characterize the entertainment.

One thinks of princely fortunes made in the course of perhaps two or three generations, and expended in a free and lavish way, that equally suggests the princely style. A wife or children have but to give expression to a wish, to find it gratified; and whilst undoubtedly the retainers of the modern prince, *i.e.* his *employés*, in all departments will find remuneration for their services kept strictly down to market value, not one penny more, benevolence is a marked feature in this social life, and every fresh appeal to pity or compassion is responded to with a charity that is liberal, in some cases almost unbounded. One thinks also of men who, flushed with success in commercial life, very naturally suppose themselves admirably adapted for political life, and possessed of an ample supply of what has hitherto been looked upon as the necessary sinews of war, fight their way by means of these into Parliament, and by their action and general behaviour there, draw down upon themselves the sarcasms of that master of letters and culture, who alludes to "Bottles," and speaks more or less contemptuously of some individuals of the great middle class as "Philistines."

We must, however, take a far wider view of the middle class. Within that class there are many "Hellenes," and, what is more important to our subject, there are thousands and thousands of gentle suffering souls with an innate refinement that would grace whatever social standing could be given. Gregariousness (if I may so speak) is a strong feature of British character. We are not a very social people, but what I mean is this; we like to act in herds or flocks, and if we see our neighbours, or intimate associates, removing from small houses into large, and launching out into a more expensive style of living, we like to do the same, and we do it, without very much consideration whether it will or will not add materially to happiness.

In a great section of the middle class, it has been the fashion of the day to live luxuriously and expensively; and throughout that class, as well as in the lower class, it is and has always been the British habit to have large families. Now what is the result? Throughout the whole class the young people are accustomed to live softly. In childhood and youth they know no hardships, yet when they themselves enter industrial life, social pressure is extreme, and how to get on is a very difficult question.

Mr. Mallock has described the case of poor colliers who occasionally are possessed of means, as he thinks, beyond their desires, and who have a spend-out which speedily restores them to their poverty-stricken state of equilibrium. But I might draw a converse picture of individuals and whole families in every direction throughout the middle class, whose not occasional but constant condition is one of possessing desires, feelings, and ideas beyond their pecuniary means; and I say this without attaching any blame. I am not thinking just now of actions which one must needs approve or condemn. I think only of human beings passively bearing a burden of wishes they cannot gratify, of ideas they have no means of carrying out, and perhaps aptitudes they must repress instead of exercise.

We must not call these families or individuals the poor. In *their* surroundings we see no squalor, no unsightly indication of poverty. But penetrate within the veil, and we find untold suffering and sorrow—anxiety about ways and means —disappointments in the search for employment suitable to the capacity—humiliations to natural and proper pride, during that search, that are almost unbearable—a slow awakening, perhaps, to the crushing fact, that this country is overcrowded, and a decision come to, that to loving mothers has the bitterness of death in it—viz. that there is no escape from the hard and cruel necessity of parting with sons, who are as the very light of their eyes, and sending them into distant colonies to rough it, when they know too well they are unfit to do anything of the kind, and that, in short, their whole nature and upbringing have rendered them mentally and physically unfit.

We must refrain from following to its often bitter end, the career of young men who emigrate. What we want to make clear is, that whilst the plutocracy is much spoken of, and wealth and luxury are the *prominent* features of middle-class life, grinding poverty is in its very midst—always as much as possible covered up and hidden—but eating nevertheless like a canker into happiness, and spoiling if not degrading the quality of human life.

Now, in addition to the problem of how, out of a narrow income (looked upon as cloth), to cut a garment that will fit the personal desires of individuals whose requirements are great and varied, there is another serious difficulty here. One social class, we must remember, hangs on the skirts of another, and our middle class has voluntarily assumed the patronage of that class we technically call the masses, or the poor. As I pointed out, the rich give freely and spontaneously enormous sums in so-called charity. Nevertheless the task is felt by many individuals (and with abundant cause) as a burden and tax upon meagre resources, almost too heavy to be borne.

What can such persons do? Starvation is a lot that British sympathy will *never* consciously allow to overtake a fellow-creature, and many a half-crown that can ill be spared passes into the hands of a collector for the poor, wrung from the sweet and tender human feeling of some already careworn widow, leaving her to pinch and scrape and sigh more bitterly than ever, in efforts to keep free from debt, and hold herself, and those she loves, from sinking irretrievably in the social scale. Half-a-crown seems but a trifling matter, but the proportions to which has swelled the practice of collecting at house doors for charitable schemes and institutions can scarcely be conceived.

In some places, spring and autumn bring constant applications, and I have heard many poor householders express the sense of irritation, disgust, and dread aroused within them by these frequent claims upon a meagre purse.

Take for example a city I know well. The number of inhabitants is about 229,000, and there are forty separate and distinct charities of various kinds, in addition to an enormous infirmary for the sick, a large asylum for the insane, eight dispensaries where medicine is supplied gratis or cheap, three ragged schools, and two huge union workhouses, which are generally full, and for the support of which the whole town is taxed.

Now think of what all this implies to straitened members of the middle class—distributors, it may be, whose businesses are declining, and who foresee that by-and-by they will collapse before co-operative stores, but who are fathers of large families —or it may be some widow lady, who, bereaved of her protector by sudden death, finds herself compelled to face the difficulties of life alone. She has children whose innocent tastes she could gratify whilst her husband lived, and an income which, when mulcted of government taxes and lawyers' large fees on the winding up of the estate, has become miserably small, and inadequate to supply her own wants and those of her children. Now this lady pays poor-rates and schoolboard rates. Society obliges her to give money for the support and education of the class beneath her own, and then expects her to give voluntarily for the further wants of that class. By post or by lady-visitor, or by a paid male agent, no fewer than forty special calls upon her purse and appeals to her benevolent feeling are yearly made. She is entreated to keep up her late husband's benefactions; she is told that this or that cause is more necessitous, more urgent, more distressing than any other; and, strange to say, not one collector ever tells of work successfully accomplished, of social evils remedied, and institutions able to wind up their work and peacefully retire ! On the contrary, all are gradually expanding. The more we do, the more remains behind for us to undertake. *That* is the keynote of the tones of every labourer in the field of destitution ; and the widow's heart, already burdened with her own cares, sinks in despair in dealing with those interminable calls upon her charity.

These forty schemes, too, are quite distinct from efforts of the Church. The widow is expected to bear her share in ministering to the poor of her own church, and upwards of a hundred churches within the city have their own district visitors, their individual schemes to elevate the masses. Undoubtedly these individual schemes cross and intersect one another, and it is not difficult for idle and designing paupers to invent a scheme, more individual still, by which they can impose and gain the point of winning aid, considerably beyond necessity as well as beyond their own deserts. But mark—this cunning that they show is partly the result of these conditions. Who can wonder that they take advantage of a clumsy, ill-adjusted system? There would be greater cause for wonder still if they did *not*, when we recall the absence of home-training and home-discipline in youth, and the surroundings that have tended to foster, never to eradicate, their vicious anti-social instincts from the first.

Our system of benevolence helps onwards in a course of vagrancy and vice. I do not say, it does no good in any form; but what I say is this: it does a minimum of good, with a maximum of effort and expense; and everywhere throughout its field of action it does some positive injury, in the way of pauperizing and further demoralizing those whom we call the lapsed among the masses.

In our city of the forty schemes, a few years ago, the fact that much injury was being done was brought before the public, and a resolve come to, to amalgamate, unite, and centralize the charities. That this was possible seemed clear, for in various cases one special section of the poor had two societies or more devoted to it. For instance, the aged poor had one for men and three for women. I give their names:—1, Society for relief of Indigent Old Men; 2, Society for relief of Indigent Gentlewomen; 3, Senior female Society to relieve Indigent and Aged Women; 4, Junior female Society for relief of Indigent Old Women.

For the class called fallen-women, there are four separate establishments:—1, A Magdalene Asylum; 2, An Industrial Home for fallen women; 3, A Rescue and Probationary Home for fallen women; 4, A girls' house of Refuge, or Reformatory.

Then for girls in danger of losing good character, two establishments, called a Training Home and an Institution.

For expectant mothers, three charities:—1, A royal Maternity Hospital; 2, A Lying-in Institution (for delivering poor married women at their own houses); 3, A Society for relief of poor married women of respectable character when in childbed.

For the sick :—1, A Society for relief of the Destitute Sick ; 2, The large Hospital or Infirmary ; 3, A smaller Hospital ; 4, A Hospital for Sick Children ; 5, A Hospital for Incurables ; 6, A Society for relieving Incurables at their own homes ; 7, A Convalescent House ; 8, A Samaritan Society ; 9, Another Convalescent Home, and so on.

The blind had two societies to aid them ; the deaf and dumb, one.

Surely it was no irrational view to take that some of these might be united, or at least that the middle class might be relieved of those tormenting counter-claims and applications at their house doors. They would be invited to send their benefactions—all they desired, or perhaps could afford to bestow in charity during the year—to one central association, and entreated to give no alms to any pauper, but to send all beggars to the association, assured that every case would be investigated and relieved if necessary, and all deception and hypocrisy exposed. The association was formed—its purpose, as appears in the title, "to improve the general condition of the poor and to discourage mendicancy, idleness, and dissipation." But behold ! After an existence of several years, with all the usual paraphernalia, an office, paid officials, etc., it is now admitted to have failed entirely in its great aim of reducing the number of separate and distinct charities. Not *one* has been found willing to retire from its own individual field of labour ; each clings to its special work of patronage, its *pet* scheme for the amelioration of the condition of the poor, and goes on as before making its appeal for support, in public and in private. *This* effort, then, resulted in adding one more to the already numerous associations ! *

The energetic thoughtful official at its head tells me, that money is sometimes sent, for distribution to the different schemes, and when marked "for the incurables," they have the trouble of ascertaining to which society for incurables they are to hand it ! "Have you succeeded," I asked, "in putting down mendicancy ? " " By no means," he replied, "and never shall, so long as wealthy people selfishly do what is easiest to themselves —give alms, instead of taking a little trouble. We know of beggars who can quite readily and regularly obtain from thirty to thirty-five shillings a week, and who consider it absurd that we should offer work and urge upon them labour as a *duty*. ' Rich men are idle, why should not we be idle also if we

* In Liverpool, I am told, this scheme has been much more successful.

choose?' is asked;" and my informer added, "to me there seems no doubt that in that class, viz. the lowest of all, socialistic ideas are gaining ground."

Now, of course, such beggars have no real knowledge of the ingenious schemes of socialism; but they exhibit a marked disregard to the rights of private property; and this practical man feels, much as Mr. Mulhall does, that there is danger to our social life in the air, *i.e.* in the mental atmosphere of the lower class, and that for our own sakes, from the instinct of self-preservation, we of the middle class must find some more effectual means of elevating the masses.

"Is there any special error in our system you can point to?" I enquired; and he replied, "Our poor-laws are a great mistake; they seem to me to foster pauperism and crush out the independent self-supporting spirit. You see the poor conceive they have a right to be supported, and population amongst that class increases far too rapidly. I know at this moment of at least one woman who has had five children born in the union workhouse, and she is a most miserable specimen of humanity—she scarcely cares a rap for any of her babes."

This system of taking women for their confinements to the workhouse seems strange, when we consider that a maternity hospital is close at hand, and two societies for helping women at these times. Nevertheless, the matron at one of the unions showed me by the house doctor's book that seven children had been born there (all of them illegitimate) during the preceding eight weeks. She also mentioned a case of a married woman who had given birth to her whole family, and that a large one, within the workhouse. This woman's husband is frequently out of work, and when a babe is expected he forsakes her. She applies for parish relief, and is sent with all her children to the union, where she is confined, remains some two or three weeks, and then rejoins her husband, until in another year or so the same thing recurs.

This workhouse, like all similar establishments, is enlarging its borders. When capable of holding 600 paupers it was filled to overflowing; and now the number necessary to provide for is computed at from 950 to 1000, and building is going on. The staff of functionaries has also to be increased, so the rate-payers may anticipate a further call on their finances. And this is only one of the two workhouses for which the town is taxed. The matron observed to me that of late the inspectors have been sending into the house an unusual number of young men, and these seemed to her of a markedly low and brutal type. This, of course, must be attributed to scarcity of work and over-population. Occasionally it happens that a mother forsakes her child, climbs the wall, and makes her escape without it. Such little forsaken ones, left a burden on the State, are boarded here and there in the country, and given the chance of gaining the love of some more kindly foster-mother, and growing up under home influences. The babes of single women are frequently healthy, but, as a rule, it is not so with those of the low class of married women who come to be confined. In most cases the women drink, and their children are rickety and diseased.

Now, cruel and heartless as it seems to be to say it—the best thing *for these children*, and in one sense *for society at large* also, would be that they should perish, and that as quickly as possible.

"We want," as Mr. Galton says, "as much backbone as we can get, to bear the racket to which we are henceforth to be exposed,"* and those who have it not, and yet must struggle for their daily bread, *will not* and *cannot* find the gift of life to be a precious boon. If we left them alone, and refrained from stretching out a hand to help them, undoubtedly the law of natural selection would speedily prevail, and these young organisms, unfit for happy, healthful existence, would be crowded out by pressure of the vigorous and strong. But social agencies are actively at work, that counteract that law of natural selection and survival of the fittest. We do not, in fact we cannot, leave these little ones to perish. Our own social nature renders that course (to us) *impossible*, although we see its salutary nature; therefore we build hospitals for sick children, and in every way oppose or antagonize the law of natural selection. To do this appears right, and certainly far more natural than to let the poor children die; nevertheless, a different view sometimes occurs to unsophisticated minds, brought into close contact with the working of the system.

"Some years ago," says a lady-friend of mine, "a restless eagerness to help my fellow-creatures took possession of me. I had taught in Sabbath-school, been a tract distributor, and, in short, tried all the occupations devised to give girls of the middle class, who are unmarried, a sense of effective life; and yet I felt a power of work in me, a surplus energy that made me miserable, so I resolved to offer my services at a

* "Hereditary Genius," F. Galton, p. 345.

dispensary for the sick poor. Our own medical man objected. He thought the risk of infection for me too great. Ultimately we compromised the matter by my becoming a member of the ladies' committee for the Sick Children's Hospital, at that time newly built. I frequented the wards, taking a deep interest in all the tiny suffering inmates, and never shall I forget the painful, complex feelings I experienced there. Many of the children had frames so small as to make me think them infants only a few months old, whilst their faces were like those of aged people—the foreheads wrinkled, the cheeks thin and withered. The eyes had none of that sweet ignorance of life which we call innocence. Rather it seemed as though a lifetime of bad experiences had been crushed into their few short years, leaving an indelible impress. And as we strove to doctor and restore these children, I kept thinking—what *will* their *future* be? We cannot give sound constitutions nor perfect health, and still less can we blot out the past and give them painless memories of happy, healthful childhood, without which, it seemed to me, no human life is ever rounded into beauty. And when from time to time I saw success follow our efforts, and the results of all the tending and nursing appear in a growing vitality and some approach to feeble health, I wondered if our office was a truly wise, beneficial one, and slowly I grew convinced that, with the best intentions, we were yet *not* doing unmitigated good. Standing one day by the bed of one of these sad cases, a little blighted bud of poor humanity, I was surprised to smell alcohol! until the nurse explained: 'Poor dear, she's one of them we call our whisky children.' Born of intemperate parents whose blood takes up alcohol from whisky, the wretched nourishment they get is mother's milk impregnated with alcohol; and they turn at first from all food that has not a similar flavour. These poor unfortunates have to be cautiously and gradually weaned from the injurious vitiated taste, and won to take the natural, healthful nourishment."

My friend goes on to say, that gravely perplexed and puzzled by the social problem, whether she was or was not engaged in useful, beneficent action, she carried all her difficulties at last to an elderly man in whose great wisdom she implicitly believed. But his replies to her inquiries upset her reverence for his wisdom. He said, "You're doing all you can for these children, and for yourself the work is excellent. Young ladies of the present day lead butterfly lives; their time is spent in flirting and folly, a poor preparation for wifehood and motherhood. These hospitals are very useful. They give to girls like you some real work, and draw out all the tender feelings of your hearts. You must not doubt that you are acting well and rightly."

It is almost unnecessary to comment on this irrational view of an important question. I have already shown how much the comfort and happiness of many individuals of the middle class are marred and frustrated by the miserable condition of the poor. No one can think that I deny what is too evident—that the interests of these classes are intertwined and closely united; but, that the poor should suffer, that little children should be kept alive to linger on a sick-bed, or to fight their way in life with odds painfully against them, merely that girls of the middle class may profit by the task of tending them, is too grotesque a travesty on human justice. It savours of the childish moral sentiments of Watts's hymns, where the spectacle of the poor is made a reason to think about and plume ourselves on our own superior position.

> "Whene'er I take my walks abroad,
>   How many poor I see!
> What shall I render to my God
>   For all His gifts to me?
> Not more than others I deserve,
>   But God has given me more;
> For I have food while others starve,
>   Or beg from door to door."

We have all outlived this way of looking at the poor, and recognize that in our modes of dealing with them we are bound to have a single eye to what is of paramount importance, viz. *their personal happiness and moral good.* If sick children's hospitals are only, or even chiefly, good as schools for unemployed young ladies, I think they stand condemned.

But I need not dwell on that point. What I am here concerned to show is this—we consciously, and of set purpose, oppose and frustrate the action of the law of natural selection and survival of the fittest. We take (so far) the destiny of these weak and sickly children into our own hands; for, *unfit* as they are, we help them to survive.

We must not overlook this fact, or turn aside from recognizing and admitting that effect of our patronage of the poor.

It was most natural for that young girl to wish to do effective work, and good that she could feel assured would prove both permanent and unalloyed. But of which of all our forty benevolent schemes could such an assurance be given? Let us consider one which is *par excellence* a ladies' charity. It helps expectant mothers, and, indeed, goes by that title in the midlands factory town of which I formerly spoke, where there is a very large and flourishing one, partly supported by the device of giving a public, ladies' charity ball! Of that kind of charity amongst our forty there are two. They differ in respect of what they give. The one supplies a nurse and a small sum of money; the other sends no nurse, but for three weeks it doles out tea, sugar, bread, meat, and soap, with one small suit of baby-linen. In the midlands these are given, and the nurse supplied as well; the ball being a successful method of keeping up the funds! and the operations of the charity are increasing and extending. Of the two others also that may be said; for one reports the number of mothers relieved last year to have been ninety-nine, whilst this year it has risen to one hundred and thirty. Now, in a factory town, where population is densely crowded, what goes on in one little dwelling is quickly known all round. A thoughtful, self-dependent, young, expectant mother has been saving her pence and sewing for nine months, preparing for the advent of her child, and often giving hints and gently kind reproofs to her poor, thoughtless neighbour next door, who is in a similar condition. But when the period arrives, behold the thoughtless one is much the better off of the two! She has been worldly-wise; for she has all she wants, simply because her hour of trial found her unprepared; whilst her more thoughtful neighbour hears of a nurse, and ladies' visits, and of sympathy and kindness shown, and she herself is left the while to the solace of her superior pride or independence! Can we wonder if the independence melts away; if, as the years roll on, that mother too applies for help, and all her prudence and *self-helping* ways are sapped and undermined by this benevolent ladies' charity?

There is nothing more certain to my mind than this—that, for years and years, these good, but foolish ladies, have been injuring the moral nature of the poor. Their patronage has helped to make them thriftless and deceitful, and has fostered the increase or propagation of the stock whose qualities tend to deteriorate the human race. It almost sickens one to think of precious sympathy and noble energy thus run to waste—nay, worse than *that*—spent to the hurt and ruin of our common Humanity, and all for lack of scientific knowledge to guide and point the way to methods that are rational and right.

One turns with a sense of some relief to education; for surely in that field no danger can exist of doing harm instead of doing good.

Amongst the earliest of the ragged schools were those founded by Sheriff Watson of Aberdeen. His heart was deeply pained by the neglected children—some two hundred and eighty—who were idling in the streets, or frequently committing crime. The parents' excuse for keeping them from school was that these little ones must make a living for themselves, and on the street there was some chance of errands for them. Now, Sheriff Watson resolved to give them education, employment, and some food, but not to clothe and lodge them. He saw the evil and danger to society of relieving parents entirely of the duty and expense of maintaining their own children, and he carefully guarded against it. Even where homes were bad, he thought them better than no homes at all; and he said the children trained in his schools were more likely to improve the tone of a bad home than be corrupted by it. The principles which he kept steadily in view were, that there should be no separation from the parents, a minimum of interference between a child and its parents, and that economy should be strictly practised. He estimated the cost of providing for each child at little over £5 a year.

Now, his grave expectation was, that when this system had improved the habits of the people, the number of children to be thus dealt with would be greatly and certainly diminished. Alas! the very opposite of this has occurred. The schools which Sheriff Watson founded have amazingly increased, both in the number of the children and in expense. The annual cost of each child has trebled. They are no longer lodged in a plain and simple dwelling, but in large handsome hospitals, erected in the country, and all the inmates are boarded, fed, and clothed within the institutions. It is no longer, then, a minimum, but a maximum of interference that takes place between parents and children; for every bad parent, and many good ones, are entirely relieved of the burden, or the duty, or the happiness—whatever we consider it—of maintaining their own children. Various influences have led up to this result. Government help has been given by means of a capitation grant on each child, and this has formed one great inducement to receive the children as boarders. Again, as numbers increase, the management falls more and more into the hands of permanent officials, whose interest it is to expand rather than to shrink, and so on.

In the reformatory and industrial schools of Great Britain there are now about twenty-three thousand children, and the annual cost is £316,000, of which £260,000 is provided by the Treasury. Parents, it is said, anxiously compete for places in the schools; and there is often ground for suspecting that a parent has caused the misconduct of his child in order to be relieved of its maintenance! The provision that they should contribute in some small measure to the cost is easily evaded, and the worst of signs is showing itself, viz. the beginning of a hereditary class, where parent and child succeed each other as inmates.

During the first few years of the existence of the system there was a diminution of juvenile crime; latterly it has not been so. The rate of diminution has decreased, and bears no sort of proportion to the increase in the numbers maintained, and consequently in their expense to the community.*

Even in this field, then, of education for the poor, to middle-class reformers there is *no* relief from a sense of failure. Sad and discouraging as are the facts, the truth stands clear before us. Our method somehow is a wrong one. The indirect effect of ragged schools, as at present conducted, is to increase the number of ragged children, and by the action of the schools upon the minds of parents and of all the adult poor around them, they foster pauperism and anti-social feelings. What they accomplish in the way of temporary good is not to be compared with the amount of mischief ultimately done in making permanent our social misery and crime.

I quote a passage from the greatest thinker that we have on social subjects: "By way of checking recklessness, and discouraging improvident marriages, and raising the conception of duty, we are diffusing the belief that it is not the concern of parents to fit their children for the business of life, but that the nation is bound to do this."† Practically the ragged schools diffuse this belief, and the evil is none the less great and charged with danger, that it is subtle, and concealed, from a superficial view.

In our city of the forty voluntary schemes we have three

* *Scotsman* of January 28th, 1882.
† Mr. Herbert Spencer, " Study of Sociology."

schools for children of the lower class distinguished by the terms—1, Original Ragged and Industrial School; 2, United Industrial School; 3, Ragged Industrial School. And besides these we have another institution called an Industrial Brigade, which in reports is spoken of as an adjunct to the ragged schools, and which, like all the rest, is steadily increasing. In 1871, eighty-two boys were sheltered there, and now one hundred and fifty is their number.

Of similar homes for girls there are two, and one of these has lately had an addition built. The matron at that home told me of a father who had one girl at the one home, and two at the other, and who refused to pay a penny. At first he paid a little, and by the time he ceased to do so the children were attached to teachers and companions. As the father threatened to remove them if a claim were made on him for aliment, the committee preferred to waive the claim rather than let the children lose the benefit already gained by their short residence there. Another case was this. A waiter in a hotel was left a widower with one little girl, a child of seven. He brought her to the home, and said he could not properly secure for her the attention she required. If she could be received he would, of course, pay her board, and feel most grateful. For a time all went well, then suddenly he disappeared from the town without giving notice. At the end of six months he wrote, gave his address at some country place, and begged to have his bill forwarded, with news of his child. This was done, but no reply and no money were ever received! After another year had elapsed a woman called once to ask about the child. She said the father sent her, but she refused to give information of his address or his employment. Again the ladies of the committee feared to lose the little one by prosecuting any inquiries; so that parent and child are separated probably, if not certainly, for ever.

This is only a sample of what takes place in all directions. Parental duty can be evaded, and human nature, which in the discharge of parental duty might be improved and raised, is injured and degraded.

Moreover, thousands in the middle class, burdened by the support of these forsaken children, have less to give to real orphans, and the public have to be urged and stimulated to meet a *necessary* outlay. I am thinking at this moment of a scene I witnessed at an English watering-place during the last summer. Children (a mixed lot of boys and girls) from a

Sailors' Orphans' Home had been brought some thirty miles by railway, and were paraded up and down and round the town, with banners flying and the infant band playing, whilst in the front of the procession marched the teachers and some clergymen. Every now and then a pause was made, the crowd encircled them, and round went little boys and out and in amongst the crowd with begging boxes. At last they all assembled on the seashore. A clergyman addressed the bystanders, and made a strong appeal to public charity, pointing to the orphans, and emphasizing, by the sight of them, their claim. The crowd became so dense that he feared some spectators might not clearly see the children. He therefore singled out a little girl, and with her in his arms mounted his rostrum. The child burst into tears and sobbed bitterly, and as he placed her on the ground again pence were thrown into her lap and showered all round about her. When the whole party had departed, carrying with them the spoil, a goodly sum, resulting from the day's display, I ventured to impugn the whole proceedings to the clergyman. I asked him if he thought it wise and right to teach young children thus to beg, to pose before the public, and play upon benevolent feeling. Could we attribute any blame if in some future time, when toiling for the daily bread proves dull or too distasteful, these children turn to begging, and fall back upon the lesson taught them by pastors and masters in their youth, viz. to parade their destitution, their necessities, and trust to human sympathy? It seemed to me that rather should the fact of their dependence on the public be concealed from these poor children, at least until they reached their adolescence and were self-supporting. His reply was simply, that the method was effective, and in *no* other way could the money *absolutely necessary* be obtained. This was an argument so far unanswerable. That orphans who are destitute must be maintained, either voluntarily or by the State, I certainly admit; and the innumerable bazaars, fancy balls, sermons, all inspired by the necessity of keeping up the funds of our unending public charities, prove to a demonstration that the burden is *too great;* the tax upon us is more than pure benevolence will bear. Hence this resort to grovelling sensationalism.

But as regards the work that most of these charities are doing, of each small scheme, the annual report is sure to say it does a large amount of good; and if we think of the enormous number of the schemes, and add them all together in our mind's eye, we are entitled to an outlook most exhilarating. Surely with all this mighty energy applied, the masses in a few short years will be elevated, and pauperism, disease, and crime be rooted out. But when we put aside reports, and go amongst the poor and into every field of action for their good, to what conclusion are we brought? To this, that on the whole, our patronage is doing infinitely more harm than good; that all these tiny streams of surface charity are simply irrigating and refreshing the field of poverty and vice, and nourishing weeds for future generations to contend with.

Not one appears to penetrate and reach the causes of destitution and misery. *These* year by year increase; and, if no better can be done, it were wiser far to withdraw from our presumptuous aim of elevating the masses, and with at least a modesty, born of our knowledge of truth, desist from futile efforts and leave the poor to elevate themselves.

CHAPTER VII.

POVERTY IN RELATION TO GENERAL WELL-BEING.

"It is the curse of England that its intellect can see truths which its heart will not embody."—LAURENCE OLIPHANT.

IN entering upon the subject of the poor I observed that we must avoid a one-sided judgment, and I used the figure of a stereoscopic slide, which by means of two representations of the same object taken at slightly different points affords in full relief a picture more truthful and complete than could otherwise be obtained.

Now, in my two last chapters I have given as full a picture as space permitted of my subject from one point of view, and here I desire to look at it from a different point, and place it before my reader under the aspect of a widely general social question.

But first let me make some preliminary observations. It is possible for us ideally to single out a special social class (as I have done with the poor) and, for the moment, to view it as though it stood alone, having sufferings and interests apart and separate from all the rest. But in reality the various classes of a social body are as closely related as are the members of a living organism, and the interests of all are intimately intertwined. Viewing society, then, as *a whole*, and bearing in mind that it resembles a living organism, it gives us no surprise to perceive that change has been the constant condition of healthful, progressive life. Whatever the future may bring forth, the past has never shown a stationary state of social wellbeing; but perpetual alteration in human conduct and feeling has characterized a growing social body. In other words, the forces that hold the units together change in accordance with

G

changes in the outward environment of the units (or the social life), and those forces that are good and useful at one time become bad and hurtful at another, and are then gradually superseded. A new adjustment takes place, which infuses fresh vigour into the society and makes it healthfully progressive. Now, it is knowledge that is the great factor in producing these readjustments; but knowledge is necessarily slow in diffusing itself throughout society, and meanwhile public sentiment clings about old customs and habits, and is invariably at first violently prejudiced against all knowledge that condemns the customs and habits of the past. Hence the pioneers of progress are often made to suffer acutely for their cause. By pioneers I mean those individual minds that, recognizing and grasping knowledge, venture to formulate or embody it in language, although they are conscious that the changes it will bring about must be difficult, to public sentiment extremely painful, but also *inevitable*, and in the end certain to prove beneficial. These pioneers put in movement the car of progress, and are themselves, if not crushed, at least bruised and wounded by its wheels.

Numerous illustrations occur to me; and as I hesitate which to choose, a picture rises in my mind of a gentlewoman, made to suffer in the cause of truth in times so recent, that there seems a double propriety in selecting it as our illustration and warning. When Harriet Martineau was engaged in writing her series of Political Economy tales, she received various cautions and hints in reference to her treatment of the Population question. "Go straight through it," wrote a friend, "or you'll catch it." Another, who seemed more an enemy than friend, suggested hesitation, professing grave anxiety lest the reputation of a lady should suffer by even touching upon a question which, nevertheless, the interests of society required her to deal with. Had she been left to herself, Miss Martineau would have felt no difficulty at all. The question was a purely philosophical one, and was certain to be regarded, at least by all free and truth-loving minds, without vulgar prejudice. It contained one of the principles of the science she was illustrating, and her simple duty was to evade no question that lay in the direct line of her undertaking. But these warnings changed the easy, pleasant task into a burden and distress to her. She was reminded of the fact that public sentiment lingers tenderly with the past, and is violently prejudiced against new knowledge that condemns the customs or habits of the past. As she sat writing her innocent little story, called "Weal and Woe in Garveloch," the perspiration many a time streamed down her face (so she tells us in her autobiography), and although she felt convinced that there was not a line in it which might not be read aloud in any family, still she feared lest the publication of what she knew to be important truth might cause her to lose credit and influence. Nor were these fears unfounded. No sooner did the tale appear, than Messrs. Lockart and Croker reviewed it in the *Quarterly Magazine*, and, taking advantage of her sex, made "gross appeals to the prudery, timidity, and ignorance of the middle classes of England," and so inflicted upon her a suffering which few, if any, pioneers escape.* This occurred in the year 1832; but now that fifty years have passed away, it is evident how great a change has taken place in public sentiment. Were we to search throughout the middle classes of England, it would be difficult in 1882 to find a single individual of ordinary education and good taste ready to cast a stone at a woman because she writes or speaks upon the population question. It is clear that the moral dignity of truth is fully appreciated when ignorance gives way to knowledge, and public sentiment (however long it takes) is always ultimately brought into harmony with *fact*.

To resume the subject of the poor. We saw that their condition is one of social misery and degradation; and if we approach them nearly enough and sympathetically, we are *compelled* to admit, that the causes of that misery and degradation lie so much in their environment or surroundings, that towards the individuals we are bound to feel only deep compassion, and to abstain from blame. We utterly repudiate and deny the statement that the condition of the labouring classes is proportionate to their faculty of desire. We believe them to be full of unsatisfied desires, and that the craving for a happiness, which they know not where or how to seek, is what often causes them to snatch at coarse indulgences and vulgar excitements, which only further brutalize and degrade. We should remember that our huge proletariat, which lives cramped up in densely crowded cities, toiling till perhaps it sweats to death at sedentary drudgery (such as cotton-spinning, weaving, tailoring, etc.), and sleeping at night in noisome, airless dens, where health and comfort are alike impossible, is the direct descendant of an agricultural class, whose whole surroundings

* "Harriet Martineau's Autobiography," vol. i. p. 199.

were essentially different. And knowing this (notwithstanding that our present knowledge of all the laws of heredity is but meagre), we have, I think, ample grounds for supposing that thousands of the members of our proletariat class are inheritors of a physical organization, with instincts and cravings far more in harmony with the life of their progenitors than with their present actual surroundings. I do not mean to assert that the lives of these progenitors were in all respects enviable. No doubt their huts or cottages were miserably poor and crowded; their lot was one of constant labour, and destitute of all the embellishments of life that education and refinement give; but when they crossed the threshold of their homes, fresh air was all around them; they could breathe a pure, uncontaminated atmosphere, and could enjoy the soothing, satisfying influences that flow from nature when unspoiled by man. Now, with a nervous system attuned to the slow and calm monotony of rural life, with aptitudes for agricultural pursuits, and an instinctive, though latent, love of animals, trees, and flowers, an infant born in a densely populated city, and destined to labour year after year amid the whirl and turmoil of industrial life in Glasgow, Manchester, or London, perpetually breathing an atmosphere made thick with smoke, is *profoundly to be pitied*. His youth and manhood must lack a condition essential to the satisfaction of his organic nature, and he will suffer discord and restlessness within himself. When to this is added the absence of right training, and the presence of poverty, a fierce competition in all the fields of honest labour, causing a lifelong struggle for existence—and very little domestic happiness and comfort—is it not marvellous that social order is maintained at all, and that a comparatively small proportion of the number, only, sink to the level of criminals and outcasts?

Satisfactory life is always the result of adaptation; and adaptation depends upon a prolonged converse between the organism and circumstances that *remain the same*. But let us for a moment endeavour to realize the enormous, the amazing change of outward surroundings or circumstance that has come to our great proletariat during the last hundred years; and we shall cease to wonder that the adaptation is as yet so frightfully inadequate, and life is felt by many human beings to be nothing but a blank, miserable existence, destitute of happiness.

We may anticipate, however, that vast and salutary alterations will slowly, surely progress in *two* directions. On the one hand, human nature will be modified and rendered fit for a complex civilization, by a new system, free from superstition, —I mean a system of direct definite training of the young to natural and worthy social life. On the other hand, still greater changes will be made in outward circumstance, as gradually men learn by experience that happiness is what they want, and that a nation's glory consists, not in its greatness of position or its wealth, not in the culture and art that flourish in its midst, not in the goodness and noble elevation of *some* of its members, but in the pure and simple happiness of each and every individual within its bounds, from the highest and most prominent to the lowest and most obscure.

As regards the latter, there are already thousands of noble hearts, on whom the burden of the poor lies like an incubus, and who would give their very life's blood to be able to stoop down, and lift that mass of suffering humanity out of its "slough of despond" and of degradation.

It is not want of human sympathy that impedes our progress in this direction; it is that *wrong methods* have been adopted, and we are only now beginning to perceive that these must be discarded and stamped out, whilst in their place must come a rational reform, which reaches to the *causes* of poverty or pauperism, and cuts away its great, outspreading, gangrened *roots*.

Mr. Matthew Arnold has told us, that on one occasion, as he and a good man were looking at a multitude of children in one of the most miserable regions of London—children eaten up with disease, half-sized, half-fed, half-clothed, neglected by their parents, without health, without home, without hope—the good man said: "The one thing really needful is to teach these little ones to succour one another, if only with a cup of cold water; but now, from one end of the country to the other, one hears nothing but the cry for knowledge, knowledge, knowledge!" But Mr. Arnold's thoughtful mind *rejected* the remedy of this good man. *He* muses upon it in this strain: so long as the multitude of these poor children is perpetually swelling, they *must be charged* with misery to themselves and us, whether they help one another with a cup of cold water or no! and the knowledge *how to prevent them accumulating* is what we want. We must let our consciences play freely and simply upon the facts before us, and we must listen to what it tells us of the intelligible law of things as concerns these children—and what it tells us is, "that a man's children are not really *sent*, any more than the pictures upon his wall or the horses in his stable are *sent;* and that to bring people into the world, when one cannot afford to keep them and oneself decently, or to bring more of them into the world than one can afford to keep thus, is by no means an accomplishment of the Divine will, or a fulfilment of nature's simplest laws, but is . . . contrary to reason and the will of God." * *This* knowledge, he asserts, ought to be habitually acted upon, as one acts upon the knowledge that water wets and fire burns.

Now, here Mr. Arnold touches the *tender spot of our diseased social state*, and points to the very nucleus or centre, from which there radiates—poverty and pauperism—social pressure and fierce competition—and disease, with all the misery and wretchedness that follow in its train.

Parental conduct has an all-important bearing upon the social state. It affects the mental, physical, and moral health of all mankind; and were parental conduct guided at all times by enlightened reason, we should ere long hail the development of a regenerated race.

I have described the influence and position of mothers amongst the poor, and have shown that as a rule they have neither ample comfort nor health. They are over-weighted by toil, and physically exhausted by child-bearing. The nervous system gives way, and they sink into peevish, discontented, hopeless creatures, utterly unfit to make the sunshine of a home. Husbands turn in for comfort to the alehouse, where stupid, heartless mirth leads to debauchery, and meanwhile the little children stand no chance of being carefully *brought up*. They *tumble up;* they are tended anyhow or nohow; and in innumerable cases, where drunkenness has grown into a habit with the father, they are left alone all day, or put in *crèches*, whilst the poor, miserable, broken-down mother stands at some factory wheel to earn the daily bread. The only useful training that they get is at the public schools, and that is of the head, —the tender feelings of the heart are never adequately dealt with there—and practically, the children are left without the smallest attempt to develop in them those qualities which are absolutely necessary to happy domestic and social life.

The fact that from delicate mothers, too large families, too crowded homes, spring many social evils, is certainly no *new* discovery. Philanthropists have long known it, and have strenuously endeavoured to obviate the evils. Some have said, society should build workmen's houses, airy and spacious,

* " Culture and Anarchy," p. 246.

and should bribe or perhaps compel the poor to leave their miserable dens, which should be pulled down at once. Much in this way has already been done by private effort, and many noble workers have followed the poor to their new dwellings, inspiring energy and hope, and striving to instil such lessons as would raise them to the level of a purer, better life.

These measures may ameliorate in a small degree some of the miseries of the poor; but when we regard them as a *remedy* or *cure* for poverty and pauperism, we are constrained to muse, as Mr. Arnold did, upon his good friend's remedy of a cup of cold water : "so long as the multitude is perpetually swelling, they must be charged with misery to themselves and us, whether they help one another with a cup of cold water or no, and the knowledge *how* to prevent them accumulating is what we want."

We may keep building up and pulling down their houses from year to year, but if *that is all we do*, it is something like Mrs. Partington's brave attempt to dry up the Atlantic with her mop and pail!

The population question has to be studied, for it lies at the very root of the whole matter; and when general conduct is widely influenced by intelligent comprehension of that question, then, but not till then, we shall possess the true foundation upon which to rear a solid edifice of social well-being.

In 1801, the population of England and Wales was 8,892,536, or let us say about nine millions. Now it has risen, as we see by the last census, to about 26 millions. Roughly speaking, we have trebled our numbers in the last 80 years, and the increase has been at an accelerated rate. I mean, that whereas for the ten years between 1841 and 1851 the percentage of increase was 12·65, from 1861 to 1871 it was 13·19, and from 1871 to 1881 it was 14·34.

This increase in our numbers is not due to the birth-rate only, but also to a fall in the death-rate. The birth-rate has been swelling, the death-rate has been shrinking, and as far as the immediate consequences are concerned, the latter is of course a matter for rejoicing. Increase in the average duration of life may, *cæteris paribus*, be taken as a fair index of material welfare; but there is much to set against this *one* encouraging feature.

By the parliamentary return of marriages, births, and deaths, registered in England and Wales in the year 1881, it appears that in different districts the percentage of marriages varies considerably. It is greater in the mining, manufacturing, and trading districts than in the farming districts, and much higher in London than in the provinces. In the district which comprises Hertford, Buckingham, Oxford, Bedford, Cambridge, the rate equals twelve persons per annum for each thousand of the population. In London it is eighteen persons for each thousand, and in the divisions which comprise Yorkshire and Lancashire, the rate is sixteen and seventeen persons to each thousand.

As regards births, somewhat similar proportions are stated. In London, there were thirty-five births to one thousand of the population, whilst in the south-eastern division, there were only thirty-one; but the rate rises again to thirty-five and thirty-six in the great manufacturing districts of the midlands and the north.

From the facts presented it is evident that the most rapid growth of our population is taking place in the great industrial centres, the mining, manufacturing, and trading districts; and the type that there prevails is necessarily widely affecting the British race.

Now, it is precisely in these centres that the greatest amount of poverty and destitution exists, and the racial type is the most degraded; and as I have (I think) clearly shown, poverty, where individuals are thoughtless and ignorant, is no restraint upon the birth-rate. It is in the poorest localities that children most abound, and prudence has *no control* over the multiplying of the lowest specimens of humanity. Prudence presupposes considerable intelligence, and some capacity for independent thought; but these poor creatures as a rule are like machines, acting, without one moment's thought, from those customs and habits which accord with natural instincts and personal desire.

Now, custom and habit are never safe guides. They are handed down to us from the past, and are often out of keeping with the present state of things. But how are social units that are unintelligent to be dealt with in regard to bad and injurious habits?

In the present stage of our civilization, social units *hold firmly together*, and *influence*—even apart from reason—is very great. The sway of intelligent minds over the unintelligent is immensely powerful when calmly exercised and wisely directed; and as yet this enormous lever has never been applied in reference to the population question.

Mr. Malthus expounded a great truth some eighty-seven

years ago, and formulated it for our benefit; since that time it has been taught by every writer of repute on economical subjects. Those authors who questioned or assailed it, such as Ingram, Alison, Doubleday, have never been able to shake its hold upon the public mind; but notwithstanding this, its action on public *morals* has as yet been *nil*, and from it our social life has derived no benefit, but rather the reverse.

Mr. Malthus' celebrated essay showed, that man tends to increase more rapidly than the means of subsistence—in other words, that population and food, like two runners of unequal swiftness chained together, advance side by side, but the pace, or natural rate of increase, of the former is so immensely superior to that of the latter, that it is necessarily greatly checked, and the checks are of two kinds. They are either positive—that is, deaths occur from famine, accident, war, or disease, and keep down the population, so that the means of subsistence are just sufficient to enable the poorer classes barely to exist; or they are *preventive*—that is, fewer births take place than man is capable of causing.

Since Mr. Malthus' day another great and powerful mind has thrown fresh light upon the subject. Mr. Darwin showed us man's place in nature, and demonstrated that the chief means used by nature in breeding a successful race is a struggle for existence; so now we know that the positive checks, that Mr. Malthus pointed out, have in the distant past been beneficial to the human race; because they have resulted in survival of the fittest.

So long as population was restrained by positive checks alone, the painful struggle for personal existence was, as regards the human stock, useful and salutary; for individuals who were physically and mentally weak were certain to succumb, whilst only the strong and able, who could overcome all obstacles, became continuators of the race through the process of natural selection. When, on the other hand, preventive checks keep down the population, the action of this force upon the *quality* of race is not so beneficial. It is the intelligent, the thoughtful, the prudent, who are swayed by the knowledge that Mr. Matthew Arnold commends, and who, reflecting that, to bring people into the world when one cannot afford to keep them, is contrary to reason and the will of God, refrain from marriage, or marry late; whilst, as a consequence of this, *their* children keep numerically weaker than the children of the thoughtless and reckless members of society, who exercise their reproductive powers to the utmost.

Another grave consideration is forced upon us on the examination of this subject. In a wealthy nation where the social units are well knit together, two powerful factors must be thought of in every social problem. I mean the force of *capital* and the force of *human sympathy*. The possession of capital signifies the possession of constant power over the available means of subsistence, and consequently the man who has ample capital, is not only himself relieved from all necessity to struggle for his personal existence, but he may at any moment lift out of that struggle the individuals who appeal to and enlist his human sympathy. Now, both individually and in his corporate capacity, man is doing this to an *enormous extent*. The poor law, our innumerable public charities, and all the efforts of private beneficence, result in placing individuals either temporarily or permanently above the field in which they would otherwise be struggling for existence. This course of human action has a twofold aspect; it affects *directly* the recipients of the bounty, and *indirectly* it affects society at large. In the first of these aspects one sees nothing to condemn; on the contrary, it is a matter of rejoicing that the great social heart of the community beats warm and true, and will not consciously permit a single member of its body, however weak, incapable, or imbecile, to perish unaided in the struggle for existence.

But in the second aspect, as regards society at large, the case is very different; for observe, no social restraints are put upon these recipients of bounty, who very naturally obey the human instinct, and perpetuate their kind.

As Mr. W. R. Greg has said: "Among savages the vigorous and sound alone survive, among us the diseased and enfeebled survive as well . . . with us thousands with tainted constitutions, frames weakened by malady or waste, brains bearing subtle and hereditary mischief in their recesses, are suffered to transmit their terrible inheritance of evil to other generations, or to spread it through a whole community."

It is an inevitable complication of our advanced civilization, with its power of capital, and power of tender sympathy, to set aside, or rather, I should say, to counteract the law of natural selection, and so to lose the benefit of its action in the survival of the fittest. We artificially select hundreds of thousands of individuals, and help them to survive, whom, if they had been left alone, hard nature would have deemed unfit ; and in our midst the race is no longer to the swift nor the battle to the strong.

But here I must call attention to a verbal point which may perplex my reader. The outcome of man's beneficence I have called *artificial*, as though his kindness to the feeble of his race was not supremely *natural*. A moment's thought will show my meaning. I am compelled to use a word that marks a difference, and shows that man's selection is not what Mr. Darwin meant by natural selection. He was pointing to the forces of the cosmos, and all opposing forces that would ruthlessly bear down, destroy, crush, or crowd out, all but the strong and vigorous of mankind; and his theory did not embrace the state of things that now prevails—a state in which wide benevolence, the exercise of all the helpful, tender qualities that characterize man's civilized human nature, combined with power derived from saved-up capital, have caused the introduction of a wholly new condition, viz. an environment that closely surrounds the feeble of the race, and gives them added strength against opposing forces.

This new condition, which affects the *quality* of life, cannot be ignored in any adequate solution of the population question.

The quality of life (and what I mean by quality is the mental, physical, and moral health of the race) is *more* important than the *quantity* of life; since what mankind is aiming at, and seeking to achieve, is happiness for all.

But let us here look at the quantity of life, whilst bearing in our minds the simile I used, of population and the food-supply, as two fast runners chained together, and advancing side by side, though of unequal swiftness. The increase of the swiftness of the second runner during the last half-century, has been unprecedented in the world's history ; and no wonder, for man has learned to put aside his self-made obstacles as well as to overcome the obstacles that outward nature placed in his path. " Forty years ago the peasants of Castile and Leon saw their wheat rotting, because it was forbidden to export it; whilst Great Britain was paying famine prices for bread rather than repeal her corn laws." *
The removal of arbitrary laws against grain has greatly told upon the increase of the world's food-supply ; and when we add to that, the opening up of new countries, with extension

* *Westminster Review*, R. P. Porter.

of railroads, the improvement of ocean navigation, all facilitating the carrying of grain from distant countries, we can perceive how, notwithstanding the enormous increase in our British population, no widespread famine has occurred to check the speed of that tremendous runner. The British runners, indeed, have been *doubly* aided to keep side by side through the discovery and opening up of new grain countries; for not only is the swiftness of the one, the food-supply, accelerated, but the swiftness of the other is checked, by the absorption (if I may so speak) of its speed—I mean the carrying away of surplus population.

A perpetual exodus is going on from England, Ireland, Scotland, and Wales. No fewer than 154,000 emigrants from our British Isles entered the United States in 1881; whilst from France, the number that took refuge there was only 5227. The fact that we so largely seek relief by emigration, shows how at home the labour market is overcrowded. The population presses upon the food-supply and strives—of course ineffectually—to outrun it. Nothing but intense suffering could uproot these tens of thousands of our fellow-countrymen, and make them expatriate themselves to seek a living in far-distant lands; and who can doubt that *many* at least in doing this have forfeited their happiness, which was bound up in friends, and all the loving ties that held them to their homes? Nevertheless, through emigration some temporary relief is given to the old country, which has its industries congested through its too rapid birth-rate; but there is a *per contra* we must look at, a very serious drawback to the benefit derived from emigration.

It affects the quality as well as the quantity of life in the old country. It withdraws a large proportion of individuals who are above average in physical strength and mental vigour; and this tells both on the actual state of things, the population of the time being, and on the stock from which the breed will be kept up. I use these terms (more applicable to the lower animals) simply to make this matter plainer and more easily understood. The physically strong are urged to emigrate. Only the other day, a clergyman in the city of the forty charities was lecturing on Canada, and eloquently advocating its claims to be considered a fine field for emigration, and an El Dorado to the British working man; "but," said he, "persons who have not sufficient stamina to stand considerable hardships at the outset had better stay at home."

Emigration, then, is a social force that sifts our proletariat and selects the finest specimens to send them to the distant colonies. It also sifts out the *men* and separates the sexes. The 154,000 contained, no doubt, many wives, and some whole families; but it is an ascertained fact that thousands of single men went out alone; and what that signifies—as *touching human happiness*—both to the men who thus departed, and to the women left behind, is sickening, nay, appalling to contemplate. Happiness is *not* meat and drink alone. At least, if it means this to *any*, they must be very few. Happiness in the narrowest sense consists in the possession of a certain amount of the requisites of existence, food, clothing, friends, love, and tranquillity, with occasional excitement. Since emigration cannot give all these conditions, but only some, it is, I venture to assert, when looked upon (and many do so regard it) as a remedy for all our social miseries, most ludicrously, yet, alas, seriously, wofully inadequate.

And moreover, as the solution of our population difficulty it fails; for great as are the numbers of the surplus population it disposes of, these numbers are not equal to the surplus numbers of the birth-rate. The Marquis of Blandford, writing in the *Nineteenth Century* for February, 1882, says: "Probably one of the least considered, though at the same time one of the most vitally important problems of the future in England, is the question of population. Matters affecting trade and the laws regarding property, are only correlate questions, which vary in their importance together with this other factor. England has practically doubled * her population since the beginning of this century, besides having kept up an immense flow of emigration to other countries. Even in the last ten years the population has increased three millions. Every temporary increase in prosperity, caused by trade or abundant harvest, can be traced by noting the fluctuation of the birth and marriage rate among the population. Nothing synchronizes more completely in an inverse ratio than the price of bread and the birth-rate among the people. That such a fact should be true is an immense indication, if any were wanted, of the improvident character of the English working class. No sooner does a small increment of wages accrue to the people during prosperous times, than it is all at once absorbed by increased cost of living through early marriage among the more improvident: but the increase of population so caused again reacts on the condition of the labourer by creating more competition in the

* N.B.—It has done *more* than this.

labour market, and therefore a lower rate of wages. The cost of production is no doubt kept down, but the profits derived from trade are absorbed by the manufacturers in the higher interest obtained on capital in consequence of the lower rate paid to labour. Capital thus rolls up with capital, and becomes more and more concentrated in the hands of the few, while the great mass of the people are only semi-maintained at what we call a food level."*

A mass of evidence which *comes to us from every side* forces upon our minds the consideration, that our general social welfare depends upon the habits of the lower classes, especially in factory towns, and all the great industrial centres; and unless we can discover how to control their too rapid multiplication, we shall continue to suffer, and that perhaps more intensely than ever, from poverty and pauperism—fierce competition and extreme social pressure—an increasing inequality in the distribution of wealth—and a lowering of the healthy vitality and vigour of the British race.

It is simply impossible for general well-being to be maintained in a society where the increase is greatest from the worst stock; and truly the individuals at the base of our society are not of such mental, physical, and moral type as to make it desirable, or even safe, that the greatest proportion of young life, the vital strength (we may call it) of the coming race, should spring from them!

But *is it* possible to regulate parental conduct, and to change time-honoured customs amongst people who are, for the most part, unintelligent, ignorant, degraded, immoral? To seek to enlighten such upon the population question in all its bearings, and to spread before them the wide-reaching evil results of their personal action, would be absurd; and it would be equally absurd to appeal to noble feeling in them, and to demand or require from them unselfish or altruistic conduct. *Unless* the method we adopt is fitting to their purely selfish, or egoistic state of feeling, *we must fail.* There is no force that we can rationally expect to influence them but one, viz. their own personal interests. But in acknowledging this fact, let us check all impulse to irrational blame. They are what they are, through the action of *outward surroundings* upon *inherited natures* they did not choose. If development has stood still with them, and left them in the stage of simple egoism, it is because society in its ignorance has not been able rightly to respond to their

* "Political Opportunism," Blandford.

claim, and mould them into better social units, *i.e.* altruistic, or, in other words, tender and considerate human beings. It is for social intelligence now to deal with them simply as they are, and by thinking *for* them, and appealing directly to their *egoism* (for that is the only force strong enough to propel them along a *somewhat* difficult path), cause them to adopt a line of conduct which will give them first greater personal comfort, and after that, in a thousand ways, will tend to their improvement, to their attaining a higher and wider mental, moral, and physical development.

The use of artificial checks to reproduction promises, in my opinion, to effect eventually all that is desirable, and seems to me the *only possible method* by which society can reach to the foundation of its miseries, its poverty, its pauperism, and check these *at the source*. If my reader does not accept my arguments, let him disprove them; but I trust that no mere prejudice, no sentiment of prudery or false shame, will deter our many warm-hearted and intelligent philanthropists from afresh examining this all-important subject, and from obeying Mr. Arnold's brave and dignified injunction, to "let our consciences play simply and freely upon the facts before us."

The facts are too momentous and too grave to trifle with; and if, as I assume, a *powerfully beneficial* policy may be adopted, responsibility must rest on all who, from ignorant bias, or from conservative feeling, which they cannot rationally defend, condemn that policy and run the risk of retarding social progress.

If the preventive checks upon population, to which Mr. Malthus looked, had sufficed to keep our rate of increase moderate, and if these checks acted alike on all minds, and not only on the wisest and the best, my argument for the propriety of artificial checks would be greatly weakened; but the restraint that signifies total abstinence from all the joys of married life during long years, and at the period of life when the domestic instincts and the human passions are strongest, is only possible to individuals who are rare amongst the best intelligences, and *nowhere* to be found amongst the masses. And after all does it seem credible, that an inexorable law of nature demands from poor Humanity a sacrifice so great? The late Mr. W. R. Greg has told us frankly that *he* at least did *not* think so. His words are: "Though man is bound, both as a condition of progress and under pain of suffering, to control his propensities and to moderate his appetites and desires, he is not bound to *deny* them. .... If he eats or drinks immoderately, nature punishes him with dyspepsia and disease; but nature never forbids him to eat when he is hungry, and to drink when he is thirsty, provided he does both with discretion. Indeed, she punishes him equally if he abstains, as if he exceeds."* Then Mr. Greg goes on to show that as regards sexual instinct the action of nature is precisely similar. If man indulges to excess, nature punishes with "premature exhaustion, with appropriate maladies," but not otherwise; on the contrary, enforced and total abstinence is punished often, if not habitually, "by nervous disturbance and suffering and by functional disorder." Finally, Mr. Greg sums up: "Providence will be vindicated from our premature misgivings, when we discover that there exist natural laws whose operation is to modify and diminish human fecundity, in proportion as mankind advances in real civilization, in moral and intellectual development." †

Now, my thesis is—that we have reached the point in civilization which Mr. Greg here indicates. We possess a power hitherto unknown; I mean a power of controlling the physical conditions of reproduction, through the application of human intelligence; and to use this power in favour of general happiness and well-being is to obey the *highest law of nature*.

The development through social life of tenderness and sympathy has led to the counteracting of the law of natural selection, a traversing of the law of survival of the fittest; and as a consequence of this there have been introduced fresh miseries and evils, relating both to the quantity and quality of human life. If we can counteract and subdue these evils (whilst retaining all the tenderness and sympathy, and all the pleasures that are unhurtful to Humanity) by action that will moderate our increase, and promote the birth of social units that are generally *most fit*, we shall unquestionably be in the path of strict moral rectitude.

Meanwhile we must not dream that in adopting a new policy our social miseries could at once be remedied. On the contrary, fresh complications are likely to arise, and the apprehension of these will strike all minds that for the first time face this great social problem. But let these not too hastily decide. A remedy is not the less a remedy because it is attended with some serious evils; and my argument is that

* "Enigmas of Life," chap. ii.
† Mr. Greg makes no allusion to artificial checks.

artificial checks to reproduction will prove a remedy, and the only remedy, for poverty and pauperism, whilst any increase in laxity of conduct that may occur, is in comparison a superficial evil, capable of control, and certain to be checked by forces of another kind, such as discipline and training.

If philanthropists, then, oppose my thesis, I think that they are bound in fairness to attack it on the main argument and not upon side issues only.

But other considerations must also come into play. The present policy of widespread benevolence and patronage towards the poor is hurtful to benefactors and to the patronized. From a material point of view patronage is an utter failure. It is no cure for poverty and pauperism. It has been patiently and persistently applied for half a century, and yet our poverty and pauperism are not sensibly diminished, whilst as a moral agent patronage is worse than worthless. It is like a Upas tree exhaling poison from its leaves. The individuals who take shelter beneath its branches, and still more the masses born and brought up below its shadow, are for the most part destitute of moral beauty; for the conditions are wholly adverse to the poor man's virtues. His dignity and pride are trampled on, his honesty and frankness never pay, and if he prefers self-help and independence to accepting favours from a patron, he must often submit to see the men who are inferior to himself, with cringing and subservient natures, advance before him in the struggle for existence, and win the prizes which *he* cannot reach.

Patronage, again, acts prejudicially upon the benefactors, thousands of whom perceive that poverty and pauperism do not decrease, and yet, continuing on the lines of patronage, because they know of nothing else to substitute, they work and labour on amongst the poor mechanically, with feelings gradually hardening, and with tenderness and sympathy depressed, through hopelessness of any adequate success.

But now, rejecting all patronage as useless and injurious, the true method in all its details would require to be scientifically thought out and acted on. It is of a twofold character, embracing on the one hand vigorous measures based upon a high and definite standard of morality, and on the other a masterly policy of *inaction*. There should be a gradual but persistent lowering of the props—a steadfast withdrawal of those charities which may prove impediments to the growth of healthy independence amongst the poor. There should be calm and forcible discountenancing of the injurious habit of recklessly multiplying—a clear indication that public opinion now runs on the lines of Mr. Matthew Arnold's dictum, that a man's children are not really sent, and that to bring people into the world when one cannot afford to keep them, or to bring more of them into the world than one can keep thus, is contrary to reason and the will of God; and lastly, there should be a firm condemnation of personal conduct in defiance of that dictum as injurious to society and *therefore* gravely immoral. On the other hand, there should be also widespread, active, and intelligent teaching of all that relates to right parental conduct, and the laws of health.

Although in recent years health lectures have been given in many towns, the population question, with all its bearings upon health, has been little touched upon, and by the clergy its gravity and importance have been untaught by precept or example. Some months ago the following advertisement appeared in the *Times* newspaper:—" Urgent case of distress. Assistance is most earnestly requested for the widow of a clergyman (whose income has never exceeded £300 per annum) left by his death with ten children, six of whom with herself are left totally unprovided for. One of the six is a daughter, age twenty-one, of deficient intellect, and five are boys. The youngest two are slightly deformed and delicate. It is desired to obtain aid to place the widow in a house of which she might let part, and also enable the elder boys to emigrate." I may be permitted, I think, to comment upon facts brought thus prominently before the public. The parents in this case cannot be excused upon the ground of ignorant degradation. They were undoubtedly refined and highly educated human beings, and the clergyman at least was not, we may suppose, wholly ignorant of the population question, yet he ignored it practically, and his personal conduct was (to quote Mr. Arnold) "quite contrary to reason and the will of God." He brought into the world diseased children and a number out of all proportion to his income. He had intelligence to understand that his command over the available means of subsistence—a command represented by £300 a year—would not be adequate to feed, clothe, educate, and give a fair start in life to *ten* children!

Must we not then conclude that here at least an over-mastering force impelled the running of all risks, even to launching paupers into an already overcrowded country? and if this be so, what right have we to look for greater self-control from individuals of the pauper class?

The fact is clear that instinct often overbears intelligence; and disastrous consequences to society will certainly prevail unless intelligence can point to a path *comparatively* easy. The poor hear from the lips of their appointed teachers that "Happy is the man that hath his quiver full." We may assuredly assume that both precept and example in this matter have hitherto been forces running directly counter to the interests and well-being of society.

Then, general benevolence has also, as a rule, given preference to the men with "quiver full;" and competition has distinctly fostered the tendency to over-populate. Infant labour in factories has been put down, and children are now protected there; but the State does not forbid children trading in the streets, and their employment as news-agents, match-sellers, etc., has of late years become so general as to call for special notice. In the city whose forty voluntary charities have been described, the evening newspapers are principally sold by children, who are frequently complained of as bold and importunate; and a thoughtful judge, within the last few months, publicly condemned the practice. He pointed out that vicious habits are contracted by these juvenile traders; and he showed the folly of a system that permitted their exposure to conditions certain to degrade them, and then established institutions to reform young criminals. Rather than cure, prevention by the State would, he thought, be easier, wiser, and more humane. In Leeds, a public meeting was held on July 22nd, 1882, to consider the very same subject. The mayor deemed it necessary to take some steps towards "the protection of the numerous children, of both sexes, who sold newspapers and light articles in the streets." One gentleman proposed as remedy to organize a Children's Aid Society, and a News-Boys' Brigade, similar to those in Manchester, and he saw "no other means, if the gutter children were to be rescued from their present normal condition, and brought up in habits of industry and honesty." Another disapproved this course of action. He remarked that—so long as they were told on very creditable authority that four-fifths of our charities promoted pauperism, it was advisable to be careful before creating more of such institutions. A third speaker urged the necessity of legal measures—"an extension of the Factory Acts so as to make it criminal for parents to send children out into the streets after a certain fixed time." He thought also, "that blame lay at the door of the newspaper offices, for that the issue of so many late editions contributed to the demoralization of children." There seemed no unanimity of opinion except on *one* point, viz. the necessity of doing something. The facts stated were, that from 250 to 300 children were regularly going about the streets of Leeds, selling newspapers and matches, and that they could make as much as 3*s.* 6*d.* in a day; but frequently parents took the money from them for the indulgence of their own drunken habits.

The account of this meeting given in the *Leeds Mercury*, is followed by a leader upon it, in which occurs this passage: "No one can watch from day to day the life which is thus growing up in the streets without feeling that with all our benevolence, and with all our charitable appliances, we have somehow or other failed to reach these poor victims of the improvident and the drunken, who are forced almost from the cradle into the gutters to earn their own living, and to supply the vicious cravings of those who should be their protectors. And yet it is doubtful whether a new organization is desirable. . . . Whatever is done should be done thoughtfully."

Surely all who were present at that meeting felt keenly with Mr. Arnold that the knowledge how to prevent them accumulating is what we want, and that the thing of primary importance is to reduce the number of the births, and to restrain the drunken and improvident from furnishing an ever fresh perennial stream of gutter children.

With the new knowledge we possess, backed by a strong opinion in its favour from intelligent and thoughtful minds— minds that have grasped the *facts in all their significance and their bearings*—philanthropists may now succeed in checking the reckless multiplication of paupers without any appeal to legal force; but certainly the State should by law protect those children who are sufferers from the customs of the past, and wrest them promptly from all trades that are degrading and demoralizing. In the interests of the whole social body, no reversing of the relations of parents and children should be possible. If individuals seek the happiness and incur the grave responsibility of parenthood, they owe it to society and to their children to support them entirely during their tender years; whilst, on the other hand, children owe no support whatever to parents during the period of infancy. And if this principle is not respected, society should teach the lesson, and in self-protection act upon it at all times stringently.

The State might also with propriety put down all charities which are subversive of this principle, and which tend to foster the immoral habit of reckless reproduction. What are called ladies' charities * are distinctly of this order. The work they do is not in any sense beneficial to society in general. On the contrary, it is materially and morally injurious. It fosters multiplication amongst units that are *socially* unfit for parental duties ; it is grievously unjust towards the prudent and industrious poor ; it saps the spirit of independence, and discourages the personal habits that ought to be promoted. As regards these so-called charities, there seems to me but two alternatives—either they should be suppressed, or wholly altered in character. As agencies for bringing educated women into contact with poor mothers who are ignorant, they might be made most valuable, both in giving knowledge requisite in the controlling of the physical conditions of reproduction, and enlightening these poor women on all that relates to parental health. If ladies who support these charities adopt this course, they will be public benefactors, and redeem, to some extent, the mischief already done by their foolish and baneful, although well-meant, benevolence. A few ladies are already carrying out this policy with marked success. The tract † they offer to over-burdened mothers has been received with gratitude, and its instructions acted on.

But now I must refrain from further comment on the many individual schemes of general benevolence. My purpose is accomplished if I have proved that in that field reform is *absolutely necessary*.

We live in a political age, and on political reforms attention is apt to be concentrated ; whilst personal conduct, with its far-reaching results for good or evil, is overlooked, or little thought of. The general mind is critical of government, and wide awake to the importance of continual readjustment there, to meet the changes that time brings about ; but it is not critical of long familiar habits and customs, although these may militate against the general happiness far more decisively than even bad government.

Our national prosperity is at this moment great, and yet

---

* The charities that assist poor women in childbed.

† "Law of Population," post free, 7*d.* Publishing Company, 63, Fleet Street, London.

we have amidst us miseries and evils so profound as to call forth in every earnest mind a sense of national disgrace. Political reform will very little—indeed scarcely at all—affect these miseries and evils; therefore political reform, although important, is of secondary importance.

The leading reforms that are of primary importance, are first—the regulation of the number of people to the available means of comfort and subsistence, that is, attention to the law of population, and wise limitation of human births; and second—the improvement of the stock of the human race by promotion of the best types, and repression of the increase of the worst, that is, attention to the law of heredity, and wise social action in accordance with that law.

In regard to the portion of the population which is so degraded as to be *incapable* of giving heed to the morals of parenthood, I believe a time must come when the State, profoundly convinced of its moral obligation to promote the welfare of posterity, will sequestrate and restrain the individuals who persist in parental action detrimental to society. It cannot be permitted that superior types of mankind should be lessened in number by the increase of the inferior.

CHAPTER VIII.

SOCIAL PRESSURE IN THE MIDDLE CLASSES.

"There are many in the world whose whole existence is a makeshift, and perhaps the formula which would fit the largest number of lives is 'a doing without, more or less patiently.'"—GEORGE ELIOT.

"SPORTS," said Thomas Carlyle, "are all gone from among men; there is now no holiday for rich or poor. Hard toiling, then hard drinking, or hard fox-hunting! This is not the era of sport, but of martyrdom and persecution. Will the new morning never dawn?"

Poor Carlyle! for him the new morning never dawned. The social pressure he was under throughout his youth, the hardness of his struggle for an honourable existence, and the accomplishment of what he thought his great mission, were too great. When his efforts were crowned by entire success, and he had won honour and wealth, his nature could not unbend; and life brought him but little of pleasant ease of mind and cheerful recreation.

Early in his career Lord Jeffrey had written to him: "You have no *mission* upon earth, whatever you may fancy, half so important as to be innocently happy; and all that is good for you of poetic feeling and sympathy with majestic nature will come of its own accord without your straining after it. That is my creed, and right or wrong, I am sure it is both a simpler and a humbler one than yours." *

But Carlyle did not think so. He strove to check his natural joyous instincts; he preached to himself continually, and wrote in his diary, "Cultivate thyself in the *want* of enjoyment, gather quite peculiar experiences therein;" and

* "Life of Thomas Carlyle," vol. ii. p. 42.

this subjective repression of his nature, added to the objective repression of his hard surroundings, combined to make him stern, unsympathetic—a gloomy, discontented man. Repression (whilst absolutely necessary in some directions) when wrongly directed, or when too great, is full of injurious consequences. Carlyle's glorious powers and noble nature were distorted and deformed by it; and what the British race requires is precisely the opposite of Carlyle's doctrine, viz. to cultivate itself in aptitudes for innocent enjoyments, and never to repress itself unduly, or in wrong directions.

I have spoken of the strain upon our factory workers, the stupid, spiritless condition their hard lives reduce them to, and how when holidays are given, and for the moment repression is removed, the dull phlegmatic temperaments will only respond to coarse and pungent stimuli, and social gatherings with them become saturnalias. But in the higher classes also the same features are clearly discernible. Monsieur Taine's description of the Derby day shows what impression our national amusements may create in the mind of a being of a different race, who is simply a dispassionate spectator.

"It is the Derby day," he says, "a day of jollification; Parliament does not sit; for three days all the talk has been about horses and their trainers." Then, after a glowing picture of the beauty of outward nature, and the gay scene on the race-course, he goes on: "The illuminated air, like a glory, envelops the plain, the heights, the vast area, and all the disorder of the human carnival.

"It is a carnival, in fact; they have come to amuse themselves in a noisy fashion. They unpack; they proceed to drink and eat; that restores the creature and excites him; coarse joy and open laughter are the result of a full stomach. In presence of this ready-made feast the aspect of the poor is pitiable to behold; they endeavour to sell to you penny dolls; to induce you to play at Aunt Sally, to black your boots. Nearly all of them resemble wretched, hungry, beaten, mangy dogs, waiting for a bone, without hope of finding much on it. They arrived on foot during the night, and count upon dining off the crumbs from the great feast. Many are lying on the ground among the feet of the passers-by, and sleep open-mouthed, face upwards. Their countenances have an expression of stupidity and of painful hardness. . . . The great social mill crushes and grinds here, beneath its steel gearing, the lowest human stratum.

"We ascend to our places; nothing seems grandiose. . . . Probably I am wanting in enthusiasm, but I seem to be looking at a game of insects." Then he describes the race, and how as the first group of riders approaches, "Hats off!" is called, and all heads are uncovered and every one rises. "The frigid faces are on fire; brief nervous gestures suddenly stir the phlegmatic bodies; below in the betting-ring the agitation is extraordinary—like a general St. Vitus's dance. Picture a mass of puppets receiving an electric shock, and gesticulating with all their members like mad semaphores. But the most curious spectacle is the human tide which instantaneously and in a body pours forth and rolls over the course behind the runners, like a wave of ink.

"There is one imposing moment, when the horses are not more than two hundred paces off; in a second the speed becomes suddenly perceptible, and the cluster of riders and horses rushes onward, this time like a tempest.

"A horse of which little is known has won, and very narrowly; the betting against him was 40 to 1; on the contrary, it was 3 to 1 or 9 to 2 against the two favourites; hence there were miscalculations and explosions. The prize with its accessories amounts to £6,775; bets included, the owner will have won nearly £40,000. We are told of enormous losses—£20,000, £50,000; last year a colonel committed suicide after the great race because he saw he was bankrupt. Several cabmen have lost their horses and their vehicles, which they risked in bets.

"To my thinking, these bets are to the palate an indispensable stimulant for heavy and rough frames; they require violent impressions, the sensation of a prodigous risk; add to that the combative and daring instinct; every wager is a duel, and every large bet a danger."

The scene after the race is now described, and further on he says: "Towards evening the carnival is in full swing. Twenty-four gentlemen triumphantly range on their omnibus seventy-five bottles which they have emptied. Groups pelt each other with chicken-bones, lobster-shells, pieces of turf. Two parties of gentlemen have descended from their omnibuses and engaged in a fight, ten against ten; one of them gets two teeth broken. . . . Gradually the fumes of wine ascend to the heads; these people so proper, so delicate, indulge in strange conduct. . . . One of our party who remained till midnight saw many horrors which I cannot describe; the animal nature had full vent. There is nothing exaggerated in Rubens's 'Kermess' in the Louvre. The instincts are the same, and are equally unbridled; only, in place of portly, overflowing, and ruddy forms, picture to yourself faces which remain grave, and well-cut modern garments. The contrast between the natural and the artificial human being—between the gentleman who, by habit, and mechanically, continues grave, and the beast which explodes—is grotesque.

"On our return there are drunken people along the whole road; up to eight o'clock in the evening they might be seen staggering and sick at Hyde Park Corner; their comrades support them, laughing, and the spectators' faces do not betoken disgust. To-day everything is allowable; it is an outlet for a year of repression." *

We perceive, then, that it is not only amongst the lower classes that holidays become saturnalias; amongst the higher classes the brutal, savage instincts are not eradicated, and there is no such wide difference as one would *à priori* expect between the British labourer and the British gentleman. There are plenty of outward distinctive marks, but the inward nature is much the same. The Derby day gives year after year a clear indication of our uncivilized human nature and our diseased social condition—a state of misery and unnatural repression which finds relief in bursts of wild excitement and brutal license. What many boast of as our "national sport," when looked at with all its antecedents and surroundings, is to the thoughtful full of cause for national shame and sorrow.

But is it reasonable to trace any connection between our diseased social state and this wild carnival, except to notice on the surface, mingling with the throng, the miserable victims of poverty, drunkenness, prostitution, and crime—these deeply seated ailments of the social organism? The throng is for a considerable part composed of the wealthy and the so-called great—the aristocracy of Great Britain, the leaders of fashion in the gay world, the lucky individuals in life's game of chance, who without much personal effort are raised above the rude necessity of labour, and may if they like make of their lives a long holiday of pleasure-seeking. Poverty at least does not approach them, save to touch the hem of their garments; so surely misery and discomfort cannot form with them, as with our factory excursionists, at once a palliation and explanation of their coarse license.

* "Notes on England," by H. Taine, chap. iv., from p. 37.

Nevertheless, when we look closely, facts are not so dissimilar as a hasty glance suggests. Amongst the wealthy as amongst the poor, the conditions of life are unfavourable to happiness, and discontent and vague unrest in literal fact prevail. A life devoted to pure pleasure-seeking seldom if ever satisfies. Humanity wants more than this; it craves to have its best and noblest powers called into play, and exercised in action that will tend in some way to promote the general good; and idle men are as a rule suffering men, however little we may think it.

The younger of the two Robert Owens tells of a deep impression made on him by an incident in his father's history, and which greatly influenced his own life. His father on one occasion, becoming exhausted by his private and public labours, went to the house of one of his partners in business to rest, and recruit his health. The partner was a man of leisure, very highly educated, and able to gratify his cultivated tastes. His marriage was a happy one, and his children were growing up around him with the fairest promise. He had a handsome town house and a country seat, six or eight miles from London, in the midst of a magnificent park. The country seat with its surroundings was the ideal of rural elegance, and its owner, who had travelled over Europe, possessed mementos of his journeyings in statuary and paintings of the best masters. Everywhere there was luxurious refinement and all that makes life enjoyable.

There Mr. Owen found his friend, a man of leisure, with no more pressing occupation than to pore over the treasures of his library or enjoy his conservatory, filled with rare plants. Mr. Owen felt refreshed, and charmed by the elegance and the complete repose. "I have been thinking," he said to his host, "that if I ever met a man who has nothing to desire, you must be he. You have health, cultivation, a charming family, every comfort wealth can give, the choicest of all that nature and art can supply."

His host interrupted him: "Ah, Mr. Owen," he said, "I committed one fatal error in my youth, and dearly have I paid for it! I started in life without an object, almost without an ambition. I knew not the curse that lights on those who have never to struggle for anything. I ought to have created for myself some definite pursuit—literary, scientific, artistic, political, no matter what, so there was something to labour for and overcome. Then I might have been happy."

Mr. Owen suggested that he might still turn into a new

course, and in a hundred ways might benefit others, whilst fully occupying himself. But the reply was: "It is too late; the power is gone; habits are become chains. For me, in all the profitless years gone by, I seek vainly for something to remember with pride, or even dwell on with satisfaction." And the effort to adopt a new life was never made; the man remained an inwardly repining *martyr to leisure*.*

Now, who can doubt that thousands of our wealthy classes are in the same position? To have no stimulus to labour, and no opening for effective labour suited to the distinctive individual capacity, are sources of misery to a portion of mankind, almost as great as are overwork and hard necessity to another portion.

As a converse to this picture of Mr. Owen's friend, let us now turn to George Eliot's Caleb Garth, the man whose happiness in his work had nothing whatever to do with its money payment. "Caleb Garth often shook his head in meditation on the value, the indispensable might of that myriad-headed, myriad-handed labour, by which the social body is fed, clothed, and housed. It had laid hold of his imagination in boyhood. The echoes of the great hammer where roof or keel were a-making, the signal-shouts of the workmen, the roar of the furnace, the thunder and plash of the engine, were a sublime music to him; the felling and lading of timber, the crane at work on the wharf, the piled-up produce in warehouses, the precision and variety of muscular effort wherever exact work had to be turned out—all these sights of his youth had acted on him as poetry without the aid of the poets, had made a philosophy for him without the aid of philosophers, a religion without the aid of theology. His early ambition had been to have as effective a share as possible in this sublime labour." †

He was, she further says, a man of a reverential soul and a strong practical intelligence, but with no keen competitive spirit, no strong self-interested views on the question of profit and loss; and consequently unfit to fight his way to fortune in an industrial field where purely selfish competition and antagonism reign. He had to give himself wholly to the kinds of work which he could do without handling capital; and he was "one of those precious men within his own district whom everybody would choose to work for them, because he did his work well, charged very little, and often declined to charge at all. It is

---

* "Threading my Way," by R. Dale Owen, paper iii. p. 79.
† "Middlemarch," book iii. p. 184.

no wonder then that the Garths were poor, and lived in a small way. However, they did not mind it."

Now, what George Eliot here depicts is a man with aptitudes for varied and useful employments, who delighted in the exercise of these aptitudes, and wanted nothing but scope to bring them into effective action. " 'It's a fine bit of work, Susan! A man without a family would be glad to do it for nothing. It makes me very happy, Mr. Farebrother,'—here Caleb threw back his head a little, and spread his arms on the elbows of his chair—'that I've got an opportunity again with the letting of the land, and carrying out a notion or two with improvements. It's a most uncommonly cramping thing, as I've often told Susan, to sit on horseback and look over the hedges at the wrong thing, and not be able to put your hand to it to make it right.' " Metaphorically speaking, many an idle human being at this moment is sitting on horseback looking over hedges at busy workers on the other side, and feeling uncommonly cramped by the dignity, the honours, the non-laboriousness of the position he holds, and is compelled to hold, for he neither had, nor has, any free choice in the matter.

My reader may incline to think that I exaggerate the position. Surely work is everywhere, and willing workers who can afford to dispense with pay, may find their mission in amateur philanthropy, if not in the sphere of politics. But the all-essential point in work is, that it should be *useful*, that it should clearly and unmistakably tend to the well-being of ourselves or others. Without *that*, it is salt without its savour. And I maintain that very few of all our philanthropic schemes for social reform can give one grain of healthy satisfaction to the mind that craves a useful outcome in the field of action. As to politics—special aptitudes and qualities are required for that, as for every species of labour; and even if it were possible (which it clearly is not) to find occupation in it for our innumerable men of leisure, we should certainly experience that too many cooks spoiled the broth, as well as that many of our cooks were bad, *i.e.* wholly unfit.

The fact is this—our present system of things does not admit of every individual finding useful work suited to his nature; and whilst it is painfully true, that "all work and no play makes Jack a dull boy," all play and no work equally makes Jack spiritless and dull. What is required is *both* ; for pleasure lies in the *alternation*, and so long as man has only one, conflict goes on in human nature. There are promptings, innocent and pure, which are repressed, and in repression the happiness of life is spoiled.

The dissipation and final ruin of many fine young men arise from no other cause than this. The outward conditions of their lives, their social standing or position, oblige them to check desires, which, if they had been able to satisfy in a normal and legitimate manner, would have given them satisfaction and inward harmony, whilst causing no injury to any human being, but which they had not moral or intellectual strength enough to hold suppressed, whilst standing upright like martyrs at the stake. Weakness, not badness, was the cause of ruin, and what was wanted was conditions favourable to happiness, for happiness makes individuals strong.

When there is in the minds of parents a scientific knowledge of human nature, and of all that is essential to the happiness of the young, conventional barriers will be broken down, and many changes take place; but there will be fewer, and in the *far distance*, we may hope, not any shipwrecks of young lives. As it is now, many middle-aged men awake, like Mr. Owen's friend, to a humiliating double consciousness. They feel a craving for useful honourable occupation, and at the same time the conviction that they could not fulfil the duties of such occupation. It is not easy to break the chain of idle habits, and whilst inwardly fretted by these, a depressing sense of feeble impotence tends further to distress and sadden them.

When health is good, individual happiness largely depends on useful activities; and a social system that fosters and encourages the growth in number of the class of the community called in the language of Political Economy " non-productive consumers "—that social system, I say, is *not* productive of widespread general happiness.

Women especially have been great sufferers from lack of satisfactory outlets for their energies. Numerous women in the middle classes, unable to marry, through no fault of their own, but simply through the force of outward conditions, have been condemned to lead narrow lives, unsuited to their nature. Have we not all recognized the truth of George Eliot's picture —" Miss Winifred Farebrother, nipped and subdued, as single women are apt to be who spend their lives in uninterrupted subjection to their elders " ?

Wives and mothers have a social position; old maids have none, unless they are wealthy. To suppress themselves is what society has required of them; to hide away in obscure corners, and to come out at intervals to fill some gap in other people's lives—to keep a widower-brother's house, perhaps, until he forms another marriage tie, when back the old maid goes into her corner, or to nurse some married relative when sickness has overtaken him—this is all the sphere that such women have as a rule enjoyed. Ought I confidently to assert that this is not enough for happiness?

All of us may readily recall some example of a life of so-called "single blessedness," surrounded by an atmosphere, a very halo of patient gentleness, and sweet, though sad, contentment. The nature has adapted itself to the inevitable, and religion has proved a solace. But approach that life more closely, and in its history read, as it were, between the lines. You will find a world of suffering, a record of continuous struggle to keep the attitude of mind which conscience approves; and you will recognize how much is lost in a society that "nips and subdues" a flower-like life, instead of offering it free scope for full expansion.

Celibacy not freely chosen, but forced upon an individual, is a *cruel lot;* and so long as to that is added social contempt (which as regards women is, in a great measure, the case), so long as old maids are more like social ciphers than social units, what wonder is there if embittered hearts and gossiping, perhaps malicious, tongues make social life discordant and contemptible?

When a pent-up river finds an outlet, its rapid course and wide outspread show the observer how great has been the force, cribbed, cabined, and confined; and similarly the "Woman's Rights" movement, with all its offshoots, shows the amount of energy that had been suffering undue repression. The last few years have seen a rapid, wide advance; and, gentle as the sex proverbially is, it has been somewhat loud in self-assertion. Men may feel surprise to see women invading provinces hitherto held sacred to themselves, demanding equal education and a fair share of what there is to gain in trades and professions, pushing their way to social influence on school-boards and poor-boards, mixing themselves up with municipal elections, and fighting for the franchise. But all this is accountable and clear as daylight to the student of past history who discerns the antecedents of those social forces that compel attention now. The lives of unmarried women, and of many who are married, have never yet afforded the general conditions of human happiness; and all the fuss and effervescence will subside when these conditions are better apprehended and to a great extent secured.

The utmost individual freedom women require, and have a right to, as much as men. Their personal development must be no longer checked; the vexed questions of what woman's natural sphere is, what she is fit for, *can* only and *will* only be solved by experience, and the free exercise of her powers in action that is tentative.

This great advance of women, whilst beneficial to the intellectual development of the sex, and certain to result in better adaptation of the laws and other outward conditions of life, carries with it in the mean time many drawbacks. It is a struggle between the sexes. Antagonism flourishes, and cordial relations between men and women become more difficult to maintain. Men of conservative nature, who love to protect the weak and who admire the gentle qualities that have been evolved during the period of the subjection of the sex, are full of fears that now the female type is altering; that Amazons, blue-stockings, strong-minded females, are alone what the future will provide as men's companions, wives, and mothers. But the type that occurs, and is even predominant in a transition epoch, is not *necessarily* the woman of the future. With a secured position on the platform of social life as elevated and a sphere as free as man's, with sex equality established and no distinct and special battle of her own to fight, the Amazon is certain to disappear. The corners that offend will be rubbed off, and the gentle female nature, for the first time allowed a full and free expansion, will get rounded into moral and emotional as well as intellectual beauty.

On the other hand, to women who are fighting their way to social liberty and equality the action of conservative men seems mean and despicable. To grudge them a fair chance of success in fields of remunerative labour; to shut the gates of high professions and bar them legally, lest *women* should enter in; to keep to themselves the privileges and rewards of great attainment in arts; and, in short, to act on all occasions on the principle that "might is right," although that is a principle not publicly acknowledged—all this is barbarous; it partakes of savage tyranny and brutal despotism. But when we look to *causes* it appears less the result of individual coarse brutality than of a social force, which, acting on individuals, induces unjust conduct.

When first a band of female aspirants to the medical profession appeared in Edinburgh, and tried to force an entrance to the sacred inner precincts of its University, one of their bitter enemies and staunch opponents, himself a doctor, quite frankly stated in public that his ground of opposition was the danger that he and his brethren might be crowded out by ladies from a field of remunerative labour. " I have myself," he said, "six sons. I purpose to make some or all of them followers of my own profession. To my certain knowledge there are plenty of young men available for all the work the medical profession affords, and we have no need of women. Why should we, then, admit them? No! let us be wise; let us resolve to keep them out, and not expose our sons to greater hindrances and social pressure." *

Here, then, is the force that, acting on men, makes them unjust and brutal in action. The social pressure *is too great*. The struggle for success in every calling is intense. The numbers are excessive, and competition is fierce and unscrupulous. The father of six embryo doctors was to be pitied, whether women entered the medical field or not; some of the six would certainly succumb to the hard struggle for existence, and bring him pain and sorrow.

He did not exaggerate the facts. Within the last eight months a dentist in Edinburgh advertised for an assistant. The salary was small, and yet there were above two hundred applicants, and half of these were young M.D.'s.

In other fields, where culture and long and careful training for the work are required, the prospects of success are quite as small. The number of pictures sent to the Royal Academy this year † has been without parallel. The average hitherto was six thousand canvasses, but fifty per cent. must be added to this year's total, and no fewer than seven-eighths of the number have had to be rejected. Let us pause to think what, in the domain of human emotions, this fact alone implies. If we could follow back to where they came from all these rejected pictures, the overwhelming disappointments, the blasted hopes, the frantic despair we should encounter would appall us. It means to many the impossibility of selling pictures, whilst on a ready sale depends the bread of loved ones. It means semi-starvation, followed, perhaps, by premature death. It means madness to some, and to all mental suffering.

The field of law is equally overcrowded. A few days ago

---

* I am not quoting verbatim, but this was the tenor of the argument.
† This was written in 1882.

two ladies were expressing deep regret on hearing of the death of a well-known lawyer, a man in the prime of life, whose death was sudden and unexpected. The son of one of these ladies entered the room, a lad of sixteen, who had been about three or four months in a lawyer's office. Hearing what had happened, he looked grave for a moment, and then remarked, "Well, mother, there is one good thing; his death is a step up for me." "What do you mean?" his mother asked; and jauntily the lad replied, "We've far too many lawyers here. The fellows tell me I'll have little chance unless a few die off, or emigrate. I sometimes think I'll emigrate myself, for, you see, it needs one to be so jolly clever to get on and make any money here!" To hear young people speaking in this tone is inexpressibly sad. How can sympathetic, generous feelings be fostered in surroundings that create a sense of gladness when a fellow-creature dies, and eagerness to fill the dead man's place? And yet what can a mother say? To chide were folly. Our lads are brought into contact with hard facts, and we must face these facts and deal with them, *not* live in a world of pure imagination, preaching unselfishness when all the time we wish our lads to achieve success in life; and that is only to be done by pushing, scrambling, and a large amount of egoistic self-assertion.

The intellectual atmosphere and moral tone in business life are so *entirely different* from the teaching given from pulpits, and in homes by refined mothers, that it is perhaps surprising that clearness of intellect is ever reached at all; and certainly it need not surprise us that a vast majority of individuals go through life confused and muddled in their thinking as to what is right and what is wrong in actual conduct. Yes, it is true, there is an utter *inconsistency* between the morals taught the young and the moral code they see acted upon by business men; and if the future is to produce noble, upright, generous, conscientious professional and business men some *reconciliation* must be found, and human conduct must be studied and then taught as a science and an art.

Some fifty to eighty years ago, as peasants multiplied and cottage homes were crowded, the problem of a career for *all* was solved by *education*. However closely packed within the home children were mostly out of doors the whole day long, and those who inherited strength of constitution had favourable conditions to healthy physical life. Those members of the group who showed superior cleverness, any mental power above average, were singled out by the parents for a higher calling than their own, and every effort was made by the whole family to give these favoured ones a good education. *That* once secured, no doubts were entertained of ultimate success in life; and meantime the social pressure at home was suitably relieved.

The picture of the peasant home of Thomas Carlyle's childhood, and his career, illustrate my meaning. The solidarity of the peasant group is striking and pleasing. Father, mother, brothers, sisters, all aim at helping Thomas to a good education; and no sooner is he earning money than *he* supports and educates his younger brother John. To that class a higher education formed the egress from a life of too great social pressure at their own level. But now there is no such escape. To rise from peasant life to commercial or professional life is only to exchange one field of social pressure for another. The struggle for success is quite as great in the sphere of educated labour as in those spheres where little or no education is required, and education makes us all more sensitive to the pain of failure.

Again, if emigration is to be the door of escape to which we all must look, then education of the higher kind is a barrier to success, and a drawback instead of an advantage. Now, what do I mean by *success?* Do I mean gain in money, and am I forgetful of the natures which, like Caleb Garth's, can make, without one mercenary thought, a joyous poetry, philosophy, religion, out of intense personal devotion to the cause of wholesome and effective work? No. But work is *not* wholesome under all conditions; and when a being, refined, cultured, social, susceptible of all the pleasures of civilized life, is forced to cut down wood in Canada, or live in isolation on a sheep-run in Australia, success to him can only mean making enough of money to return to his old home before capacity for enjoyment has wholly left him. Meanwhile the qualities that rendered him fit to enjoy and embellish social life in intellectual circles cause him in solitude to pine and suffer acutely. I can conceive no greater act of cruelty than to call out the higher powers of individual human nature, and then thrust the individual into conditions where none of these powers can meet with satisfaction. Yet *this* is what many parents do. If money can be made in distant colonies, the real fitness of the workers for the work is little thought of, and present happiness is wholly sacrificed to a possible future good. Moreover, when the future good arrives and money has been gained, how often does one see that all the sap and pith of sweet humanity are gone! "The human heart finds nowhere shelter but in human kind," and when removed for years from its true shelter it is apt to fossilize or petrify.

A young man writing from Manitoba, on March 31st, 1882, says: "I have been asked more than once if there are many young ladies here, and with the exception of a few of these fair creatures at the town of Birtle, we are all out in the cold here. But for my own part, I have become wedded to my pipe, and as tobacco is cheap, I find it suits me better than a wife with expensive tastes!" No doubt this man will marry when he has made a fortune and can afford both the luxuries he thus classes together! His future wife, methinks, is to be pitied.

The British nation is intensely proud of all her colonies, and perhaps with good reason; but after all, to thousands they have proved the grave of hopes, the death of happiness, the birthplace of despair. Social pressure at home makes emigration necessary, but it is not a charming and sublime exit from all our difficulties, and as a permanent relief from our too rapid increase of population it is failing.

The annual excess of births over deaths is larger than the diminution of our numbers by emigration, and therefore social pressure continues. It is for parents to take wider views of success in life, and weigh well the prospects of their children. Happiness is the great object of life; and individual happiness depends upon conditions suitable to the individual's nature, both the inherited nature, and what it has become by training, and by the action of outward surroundings upon it, during childhood and youth. Many a child is educated into physical unfitness for enduring hardships; and emigration is simply the old country getting rid of surplus numbers by the *barbarous, cruel check of premature death.*

The remedy is to regulate the birth-rate by thoughtfulness and care. Surely no parents are incapable of forming an estimate of their own social position, and what number of children they may reasonably hope to rear, train, educate, and put out into the world with fair prospects of achieving a happy, useful, satisfactory life. If parents exceed this number, and run all risks of misery for their children, an enlightened public opinion should condemn it as a social crime.

That public opinion is moving in this direction is certain; but that indications of this should appear amongst the young is humiliating to our adult population. A few years ago an educated mother of the genteel but poor portion of the great middle class was deploring her hard life before her children. Her large family, she said, gave her constant care and toil, and left her neither time nor money for amusements. To her surprise, a boy still in his teens, quite gravely put to her the question : "Why did you bring us here, mother, when you had no fortune and no good looks to give us?" This is a striking sign of what is going on around us.

Up to this epoch the social organism has been *unconscious*. Blind impulses have been obeyed and reflex or spontaneous action has controlled our destiny. By the law of natural selection, however, which (when uninterfered with by man himself) brings about survival of the fittest, and by the slow uprise of *duty*, viz. that moral sense in man which has its origin in the tendency of right action to promote the general welfare, *progress* has actually resulted, although at fearful sacrifice of individual life and happiness, and, as Mr. Greg expressed it, by " marching forwards in an exceedingly piecemeal fashion !" It was the infancy of the social organism, and stumbling steps are natural to infancy.

A *new* era is dawning for us now—the era of self-consciousness. The social organism has developed *intellect* as well as a new power of sympathy and deep emotional feeling. It demands *reasons* for all things. It boldly asks, " Is life worth living ? " and will take no answer that does not commend itself to able reasoning powers, to a rational and calm intelligence.

It finds its sympathy at variance with intellect ; for whilst sympathy prompts it to indiscriminate charity, and to the cherishing and keeping alive of individuals who are feeble and diseased (so that they multiply their kind), intellect condemns such action as morally and physically injurious to the race. It sees distinctly that in acting thus, it frustrates the law of nature by which survival of the fittest is secured, and makes more permanent the diseased state of the whole society.

Again, it finds its sexuality at variance with intellect ; for whilst the purely natural impulse leads to the rapid birth of children, these children are often mentally or physically diseased—or have too feeble a vitality to pass through life with healthy, joyous vigour—or find the fields of labour they must occupy, too crowded to admit of ordinary capacity gaining success—or in some other way the lives the parents give are not surrounded by general conditions favourable to human happiness.

Without these conditions society begins to judge that life is not worth living, and to condemn those parents who act as though *they* thought otherwise. Blind impulse in this matter will soon no longer be tolerated; for human intellect has discovered how it is possible to dissever the connection between sexuality and the law of population, and to bring about a state in which, without undue repression of that natural function, sympathy and intellect in the social organism can harmonize and *be at peace*.

Hitherto there has been a groping after social regeneration, through political reform, reforms of government, of law, of classes, and so on. But *now* eyes are turned in the true direction. A change in *individual conduct* throughout the *whole social body* is required. A change which will in time procure the advent of a healthy, pure-blooded race, in numbers that will not prevent their full development, their ample possession of all the comforts, and many of the amenities of life; and in the end, will bring them euthanasia painless and unregrettable, instead of death when immature or by acute disease.

In artificially checking the birth-rate intelligence controls the law of population, and *heredity*, with all its deep significance, becomes a social force, tending to race improvement and to general well-being. Irrational breeding stands in the way of rational training, and tends to social misery of every kind. But as society slowly but surely attains self-consciousness, rational breeding will be followed by rational training, and individual conduct will take the form which best accords with all the highest interests of the whole social body.

There will no longer be a crushing social pressure which cramps all our activities, and renders it well-nigh impossible to force a way to happy, healthful, free expansion. Political economy, with competition as its great factor of advance, belongs to the spontaneous, unconscious epoch of our history. By all economists before John Stuart Mill it was assumed that when that science showed what course of conduct tended to the most rapid increase of wealth, all individuals who accepted the arguments would certainly adopt that course. But Mr. Mill repudiated such an inference, and held that it belonged to every individual to decide for himself, whether the *end* to be gained is equal to the cost of its attainment. There are many now who recognize that wealth is *not* the highest good for human beings, and that when much wealth is ill distributed it is the direct cause of discontent and many still graver social evils.

A science of society in its conscious, rational state is sure to be developed and to supersede political economy; a science in which sympathy and co-operation will take the place of competition, and not wealth alone but happiness will be the object that every social member seeks to attain. The misery of too exhausting and laborious work, and the misery of idleness, will be alike condemned and shunned; whilst active employment, suited to the capacity of each, will be embraced by all; and to that happiness which springs from exercising all our faculties and then reposing them, there will be added all the higher joys of mutual helpfulness and sympathy in co-operation.

CHAPTER IX.

THE EVOLUTION OF MODERN SENTIMENTS.

"The facts in moral history which it is at once most important and most difficult to appreciate, are what may be called the facts of feeling."—LECKY.

> "He roved among the vales and streams,
> In the green wood and hollow dell;
> They were his dwellings night and day—
> But nature ne'er could find the way
> Into the heart of Peter Bell.
>
> *   *   *   *   *
>
> A primrose by a river's brim
> A yellow primrose was to him,
> And it was nothing more."

THERE were no tender chords of feeling thrown into vibration within Peter Bell, no delicate shades of pleasurable sensation called into his consciousness by the sight and scent of the lovely primrose; yet no one would deny that the poet Wordsworth in his picture is just to human nature. There are plenty of Peter Bells amidst us even now, and there are also tens of thousands of human beings of similar clay, so far as one can judge by the exterior, who yet in point of feeling are cast in a wholly different mould—unlettered men it may be, without one spark of the genius of a Burns, but whose natures can be thrilled to sudden sweetness and pathos by the mere sight of some wild flower, and who, had they the gift of song as Burns had, would give a fit expression to their rich emotions in rhythmic measure as full of charm as are his sweet lines to the mountain daisy.

If it were possible for us to follow accurately step by step the intellectual development of a man from infancy to age, and mark the process, we should see how every outward change, however slight, in his environment, is instantly responded to by change within—how one by one his crude ideas fade away and are dropped off to be replaced by thoughts of wider range, and truer to the facts of life around him—how with the larger grasp, the outward seems to grow to his new powers of comprehension, and steadily intelligence ascends from simple perceptions to the more compound, from simple reasoning to reasoning more and more complex and abstract; till thus, along a path of infinitely small gradations the force of evolution has borne the child to manhood, transformed the germs of thought within him to comprehensive and wide cognitions, and given him ample powers for new attainments in the sphere of intellect and knowledge.

Now, a similar process takes place in individual feeling, or the sphere of the emotions. A child's feelings, like his thoughts, are simple, not compound. They are for the most part vague and indefinite, and always fleeting or evanescent; but as he grows, his powers of feeling grow also, and the feelings themselves alter in character. Their childish simplicity passes away, they augment in mass, they become compound and complex, more permanent and coherent in their nature, and far more delicate in susceptibility. Consequently the breadth of range, the depth and richness of emotion possible in an adult, as compared to the emotions of a child, are as the music of an organ to the sweet notes that lie within the compass of a penny whistle.

The group of complex emotions that we call æsthetic were absent in Peter Bell. He had no sensitiveness to beauty of form or colour, as shown in cloudland, mountain-top, or lowly primrose at his foot.

> "The soft blue sky did never melt
> Into his heart, he never felt
> The witchery of the soft blue sky."

Yet, bad man as he was, victim to all demoralizing effects of a vicious city-life, he trembled in the presence of a woman's sorrow, and showed himself capable of a wealth of sympathy, a deep emotion full of mingled elements—compassion, fellow-feeling, impulse to comfort—until as the widow sat—

> "In agony of silent grief—
> From his own thoughts did Peter start;
> He longs to press her to his heart,
> From love that cannot find relief."

And so it is with millions of our countrymen whose hard lives make them outwardly seem dolts or clod-hoppers, blunted to all fine feeling. When the spring, the hidden chord that touches their most tender vital part is played upon, the emotion stirred, the sentient state the man experiences and exhibits is unlike the simple feelings of a savage and inferior race; it shows what evolution effected through a long course of years by action within and on his ancestors—it shows emotion far removed from the rudimentary stage and far advanced towards a stage of rich, full, mellow maturity.

Now, it is this *evolution of sentiment*, not in the individual but in the race, with which I have here to do; and my purpose in this chapter is to point out some of the modifications certain to take place as social life progresses in the path that leads to general happiness. My reader is probably quite alive to the fact that distinctive sentiments and modes of feeling characterize the different races of mankind, as well as distinctive outward features, and that the impressing upon a plastic race of these divergent states of feeling is mainly, though not entirely, due to external conditions, by which I mean, not climate, geographical position, and such like *only*, but also the form of civilization that has taken root, and moulded the habits and customs of the race. For instance, the Greek civilization tended to develop largely the æsthetic group of feelings; whilst on the other hand in Scotland (through outward influences which I must not pause to trace) these feelings have been stunted in their growth, and moral sentiments have had a far deeper and more firm development. Amongst barbarous tribes of men, the simple but more violent emotions, such as anger, fear, jealousy, revenge, generally hold sway; but there are also uncivilized communities where these fierce passions are little known, and where, in consequence of the absence of all warlike surroundings, the gentle, tender sentiments, that have for their foundation family ties and peaceful social life, prevail, and are considerably developed.

Again, the sentiment of love of property is clearly of slow growth, dependent on the growth of intellect in the race, with which it seems to correspond; for mark, a savage only cares to obtain the things that he *immediately* requires. He lacks imagination wherewith to picture to himself what he will want to-morrow; he lacks intelligence to meet remote contingencies, and sympathy to give a motive for providing for others; consequently improvidence is with him the rule. But in a

civilized community, where an established government gives
security of possession, prudence and forethought meet with
due reward. The man who steadily accumulates, surrounds
himself with comforts and inspires the like desires in others.
Secure in all his comforts, his energies are free for self-develop-
ment, and the man himself becomes more complex. He has
intellectual tastes that bring him pleasure; he has love of
travel; he delights in seeing and possessing works of art ; he
has keen sympathies which give him exquisite enjoyment when
he finds means to minister to others' happiness; and for all
these new gratifications *he must have property*. As he goes
on accumulating, he recognizes with his expanded intellect the
varied and wide result of all his labour, and his acquisitiveness
becomes a compound sentiment, made up of elements in great
part intellectual, but which in his descendants may appear as
the pure, unconscious, yet complex emotion, which we call
love of property.

Now, is this love of property capable of abuse? Are we
bound to admire it in all circumstances because we see its
genesis to be respectable, nay admirable, and clearly beneficial
to mankind? *A priori* we should expect that in all circumstances where it is wholly absent, there would be little or no
development of æsthetic tastes and pursuits, although there
might be (as in effect we see amongst the Dhimals and Santáls
and other tribes) a wide development of sympathetic emotions
and of all the amenities of peaceful social life. This forecast,
however, has been lately proved a false one. Dr. Le Bon has
laid before the French Geographical Society an account of a
primitive unsophisticated group of people, numbering several
hundred thousand individuals, dwelling in a remote region
high up among the Carpathian Mountains of Galicia, of
whom he says, "Riches have no charm for them." They are
poor, living principally upon oats, boiled or ground, and
converted into cakes, and on goats' milk whey, of which each
man consumes three or four litres daily. They enjoy perfect
health and live long lives. They are frank in manners and
extremely quick of apprehension. They are fond of dancing,
and when the labours of the day are over they meet to indulge
in their favourite pastime. But mark this point. Dancing is
not their only art. They are poets and musicians also—
"born *improvisatori*," Dr. Le Bon calls them; and many of
them sing their own songs set to music of their own com-
position. Their poetry is tender and artless in sentiment,

generous and elevated in style; and he attributes these qualities to the wealth of spontaneous resources possessed by natures which neither know violent passions nor unnatural excitements.

We see, then, that where there is no development at all of this sentiment which we call *love of property*, there may, nevertheless, be a considerable development of the group of emotions which we call æsthetic, and which we look upon as evidences of refinement and culture. Therefore we *are not* entitled to allege that were the British love of property and instinct to accumulate vastly moderated, there would result from this the gradual decay of art and a reversion of our race to a more barbarous, uncivilized condition.

Looked at *en masse* the British race presents a wonderful contrast to the Podhalians; for not only have riches *some* charm for it, but they have a charm out of all proportion to their true value, which is their power of helping man's advancement and ministering to his happiness. Throughout society we note this aspect of the general mind. A man's worth is very commonly estimated by his wealth. The ability shown in building up a fortune is conceived as fitting any man to enter Parliament, whether or not he has had the intellectual training and the culture absolutely necessary to make him useful there. Millionaires are subjects of universal envy. The successful merchant goes on from year to year accumulating, long after he has amply provided for himself and family, and for future contingencies, so far as that is possible; while the improvident poor man's state of mind relative to money is as a rule extremely similar to that of the lowest savage towards his fetish. He feels a reverence for money, compounded of mingled admiration, awe, rising almost to fear, and desire, with a sense of distance from his god which makes him helpless, hopeless, abject, and despairing.

Now, the Podhalians notwithstanding, I think we may with certainty infer that without the outward accumulation of wealth, and the inward corresponding feelings, viz. respect for property and individual acquisitiveness, no nation could ever reach a high state of civilization and be in the van of the great army of Humanity, as it pursues its conquering course, and with bloodless weapons masters Nature's forces, wrests her grand secrets from her grasp, and in the glorious and manly strife expands its own intellectual powers, and shapes its destiny in accordance with virtue, happiness, and peace.

But Mr. Herbert Spencer tells us that in some established

societies "there has been a constant exercise of the feeling which is satisfied by a provision for the future; and there has been a growth of this feeling so great that it now prompts accumulation to an extent beyond what is needful."* Undoubtedly *we* are a case in point. Great Britain as a nation has the impulse to accumulate beyond what is needful, and this is true, in spite of the apparently contradictory fact that throughout the nation improvidence is largely displayed, and there are in it literally hundreds of thousands of individuals with no property at all, and with no desire or capacity for accumulating property. Capitalists have often said, and with reason, "Why should we raise the wages of our men? They never save, they never make provision for the future."

With, of course, many honourable exceptions, the individuals of our great working class have far too little of the sentiment of acquisitiveness, whilst our capitalists have far too much; and the problem for the future, *if* as a nation we set ourselves to work out general and universal happiness, is, how by wise institutions and by wise and enlightened training of the young, to stimulate this sentiment where it is deficient, and to repress and carefully subdue it into proper proportions where it is too largely developed.

The terribly unequal development of acquisitiveness throughout the community is due to perfectly definite causes, although causes so complex that only a student of evolution can trace them out by studying the history of our extremely rapid growth in industrial activities, and watching the action and interaction of the forces these activities have brought into play. The inquiry, however, is apart from our subject. The inequality exists, and is evident on every side; and what it deeply concerns us to know and understand is, how far this inequality hurts our social well-being, and checks our intellectual and moral development.

For happy social intercourse amongst individuals, for mutual respect and confidence, and sympathetic emotional enjoyment, some affinity of sentiment is absolutely indispensable. A wide disparity of feeling is as obstructive to the flow of sympathy as a wide disparity in intellectual thought and knowledge. If in the play of circumstance an individual who resembles somewhat an inarticulate Burns is thrown in daily, hourly contact with a Peter Bell, there can be no enhancing of the joys of life through interchange and sharing

* Volume of Essays, 2nd series, p. 132.

of æsthetic emotions. In that direction Peter's mind is blank, his nature dumb, and both are losers by the want of equality in development of æsthetic feeling.

Now, what we see between these two, let us extend to a whole group of individuals. In a community, whether great or small, the steady growth of social qualities and of the happiness that flows from these, depends upon the opening up of channels of sympathy throughout the whole group, and the forming of *community of interests*. I do not use the word interests in the business sense of material profit, nor in the vulgar sense of idle curiosity as regards one's neighbours, but in the purely intellectual and emotional sense. There can be no such channels of sympathy, however, with a great mass of heterogeneous social units grouped into classes that are wholly different in feeling, and therefore in action, for it is always feeling that directs and controls our action.

I might, of course, enlarge upon the differences in culture and intellectual thought that impede sympathy amongst us; but others have spoken far more eloquently on that theme than I could do; and relatively to general well-being I think it much less important than the difference in this one emotion of acquisitiveness and love of making provision for the future. Its strength and potency with a vast number of our social units can hardly be over-estimated; it forms the pivot of their daily revolutions. Their thoughts and feelings are centred *in* and circle *round* property, and the relations of property to themselves and others; and to acquire property (even after they appear to neighbours and friends to have infinitely more than they know what to do with) they will risk serenity of mind, health of body, and even life itself.

As Mr. Howells truly says in his graphic description of the business aspect of New York on a hot day, when the mortal glare of the sun blazes in upon heart and brain, "The plethoric millionaire will plot and plan in his office till he swoons at the desk." But such a man is essentially a victim to the over-development of acquisitiveness, in the same way as a drunkard is victim to the unhealthy development of the instinctive impulse to obtain the pleasurable sensation given by stimulants. The latter has a depraved taste, a physical craving that is injurious to himself, and creates in others coldness and contempt. The former has a depraved taste also, a mental craving that is injurious to himself, and produces in others estrangement and a lack of sympathy which in some takes the form of envy, malice, and all uncharitableness of feeling; in others comes out as cold disdain, and shows a moral estimate (mingled with self-approval) of a fellow-man as little else than a fool.

We too must form a judgment or moral estimate, less of the man, however, than of the quality, which, over-stimulated and over-developed, becomes a mental disease, an aberration of the mind, a social vice to which he has unhappily fallen victim. This quality, then, of acquisitiveness is a comparatively new emotion. In the barbarous stage of its existence the British race had none of it, and our infants or young children show no signs of it now. In feudal times it was rudimentary, but quite subordinate to feelings of a different and less complex character, such as courage or bravery, hatred, warlike ambition, etc. In our industrial age it has been rapidly, enormously, but very unequally developed; and its results are mixed—tending to good, but also to unmitigated evil.

These results are both subjective and objective. The subjective, that is the results within the individual, are extremely good up to a certain point. Every adult ought to care for his own comforts, and provide for his own future, otherwise he lacks one of the distinctive, noble characteristics of civilized man. He is not a self-dependent, rational creature, but a feeble one, certain sooner or later to be a burden and drag upon society in general.

This self-regarding quality, exercised with scrupulous honesty, with manly justice, and with delicate consideration for the happiness of others, conduces to the healthy expansion of the mind and heart of man; but only so long as the direction of the force is towards a worthy goal. When that goal is reached, when the man can no longer say and feel, " my purpose in acquiring property is to maintain and educate my family, and to make certain that the sum of life to them and me shall be as great as possible ;" *then* a different state occurs, a new condition of mental phenomena sets in. The man cannot intellectually justify his continuing to accumulate. If, nevertheless, he goes on and on amassing wealth, his conduct is from reflex action (like an automaton), and not from motives that are rational, wide-reaching, and elevating. The mental quality that now directs his course is a mere passion, an appetite, which, to indulge *ad libitum*, will slowly but surely degrade the man, and sink him to the level of a poor drunkard, or an unreasoning brute.

A miser gloating on gold has long been looked upon as a despicable creature, and in literature his figure is held up to public ridicule and scorn. But the merchant prince, rolling in wealth, who yet devotes his days and years, his whole mind and heart to business, and *to that alone*—the capitalist who, in employing thousands of his fellow-creatures, reaps golden harvests, and yet keeps down wages, and strains every nerve to magnify his profits—the bold speculator whose game of chance has so enriched him that his appetite for gain is infinitely whetted, and to indulge it he will hazard all that nobler men deem the most precious things in life, and even that honour which *they* would rather die than lose—these men, I say, have never yet in public esteem taken their true places at the miser's side; yet they are *social deformities*, and as mentally diseased as he.

The social or objective results of this love of accumulating are also twofold. When a young man enters upon industrial life, and steadily pursues the path that leads to honourable success, his example and the tone of mind which he conveys to others inspire social virtues, or we may call them qualities, without which no civilized community could long exist. These qualities are prudence, forethought, enterprise, patience, perseverance, etc. But when acquisitiveness has grown to a deformity within the individual, and passion for money has, like a poison, corroded the fine tissues of his emotional and moral nature, the atmosphere that surrounds him is full of evil germs, that root themselves in other minds, and blossom into antisocial qualities and feelings which disturb the general happiness and degrade the moral elevation of our social life. Envy and jealousy are amongst these, a fatal tendency to discontent with equitable returns for labour expended, a blind ambition to make wealth at any cost, a mental state of feverish unrest, which strains the moral fibre of the whole community, lays it open to manifold temptations to dishonesty and injustice, and in short keeps it (if I may so speak) for ever hovering on the very verge of crime.

Now, to my reader's mind there is almost certain to occur the thought, that I am covertly condemning the laws of property that tend to inequality in wealth, and that my hopes of reform lie in some socialistic scheme, some measure for arbitrarily despoiling a whole class, such as, for instance, Mr. Henry George sets forth in his attractive work, "Progress and Poverty." But it is not so. In the first place, I believe that the moral sentiment or tone diffused throughout a community affects the general happiness much more than even the system of its associated life; and any *forcible* despoiling of landlords or any other class would create, in some a sense of gross injustice and personal injury, in others of indifference as to justice, which, looked at from the point of view of general well-being, would be a greater moral injury or *loss*, than any *gain* that could arise from giving more equality in wealth. In the second place, outward equality produced by external force or authority, and not by impulse from within, could never be maintained.

The measure Mr. George suggests would prove *no* remedy. After seizing landlords' rents we should find ourselves compelled to seize capitalists' profits, and so on *ad lib.* The course would be a vicious one, a terrible relapse from our at present tolerably advanced position in the scale of civilized humanity. We stand possessed of a considerable amount of liberty and fraternity. Are we to sacrifice *these* in pursuit of what we as certainly require—equality? I answer, no. We have to solve the problem, how, without recourse to any measure that savours of barbarous tyranny and despotism, we shall maintain our liberty, enhance our fraternity, and by spontaneous generation give social birth to a glorious equality, not artificial and mechanical, but natural and intensely human.

Our course, as I conceive it, is a progression step by step, through greater knowledge as to the true conditions of happiness, through change in public opinion and estimate of wealth, and through modification in individual sentiment finally, to a conscious, purely voluntary adjustment of outward system to suit the changed condition of the social units.

But now observe, this complex emotion of love of property or tendency to accumulate has not yet reached its final stage. It has to be weighed in the balances of right reason, and publicly pronounced upon. It has to be treated educationally, and inpressed upon the plastic young in lines that will beautify and not deform the character; and throughout society it has to undergo a process which we may describe in terms of clear, though inelegant political slang, as "levelling up and also levelling down."

Of jealousy I have spoken as an anti-social feeling. I may add, that it is essentially a semi-barbarous one. Like every other sentiment, it has a long, interesting history of its own, and its pedigree is much more ancient than that of the emotion we have just been considering. At one period of man's deve-

K

lopment it was useful to the welfare of the tribe, therefore it was *then* a social virtue. That period, however, has long since passed away, and it survives amongst us as a social vice. From acquisitiveness it differs essentially. That emotion can be exercised sympathetically and made to minister to the happiness of all; whereas jealousy cannot be so exercised. If I am jealous, it means I have a happiness, a joy that I would guard *from* others and would *monopolize*. The happiness may be self-produced or it may rest upon a being whom I love, but in any case it causes in me fears of interference, accompanied by suspicion of my fellows, and a general tendency to dislike, nay, even to hate them, if they dare to meddle with my source of joy. It is clearly then anti-social in its essential nature, and if a human being becomes sympathetic *all round*, he is incapable of, or he has risen above, the egoistic passion of jealousy.

But are there any such human beings? or is it but the "stuff that dreams are made of," when we picture to ourselves a far-off social state, in which that base emotion is only conspicuous by its absence? I maintain that here and there amongst us, even now, are individuals free from it *wholly*, and that in them, as well as through other evidence of which I shall shortly speak, we have a promise of what *shall be*—a vista of human happiness such as the world has not yet seen.

It is unnecessary for me to enter fully into the question of how jealousy has done useful work in the past. The mere fact that it has been developed more or less in all races that have passed from barbarism to civilization proves to an evolutionist's mind its usefulness. But we can place against that fact this other fact, that it is useless now, and not feel that there is between the two a single shade of inconsistency or contradiction. What Mr. Darwin says of jealousy is to this effect. "Amongst savages, intemperance, utter licentiousness, and unnatural crimes prevail to an astonishing extent. As soon, however, as marriage, whether polygamous or monogamous, becomes common, jealousy will lead to the inculcation of female virtue." Here Mr. Darwin gives the key which opens to us the secret of the past usefulness of jealousy. It has in a great measure swayed the destiny and shaped the tenderly emotional nature of woman; and its history is inextricably intertwined with hers, in all the varying degrees of servitude that mark her slow advance from the condition of absolute chattelism to that of rational equality with man. He, from his great endowment of superior strength, has from the first, and that unconsciously, adopted and acted on the false theory, that "he was made for God alone, whilst she was made for him;" and in the struggle to make good that theory, jealousy has evolved and formed a powerful factor in producing a very different emotion, viz. constancy, a social virtue for all time, and one as certain to wax and grow, as jealousy is to wane and slowly disappear.

When one turns to literature, in search of a reflection of the history of jealousy, there comes before one's mind at once the play of *Othello*, and perhaps nothing could better show the demoralizing anti-social nature of the passion, and the subject position of woman and general barbarous social conditions in which it flourishes. First Shakespeare reveals the lack of equipoise in the jealous mind. The merest trifle upsets its equilibrium; and although the Moor says bravely enough to Iago: "I'll see, before I doubt; when I doubt, prove;" that villain knows better. He says to himself aside:

> "Trifles, light as air,
> Are to the jealous confirmations strong
> As proofs of holy writ;"

and in effect Othello is incapable of sifting evidence, and the poor device of the stolen handkerchief seals the fate of Desdemona. Shakespeare exposes, too, the very foundations of the passion, viz. gross masterhood and pride of power. Othello's wife must be his slave and puppet. "Out of my sight!" he cries, and patiently she goes. Then the next instant: "Mistress," he calls, and she returns. "You did wish" (to Lodovico) "that I would make her turn." Poor Desdemona is the very type of patient, gentle, subject womanhood; the ideal woman of a rough, barbarous age. Her father describes her as

> "A maiden never bold;
> Of spirit so still and quiet, that her motion
> Blush'd at herself."

She despises jealousy, and perceives its low nature, for she says to Emilia:

> "My noble Moor
> Is true of mind, and made of no such baseness
> As jealous creatures are."

Nevertheless, when the base passion transforms her "noble Moor" into a very monster of injustice and savage cruelty, she is without one spark of the dignity of a free creature, and resembles nothing higher than a pet dog that fawns upon and licks the hand that strikes him. Her only dignity is shown in her refusal to display her sorrow to Emilia.

"Do not talk to me, Emilia;
I cannot weep; nor answer I have none."

But to *him* her very mind is subject and prostrated by the ignominy of his treatment of her. She makes her patient moan:

"Tis meet I should be us'd so, very meet.
How have I been behav'd, that he might stick
The small'st opinion on my least misuse?"

I cannot but think that even the haters of the typical strong-minded woman of the present day and of the still more objectionable "girl of the period" would, in condemning them, never take Desdemona as their ideal type. No! we have no admiration for her now; only pity or compassion, and a clear apprehension that the whole play, with its warlike men, its coarse ideas, its fierce passions, its bloody deeds, its uncontrolled hatred, revenge, and jealousy is simply barbarous—of a past age to which, happily, we shall never return.

Cold-blooded murder as a result of jealousy, fiction has, perhaps, rarely depicted. If it (that is murder) appears now in the pages of a novel, the sentiment of *love of property* is much more likely to be concerned in the motive. But since Shakespeare's time jealousy and duelling combined have had a long period of real and representative existence. Our growing civilization condemned and put down duelling, since when jealousy has received less favour at the hand of authors; and in the present day a first-class novelist would hesitate to draw a hero vile enough to pick a quarrel with a rival for a woman's love.

Nevertheless, George Eliot represents her Adam Bede (fine fellow as he was) as capable of this, early in this century; and she shows that jealousy in him is partly a sense of property in Hetty—the wounded pride and dignity of a *too* despotic nature which finds relief in fighting Arthur Donnithorne. In another of her great works we have a different picture—a noble woman wrestling with jealousy in the secrecy of her own chamber, rising above all anti-social feeling, and going forth to the help of her rival in such a glow of generous emotions, that the force is irresistible—she not only wins the confidence of Rosamond Lydgate, but, for the time, she draws that selfish, narrow nature up to the level of her own.

Turning now from works of genius, I take up a commonplace second-rate tale of the present year, sold at railway stations, and, it is said, widely read.* The plot is certainly original, and, as I think, absurd. Elaine, a lovely girl, lay dying—so every one thought. Doctors and physicians had withdrawn. Her case was hopeless, and she could only live a few hours. She calls to her side another lovely girl, her cousin Madolin, and reveals that she had given her heart away to a Sir Aldewin, who, she thinks, also loves her. Now she desires —intensely longs, indeed—to be his wife, die in his arms, and have above her grave a marble cross, with these words inscribed: "In loving memory of Elaine, wife of Aldewin Chesleigh." But Madolin has herself only a few days before exchanged lovers' vows with Aldewin, and promised to become his wife. What will she do? She hesitates. Her heart is well-nigh broken; nevertheless, the pressing sympathy of the moment prevails. She goes to her lover, and actually persuades him to consent, against the dictates of his better judgment; for he says, "I cannot. I should loathe myself for being untrue." The marriage takes place, and, of course, Elaine recovers to confound the doctors, and create an involvement sensational enough to carry on the reader. Now, surely here, if anywhere, is a fair field for jealousy—nay, one might half suspect the author meant to make the interest of the story hang upon the phases of that emotion. But it is not so. The girls, and also the young man, are drawn as quite incapable of jealousy developing to hatred and revenge. Aldewin, in a burst of sorrow, cries out to Madolin, "I cannot even dislike Elaine. I cannot grieve that she lives, even though the price of her life be my happiness." And of Madolin the author writes, "All her life long she loved Elaine, and that which had happened, so far from lessening her love, had rather increased it." After two years the young wife grows restless. She dimly feels that *her* love for her husband is greater than *his* is for her. She questions him, and when he answers, half reproachfully she exclaims, "I am not jealous; I have never been jealous. If I were to grow jealous of you the pain of it would kill me!" The truth comes to her, however, and in Madolin she recognizes the woman to whom her husband had given his heart—his whole heart—previous to the marriage forced upon him. How will she act and feel now? She suffers keenly, but the suspicions, the pains, wither away "in the clear light of

* "The Love that Lives."

Madolin's sweet presence." She clung to her and cried, " Keep my faith in you alive, Madolin ; never let me mistrust you." And Madolin replies, " You never shall, darling, if I can help it ! See how I love you. It seems to me always that I am your mother and sister, cousin and friend, all in one." The author of this sentimental tale understands something of the human nature that surrounds us now, and has grasped the truth that love will cast out jealousy, that as sympathy expands, its tender consideration for the happiness of others will make jealousy simply impossible, and that at the present day it is disgraced, and must not even wear the semblance of virtue. Had the author only elevated and put into its true place the modern sentiment of love of truth, as well as she has debased and relegated to its true place the semi-barbarous one of jealousy, we should have heartily admired her work. But then, alas ! for her purpose, the *plot* had proved " the baseless fabric of a vision ! "

We must not look to fiction for our *ethics ;* but we may see there the standpoint average humanity has reached, and the confusion of ideas upon the subject of right conduct, natural to a transitional epoch. Now, although I assert that jealousy is passing away, its usefulness is long since gone, average human nature despises it, select individuals are free from it, public opinion condemns it—nevertheless it exists, and is rampant in our midst, a powerful anti-social emotion causing alike to those who are the subjects of it and to those who are the objects of it, a weight of suffering, sorrow, and shame, that lies like a vast incubus on human happiness.

Have we, as evolutionist reformers, no other than a passive *rôle* to play in face of this great source of misery ? There are many earnest and devoted people who note the tendencies of modern thought, and fancy evolution principles almost preclude human action ! But they mistake. What evolution means is this : that there is a predicable order in the sequence of phenomena, a discoverable method in the march of events. Man is within this order. He is *of* the forces, not simply *under* them. The *human intellect is an enormous cosmic force*, and on its exercise by myriads of individual units depends all social progress. Far from stopping action, evolution principles, if they are rightly understood and firmly embraced, will stimulate effort, and call into play the noblest powers that man possesses, whilst guiding and directing them towards the full accomplishment of worthy aims.

Now, in this sphere of human sentiments or emotions we stand in need of action, and none of us are justified in being passive. There is scope for vast improvement in character and conduct by fostering the sympathetic and repressing the anti-social feelings within and without ourselves ; and if such action were general, Mr. Herbert Spencer would cease to tell the world that emotional readjustments to the needs of life are tardy in the extreme. To rid ourselves as *rapidly* as may be of jealousy is *essential*, otherwise the great movement in favour of equality of sex will necessarily meet with checks and grave obstructions.

But this subject I shall enter into fully in another chapter; meantime my object is to indicate the method of our action. Changes within ourselves occur in similar fashion to changes without. Old laws give way to new; old institutions are replaced by institutions of a better kind; but nowhere do we see a clean sweep (if we may call it) of what is old, leaving a blank and arid waste unfilled with what is new. And so it is with human sentiment. The changes are *displacements;* old emotions make way for those of wider range, of nobler, finer issues; and jealousy will be subdued and finally expelled by entrance of the sentiment of truth. This sentiment shows itself objectively in candour and transparency of word and deed, subjectively in honesty of mind; and, take it all in all, it is a royal emotion, as sure to elevate humanity as it is certain on the other hand to regulate the whole social life, and make it a fit abode of peace and happiness.

By obeying our own instincts to be true, and helping the impulse in others, above all by training the young in habits of truth, and so implanting in them the love of it, we shall accomplish a great end in reference to jealousy. But let me illustrate this point by simply relating an incident as it occurred in real life. A girl had been engaged to marry a gentleman, whom I shall call Edgar, for eight long years. During the greatest part of that time the intercourse was kept up by correspondence only, for he was in India, she in England. Meanwhile, she formed relations of another kind with one whom I call Roger. For a considerable period the girl was in the presence of the latter daily; and thoughtlessly, but very guilelessly, she let her heart go out to him. She called this new relation friendship, but in reality it was a fuller, deeper sentiment than the love she bore to Edgar; and, strange as it may appear, she never spoke to Roger of her engagement, till suddenly the prospect of its fulfilment was brought vividly before her. Edgar wrote of an immediate return, and of his joyous hopes of making her his wife. And then the disillusion was complete, for Roger made known that he had dreamed she understood his love! Ignorant of any obstacle, he had been courting her affection in the deepest sense, and his avoidance of a frank avowal arose from scruples of a conventional nature. Would her relatives think him rich enough? The girl was bitterly distressed. Her mind, distracted, perplexed, could see no issue to her false position. Roger urged: "You know your own heart now. Can you deny your *warmest* love is mine?" And because she could not deny that fact, it seemed to her as though it would be selfish and mean to reject the absent Edgar, and consent to marry Roger. Duty, she thought, "is always difficult, and I am called on to choose the hardest path, and turn away from Roger." But ever as she pondered thus, fresh doubts arose, and fears were mingled with her sense of shame. What if masculine jealousy brought fresh disaster? Poor girl! her misery was complete, and her emotions overwhelmed her judgment. Fortunately for her, she felt her own incapacity, and knew she could depend upon the reasoning powers and moral nature of a clergyman within reach. To him she went, and bravely laid bare the secrets of her heart and mind. He gravely said, "Now let us for a moment reverse the position. Edgar, we shall suppose, has written to you to come out to India and marry him; the fact being that he has fallen desperately in love with another girl—*innocently*, remember. He could not help it, he did not mean to do it, but unconsciously his heart has slipped away from you to her. He is shocked, he blames himself (and not without *some* cause) for falseness, and with a martyr's spirit, a heroic sense of honour, he resolves to keep his engagement, and, 'so help him God,' to make you happy. He meets you then with a lie in his heart, not on his tongue, for he will carefully avoid the subject of your rival. But tell me, Is he right? Would you be happy, think you? Would you miss nothing? Would you feel, if you came to know the truth, that his generous action was *just to you?*" "Oh no!" she cried, "I could not bear it; he ought to tell me all, and I would never wish to sacrifice in that way *his* happiness to *mine.*" "Just so," her monitor replied; "*truth is always best;* but I see you think that Edgar is not so unselfish as are you yourself. In this are you quite just to him?" "I cannot tell," she murmured; "but are not men jealous?"

"Jealous!" he answered; "ah, well, we men *are* frail, no doubt! But were I Edgar, I tell you frankly it would not mend matters to me, that I had won my wife without the priceless jewel of her love. No, no, my young friend, be *true* to yourself, and *just* to him, and fling to the winds the fears that cause deflection from the path of rectitude." Strengthened, enlightened, comforted, the girl went through her painful task. Relinquishing her false, heroic mood, but in no selfish or self-sparing one, she met Edgar, and later she became the wife of Roger. To her wise mentor she expressed her mingled sense of thankfulness and sorrowful contrition. "He freed me at once," she said; "he grieved, but yet he thanked me for my candour. He would never wish my hand without my heart; he'd rather keep my friendship pure and sweet, than my love with any tinge of falseness through it. But oh! that I could cure the pain and sorrow that I have caused him." "Hush, hush, my child!" he gently said; "don't let your self-importance grow inflated; his own true heart is his best comforter at present, and by-and-by it will be, shall I say, born again. It will send out fresh shoots and leaves, aye, and blossom into fruit all in good time. You have pained but *not misused* him. Let the past be past for you, for him, for all."

My optimism, or tendency to be hopeful in respect to the prospects of the human race, has often been opposed by dreamy evolutionists, who refuse to look vigorously into the causes of social misery. "What folly," it has been said to me, "to think that sin, pain, sorrow, can ever pass away!" As regards sorrow, death and dying are certainly sorrows we shall never be able to surmount;* and as our powers of loving expand, these sorrows will be even more deeply felt. But if we reflect on what is around us, and look back upon our own inward histories, we shall perceive that these are not our only sorrows—far from it. There are emotional trials, in which death has no part, which, nevertheless, prostrate us more completely than even when we mourn the loss by death of our best beloved. These sorrows will be mitigated to an enormous extent in two ways. First, the bitterness will be removed when we can lay no blame upon our fellow-creatures, and feel no sense of personal injustice. The trial may be of an entirely different order from what overtook Edgar; but whatever the nature, if the subject of it knows he has been treated without deception, and with scrupulous justice, the sorrow will have

---

* Premature deaths, however, will undoubtedly be much fewer.

lost its greatest sting. The second mitigation lies in the greater sympathy the social units will feel and show.

Which of us who has lived and suffered long, does not know that the universal consolers are time and human sympathy? And to the extension of the latter an evolutionist, at least, can set no bounds. But *justice!* Alas! when will that sentiment hold universal sway? Of all emotions it is the one demanding *most* in the way of human capacity. It presupposes intellect, viz. reasoning powers and imagination; also knowledge, sympathy, and a delicate susceptibility to quantitative relations, to a sense of equalness, or a want of equalness in every possible position. So complex an emotion could only be of slow development; but step by step as man progresses in education, culture, and true refinement, *it* will progress, and make its influence extend to ever-widening circles.

That it is *growing* steadily is too obvious to need proof. One trifling indication, however, I may here point to. In will-making, or the disposal of personal property, justice has greatly modified custom. I do not allude specially to the upper classes and the old law of primogeniture; but in the great upper middle class, where fortunes have been rapidly made, and property does not mean land, the father of a family would as a rule at the beginning of this century dispose of his estate in unequal portions, leaving more to sons than daughters. At the present day a father makes his will on lines more just and equitable. "Let all share alike," is his principle; and there are even some who feel and try to act from a still finer sense of justice—a desire to weigh in delicate scales the relative position of sons and daughters, and to make the balance good by leaving most to daughters. These men say, "My daughters may not marry, and alas! for them, marriage seems year by year to grow more difficult. If they do *not*, and if at my death my property is not enough for all, their lot in life must necessarily be more precarious than that of my sons; for even if I train them to a trade or a profession, their sex will prove a stumbling-block, a disability; therefore, I must put this personal disability into the scale that I may act with impartial justice."

But now, turning from all the *deep* emotions of humanity—the powerful sentiments that lie at the foundation of our social life—let us merely glance for a moment at the changes taking place in those emotions that play upon the surface of that life; I mean the sentiments that have to do with conventional propriety. We have only to read carefully Lord Chesterfield's "Letters to his Son," written about the middle of the last century, to be struck with the alteration in the habits and customs of polite society, and in the state of feeling thought proper to polite society, which has taken place. It is truly marvellous. Imagine a father gravely writing to a son : " If you love music, hear it ; go to operas, concerts, and pay fiddlers to play to you ; but I insist upon your neither piping nor fiddling yourself; it puts a gentleman in a very frivolous and contemptible light, and takes up a great deal of time which might be much better employed. Few things would mortify me more, than to see you bearing a part in a concert, with a fiddle under your chin." Or again, " Having mentioned laughing, I must particularly warn you against it ; and I could heartily wish that you may often be seen to smile, but never heard to laugh while you live. . . . In my mind there is nothing so illiberal, and so ill-bred, as audible laughter. It is low and unbecoming, not to mention the disagreeable noise that it makes, and the shocking distortion of the face that it occasions. I am neither of a melancholy nor a cynical disposition, but I am sure that, since I have had the full use of my reason, nobody has ever heard me laugh." Poor man ! we can afford without any cynicism to pity him as a stiff-jointed puppet in an artificial society, when we reflect upon the freedom, the ease of future generations, whose laughter will be as natural and guileless as a child's—the rippling overflow of joyous life. The changes, too, in woman's position, and the sentiments she calls forth, are equally marked. " Women," he writes, " are only children of a larger growth ; they have an entertaining tattle, and sometimes wit ; but for solid, reasoning good sense, I never in my life knew one that had it. A man of sense only trifles with them, plays with them, humours and flatters them ; but he neither consults them about, nor trusts them with, serious matters. Here dissimulation is very often necessary, and even simulation is sometimes allowable ; it pleases them, may be useful to you, and is injurious to nobody ! Women," he goes on, "in a great degree establish or destroy every man's reputation for good breeding ; therefore you must overwhelm them with small attentions. You must be sedulous, and rather overofficious than under, in procuring them their coaches, their chairs, their conveniences in public places." "The company of women of fashion will improve your manners, though not your understanding," he writes. But in another letter he guards him against wasting his evenings frivolously in the tattle of women's company. Yet Lord Chesterfield was for his period a worldly-wise and affectionate father. He condemns the barbarous sentiments of anger and revenge, and exhorts his son to gentleness and generosity. The lesson for us is clear. Throughout the whole range of public and private social life change in sentiment is *rapidly* going on. What is thought *proper* now will very probably not be thought proper in the year 1919; *therefore*, let no mere conservative bias thwart our judgment. All feelings are open to criticism, and it is by permitting the searching light of intellect to penetrate into the dark corners of human emotion, that we shall best secure or aid progress in the right directions. The subject is a very large one, and it may seem to my reader that I have treated it superficially; but my object is not psychological, but purely sociological. It would not be relevant to my purpose to analyze and classify human sentiments. The order of evolution is what I have to do with, and it only falls within the scope of my work to make a few general observations, and note one or two important points. These are, first —that in this sphere of feeling the path of advance towards greater happiness, lies in fostering the sympathetic, and repressing the anti-social emotions; second—that love of property must be modified and subjected to reason; third—that jealousy is anti-social and must die out; fourth—that love of truth and the sentiment of justice are of recent growth, and demand general attention and aid in their development; and fifth—that the sentiment of what is proper or improper in conventional society, is no true guide to right conduct.

CHAPTER X.

THE PERIOD OF YOUTH.

> "From these high windows behind the flower-pots young girls have looked out upon life, which their instincts told them was made for pleasure, but which year after year convinced them was somehow or other given over to pain."—J. H. SHORTHOUSE.

THERE appeared in the *Saturday Review* for March, 1868, an article entitled "The Girl of the Period," written by a woman who certainly possesses what Lord Chesterfield called "solid, reasoning good sense," the type of woman which, according to him, had no existence in *his* day, but of which we have at least a few specimens now.

This writer holds advanced opinions upon many subjects. She has given to the world works that show knowledge, high culture, experience of life, and wide human sympathies. It is impossible for us to impute to her any prejudice against her own sex, any want of true sympathy with girls; nevertheless, in this article her teaching is conservative. She compares a past state of things with a present state of things, and gives a distinct opinion in favour of the former.

"Time was," she says, "when the phrase—a fair young English girl—meant, a creature generous, capable, and modest. It meant a girl who could be trusted alone if need be, because of the innate purity and dignity of her nature, but who was neither bold in bearing nor masculine in mind; a girl who, when she married, would be her husband's friend and companion, but never his rival; a tender mother, an industrious housekeeper, a judicious mistress." "Of late years," she goes on, "we have changed the pattern, and have given to the world a race of women as utterly unlike the old insular ideal as if we had created another nation altogether. The Girl of the Period is a creature who dyes her hair and paints her face; . . . whose sole idea of life is fun; whose sole aim is unbounded luxury; and whose dress is the object of such thought and intellect as she possesses. Her main endeavour is to outvie her neighbours in the extravagance of fashion. The Girl of the Period has done away with such moral muffishness as consideration for others, or regard for counsel and rebuke.

"No one can say of the modern English girl that she is tender, loving, retiring, or domestic. Love is the last thing she thinks of, and the least of the dangers besetting her. Love in a cottage—that seductive dream which used to vex the heart and disturb the calculations of the prudent mother— is now a myth of past ages," and her only ideal of marriage is " the legal barter of herself for so much money, representing so much dash, luxury, and pleasure." She concludes with a grave warning to the young of her own sex, in reference to marriage. " The Girl of the Period," she tells them, " does not marry easily. Men are afraid of her; and with reason. All men whose opinion is worth having, prefer the simple and genuine girl of the past, with her tender little ways, and pretty bashful modesties, to this loud and rampant modernization, with her false red hair and painted skin, talking slang as glibly as a man, and by preference leading the conversation to doubtful subjects."

Now, I entirely agree with the writer of this article in her condemnation and denunciation of the girl of the period whom she describes. The picture is a truly odious one, and I suppose we must admit that there are amongst us a considerable number of the type. But the article seems to me to carry two inferences with it, which I regard as erroneous. First, she assumes that if we reject her type of girl of the period, we are shut up to the type of the past; and because we cannot admire or tolerate the loud and rampant modernization, we must revert to the girl whose chief attractions lie in "tender little ways, and bashful modesties." Second, she infers that the proper standard of feminine excellence is masculine desires. Her principle seems to me to be that woman is made for man. At all events, in exhorting the girl of the period to lay aside her absurdities, she appeals to no higher motive than that men do not approve, and do not want that kind of thing in her.

I do not for a moment deny that some men are excellent

judges of what is good in woman; but that man's opinion is to be her guide and rule, that she is to find out, in short, what *he* requires and wishes for in a wife, and shape herself accordingly, is to my mind teaching of a misleading, mischievous character. And in reference to the spirit of the age it is like a false note, a jarring discord in the music of the future. The true harmony lies in a wholly different view, viz. this—woman exists as man exists, for the development of her varied faculties, for the achievement of the highest work within her power, and the attainment of the greatest goodness and happiness possible. That her happiness is deeply involved in her relations to man, and that her goodness must be compatible with, and in no way obstructive of, *his* happiness, are undoubted facts; but the true measure of her standard is *not* his requirements, and her mental and moral calibre are not to be controlled by, and fitted to, his personal tastes.

But is this question about our ideal in womanhood, with its necessary discussion of the frivolous lives of spoilt girls, a fitting subject to introduce here? Am I justified in asking for it the attention of minds that may be occupied with such grave matters as capital punishment, our criminal code, the reform of drunkards, the land laws, church disestablishment, etc.? In my opinion, the question is an extremely important one. It forms part of a wider subject, which in relation to general welfare and social advance, comes next in importance to that of poverty and increase of population—I mean the subject of the *Period of Youth* in both sexes. Let us put it in this way. Given the prevailing disposition, character, habits, and customs of the young men and women of the present day, and given the conditions of their life—that is their environment or surroundings—what result may be rationally looked for as regards present happiness and as regards the moral elevation of the next generation, of which they are the potential parents?

In dealing with poverty, we found the pauper a blot upon our nineteenth-century civilization and a reproach to the whole British race. We saw clearly, that to abuse and attach personal blame to the pauper was irrational and cruelly unjust. His antecedents and his surroundings explain his existence; and it is by enlightened modification of these conditions alone, that we shall compass the great end of securing his gradual and total extinction. Similarly with the *Saturday Review's* girl of the period. She is a blot upon our civilization, and we may place her figure side by side with that which Mr. Lecky calls "the most mournful and in some respects the most awful upon which the eye of the moralist can dwell. That unhappy being whose name is a shame to speak . . . who is scorned and insulted as the vilest of her sex, and doomed, for the most part, to disease and abject wretchedness. . . . That degraded and ignoble form. . . . She remains while creeds and civilizations rise and fall, the eternal priestess of humanity, blasted for the sins of the people."* This latter figure is more despised than the former, more publicly and popularly recognized as ignoble; but the *girl of the period* form is in reality not *one whit* less degraded. Both are vicious, unseemly outcomes of our social system, and point to something radically wrong in our nineteenth-century civilization. Neither should be treated superciliously or with abuse. Compassion is the tone of mind they should inspire, and on the part of all reformers, persistent, earnest endeavour to remove the causes that perpetuate their number, or give them birth. As regards the prostitute, the fundamental and primary cause of her existence is poverty, with its antecedent, a too rapid birth-rate. But there are other secondary causes on which we shall elsewhere touch; meanwhile we turn again from her, to her companion picture—the objectionable *girl of the period*.

And let us glance first at the essential nature of youth. It is a period of effervescence. There is in youth, if it is healthy, an exuberance of life, which is ever ready to bubble up and overflow. It tends to movement or to action in all directions, and the movement is threefold. It may be muscular or physical, it may be emotional, or it may be intellectual; but if the latter, the activity is not of the judgment or reasoning powers, but rather of the imagination and the faculty of quick perception. Now, society is bound to find safe and ready channels for all this surplus energy. Young people of both sexes require a sphere for useful activities as well as recreational activities, a wide choice of intellectual pleasures, and a field for the free play and exercise of all the spontaneous and varied emotions that are not of an anti-social nature.

In the novels of the present day, as I observed in my last chapter, we must not expect to find guidance to right conduct; but what we do find there is a true representation of the inner and outer life that surrounds us. One of these novels gives a striking description of the position and feelings of a young

---

* "History of European Morals," vol. ii. p. 299.

girl, charged with life, as a thunder-cloud is charged with electricity, and who for lack of outlets that are pure and safe shows herself up to a certain point in the novel a heartless mischievous flirt—one of the loud and rampant kind, ready for fun at any cost. But suddenly this girl, Fanny Dover, finds herself in a new position. In a house full of helpless women and a helpless man, *something* is expected of *her*—something *depends upon her*. She is transformed at once! But we shall let her speak for herself, and in transcribing, whilst in one sense the words I write are fiction, I also feel that they are fact.

The hero of the tale, who is a woman-hater, or thinks himself so, says to Fanny Dover, " Please explain a remarkable phenomenon. You were always a bright girl, and no fool; but not exactly what humdrum people would call a good girl. You are not offended?"

"The idea!" says Fanny. "Why, I have publicly disowned goodness again and again. You have heard me."

"So I have, but was not that rather deceitful of you? for you have turned out as good as gold. Anxiety has kept me at home of late, and I have watched you. You live for others; you are all over the house to serve two suffering women. That is real charity, not sexual charity which humbugs the world, but not me. You are cook, housemaid, butler, nurse, and friend to both of them. In an interval of your time, so creditably employed, you come and cheer me up with your bright little face, and give me wise advice. I know that women are all humbugs; only you are a humbug reversed, and deserve a statue—and trimmings. You have been passing yourself off for a naughty girl, and all the time you were an extra good one."

"And that puzzles the woman-hater; the cynical student who says he has fathomed woman! My poor dear Harrington, if you cannot read so shallow a character as I am, how will you get on with those ladies upstairs—Zoe, who is as deep as the sea, and turbid with passion—and the Klosking, who is as deep as the ocean?"

She thought a moment, and said, "There, I will have pity on you. You shall understand one woman before you die, and that is me; I'll give you the clue to my seeming inconsistencies —if *you* will give *me* a cigarette."

"What, another hidden virtue! you smoke?"

"Not I, except when I happen to be with a noble soul who won't tell."

L

Vizard found her a Russian cigarette, and lighted his own cigar, and she lectured as follows :—

"What women love, and can't do without if they are young and healthy, and spirited, is—Excitement. I am one who pines for it. Now, society is so constructed that, to get excitement you must be naughty. Waltzing all night, and flirting all day, are excitement. Crochet, and church, and examining girls in St. Matthew, and dining *en famille*, and going to bed at ten, are stagnation. Good girls—that means stagnant girls; I hate and despise the tame little wretches, and I never was one, and never will be. But now look here ; we have two ladies in love with one villain—that is exciting. One gets nearly killed in the house—that is gloriously exciting ; the other is broken-hearted. If I were to be a bad girl and say, 'It is not my business ; I will leave them to themselves, and go my little mill-round of selfishness as before,' why, what a fool I must be ! I should lose excitement. Instead of that, I run and get things for the Klosking—Excitement. I cook for her, and nurse her, and sit up half the night—Excitement. Then I run to Zoe and do my best for her, and get snubbed—Excitement. Then I sit at the head of your table and order you—Excitement. Oh, it is lovely !"

"Shall you be sorry when they both get well, and routine recommences ?"

"Of course I shall ; that is the sort of good girl I am. And oh, when that fatal day comes, how I shall flirt ! Heaven help my next flirtee !"*

Now, the Fanny Dover type is as common as possible amongst English girls. They are neither good nor bad in any strong sense. They are simply shallow natures, bubbling over with exuberant life, and ready to be good or bad according to surrounding conditions. In themselves they have little instinctive sense of what is right or what is wrong, no moral sentiments strong enough to be a motive power, holding them back from one course, and impelling them to another, and all their motives, therefore—the forces that direct their energies—are sure to be derived from the predominant tone of thought and feeling holding sway in the social circle of which they form a part.

Society, then, has here and there and everywhere a Fanny Dover of neutral tint and shallow nature. If society can utilize her surplus energy, leaving only enough for recreation ;

---

* "A Woman-Hater," by Charles Reade, vol. iii. p. 33.

if it can give her a useful honourable career, and lay on her young shoulders a sense of responsibility, not *burdensome*, but *perceptible*, no more outwardly is wanted. The simple lapse of time, the discipline and experience of life, will check her levities, give depth to her nature, and colour to her character. But if, on the other hand, all that society requires of Fanny Dover is to amuse herself and obtain a husband, the girlish innocence will run to waste, and from a few years of frivolous and mis-spent life there will emerge the heartless, hardened coquette, the loud and rampant girl of the period, who dyes her hair, screws in her waist, paints her face, and carries with her a moral degradation which will taint directly all the young of her own sex, and indirectly all the young of the opposite sex who come within her reach.

Now let us look at the quiet girls, of whom Fanny says, " I hate and despise the tame little wretches ; " good girls but stagnant, she calls them. Stagnation, however, is never an element of healthy youth, and in all probability these girls differ from Fanny, not in amount of force, but in the direction the force has taken. Hers overflows at the surface, theirs runs in inward channels ; and still waters usually run deep. Crochet, or nowadays Russian embroidery, lawn tennis, church-going, Sabbath school teaching, and a little mill-round of very trifling occupations, is all their visible life, and to Fanny it looks tame. She thinks contentment with a very humdrum lot is the key-note of their position, but she is wrong. They have their surplus energy as well as she. Timidity, innate refinement, or some high moral instinct, prove forcible enough to control the outward conduct, but the interior ebullitions are strong and deep. Vague longings for a wider life, wild dreams of what *may be* perchance, vain flights into a world of purest fancy, alternate with great gusts of passionate emotion, followed by numbness—a dull sense of pain and self-reproach—because they cannot feel the resignation which they think is right, and inwardly repress their hot desires, their restless state of juvenile impatience. To the outwardly good girl as well as the outwardly bad girl, the conditions of her life are thoroughly unsatisfactory.

I do not expect many educated women to admit this. We have all a happy power of forgetting our past sorrows, and it is marvellous how little real sympathy is felt by even tender-hearted mothers with the troubles of the young. Their own lives are filled to overflowing with duties, cares, and pleasures, and they have no surplus energy to expend in recalling the inward history of their own youth so that they may truly apprehend the natures of their children, and the griefs, that to a practical and busy woman of mature age, are apt to seem fantastic.

But we have the direct evidence of some women of mature age to the fact, that the conditions of their youth were unfavourable to happiness and goodness, and in novels the theme is harped upon perpetually. Who does not sympathize with the girl who exclaims, "I am weary of theories, I want flesh and blood, I want to feel myself a part of this striving, eager, anxious humanity, on whose labours I live in comfort, by whom I have been educated, to whom I owe all, and for whom I have done nothing"?*

Madame George Sand deliberately tells us that she was shut up in egoistic sorrow till she was thirty, and Miss Martineau gives a clear and definite account of the misery she endured during the whole period of her youth. "My life," she writes, " began with winter. I did not know ten years ago what life might be in regard to freedom, vigour, and peace of mind ; the spring, summer, and autumn of life were yet to come. At past forty years of age I began to relish life, and for ten years I have been vividly conscious of its delights."

Now, Miss Martineau's youth was spent in a comfortable English home. She was surrounded by love and care. Food, clothing, education, all were hers ; but some condition, in spite of all these blessings, made her miserable ; and observe, whatever it was must have been *outside* of herself, because later, she proved that she possessed the power to relish and enjoy existence. Probably her mother would have said that two conditions were lacking to make her happy, viz. health of body and contentment of mind ; but in reality both of these, one partially, and the other wholly, were due to outward pressure. The girl's whole nature was cramped. She had no freedom to expand and develop according to her own inborn pattern, and all the forces of her organism turned inwards, fretted and corroded both her mental and physical health.

What to others might prove a cause of misery—the failure of a father's house of business—was in her case a blessing, for it opened the gate of freedom, and gave an outlet to her energies. "My mother and her daughters," she says, "lost at a stroke nearly all they had in the world. I, for one, was left

* " All Sorts and Conditions of Men," Walter Besant.

destitute; that is to say, with precisely one shilling in my purse. I rather enjoyed it, even at the time; for there was scope for action; whereas, in the long, dreary series of preceding trials, there was nothing possible but endurance. I had henceforth liberty to do my own work in my own way; for we had lost our gentility. Many and many a time since have we" (*i.e.* her sisters and she) "said, that but for that loss of money, we might have lived on in the ordinary provincial method of ladies with small means, sewing and economizing, and growing narrower every year; whereas by being thrown, while it was yet time, on our own resources, we have worked hard and usefully, won friends, reputation, and independence, seen the world abundantly, abroad and at home, and in short have truly lived instead of vegetated."*

Do I desire, then, for young girls *unlimited liberty?* By no means. On the contrary, I know many girls who have far too much liberty in *some directions*. As for instance in the spending of money, the ordering of servants, etc. However wealthy a girl's parents may be, to acquire the habit of lavish, thoughtless expenditure is hurtful to character, and may in the future render her life miserable; and authority given to the young to control their fellow-creatures, and command obedience to their slightest whim, is certain to foster the *love of power*, an emotion which ought to be very carefully and judiciously repressed.

No, it is not less interference, in any general sense, that is required. It is a thorough understanding of the nature and position of girls; and whilst breaking up the purely conventional system that surrounds them, and loosening all the bonds that hold in check their innocent activities, it is the bringing them into contact with external forces that will discipline the mental and emotional nature, and fit the character to the true needs of life.

The "shut up in egoistic sorrow" state is not confined to girls of an inherited superior capacity, such as George Sand and Harriet Martineau. Many a simple girl of ordinary powers suffers in the same way, although it may be in a less degree. She finds on her return from school, that in the busy world around, the eager, striving, anxious crowd, she is, and is compelled to be, a *trifler*—nothing more. In her home, the necessary domestic duties are all efficiently performed by servants, or it may be by older sisters, and if a few are yielded

---

* "Autobiography," vol. i. p. 141.

up to her, it is a make-belief of usefulness, a sham that does not satisfy her craving.

Brothers' careers are carefully canvassed and considered, active measures are taken to give them a good chance of making life effective; but for her, *effective* life seems never thought of. She feels herself a surplus commodity, the sense of present uselessness depresses her, and when she ponders on her future, anxious fears take possession of her mind. There is a tacit understanding that the *rôle* for her is marriage, and that if she does not marry she will disappoint her friends and be a social failure; and inwardly the girl shivers at the thought that some day she may find herself that evil thing which forms the very bugbear of her young imagination—a sour old maid! What shall she do, what *can* she do, to escape that awful doom? Nothing at all but wait, and vainly strive to possess her soul in patience.

The strict, conventional law in reference to marriage, she clearly sees engraved on all the minds around her. It is the *man's* prerogative to seek, and actively to create the warmest of all relations. It is the woman's duty to sit still, be *passive only;* and if she feels emotions stirred within her, she must conceal them, until the moment comes when she is asked to speak a final yes or no. The position does not differ in essentials from the Babylonish marriage market, which Mr. Long has placed so powerfully before our eyes. Rows of women, ranged according to their external points, from loveliness to plainness, and down to ugliness, with every shade between, sit waiting in the market for men to bid, to purchase, and to make themselves the masters of their destiny.

Now, there are many nineteenth-century girls who rebel at this enforced passivity. I do not say all or nearly all, for in a transitional age we have of course a mixed inward as well as a mixed outward condition; and the attitude of the female mind is in some cases fitted to a past stage of civilization—the stage in which barbarous surroundings rendered the entire subjection of woman natural; whilst in other cases the state of thought and feeling is fitted to the rapidly advancing epoch of *perfect equality of sex.*

The girls I speak of have great *spontaneity* of heart and mind. They are not simply responsive. It is their nature to throw out tender tentacles of human feeling, and to follow these with candid word and generous action. Now, there is nothing culpable in this, nor is it even dangerous, where refinement exists, and where there have been inculcated the two modern sentiments—love of truth, and justice, *i.e.* a conscientious regard for the feelings and happiness of others. But in a society, where the question of marriage is made too prominent, such girls are never able to be free and natural. They may obey their impulses to some extent with their own sex, but with the other sex, they must be carefully upon their guard, they must be formal, cold, unnatural, or run the risk of being taken for husband-hunters, and condemned as wanting in the very quality they eminently possess—delicate womanliness.

Cramped, then, on every side, with no sphere of useful activity, and no freedom to seek for themselves a life filled with emotion, they either turn to religion for comfort, and spend their youth in a piety which is morbid, or they draw in all their tentacles like a sea-anemone (out of its native element) and live shut up in "egoistic sorrow."

But the conditions of a young man's life are in marked contrast to all this; and naturally the tone of mind engendered, *is a sense of personal superiority.* And here again let me resort to fiction, for Mr. Black has excellently portrayed the boyish spirit of masculine contempt with which a brother gazes on and spurns the narrow life his sisters are compelled to lead.

In the tale called "Beautiful Wretch," a young lad exclaims: "I see the puzzlement you girls are in who haven't got to earn your own living. You don't know what on earth to do with yourselves. You read Ruskin, and think you should be earnest; but you don't know what to be earnest about. Then you take to improving your mind, and cram your head full of earth-currents, and equinoxes, and eclipses of the moon. But what does it all come to? You can't do anything with it. Then you have a go at philanthropy—that's more practical: Sunday school teaching, mending children's clothes, and generally cultivating pauperism. Then, lo and behold! in the middle of all this there comes by a good-looking young fellow; and phew! all your grand ideas are off like smoke." The girls drop their philanthropy at once, and vie with one another to obtain that greatest blessing to the sex—a lover!

Now, in view of the permanent and natural relation of the sexes to one another, I think my reader will allow that the disposition necessary to cultivate in lads is *respect for womanhood*, with tendencies to judge it *fairly*, and to value it according to its real and not its superficial merits. The conventional surroundings of youth, however, bring about an opposite result. Contempt for girls, for their frivolous, or at least comparatively useless lives—self-importance and a sense of power and masculine superiority—impossibility of knowing an educated girl's genuine nature and disposition, and therefore liability to be attracted by personal beauty only, or by fictitious charms, these are what a young man feels, and what the best young men both suffer from and bitterly deplore; whilst on the worst they act injuriously, and slowly but surely degrade the character.

A lad at seventeen is as a rule set free from trammels. His career is decided. He enters an office, or at a University prepares for some profession, and takes into his own hands the guidance of his private life. He finds himself at liberty to strut along the street, cigar in mouth and cane in hand, at any hour, and stare at girls of every order, class, or social station. He explores in all directions, music-halls, dancing-saloons, billiard-rooms, gambling-houses, and it may be spends hours in chaffing (as he calls it) girls of the counter, at the bar, or in his favourite resort, the cigar-shop. Nothing is forbidden to him. It is the spirit of the age to give young men in some directions unlimited freedom, even to license. As proof of this, I need only point to the complaints of students one listens to in every University city. They interrupt business at political meetings. They sing and shout and are so noisy in theatres, that unless play-goers have an unusual amount of nerve force, they feel compelled to stay at home on students' nights. When snow is on the streets they will obstruct a thoroughfare, and take their own amusement at their neighbours' cost. They treat professors occasionally with open rebellion and derision, and they have stopped a grave Lord Rector in the act of delivering his inaugural address and forced him to retire.

All this they look upon as *manly*, and as a self-assertion, only *proper* to their youth and grade. And public opinion tolerates it! Professors wink at it; fathers laugh at it; and sufferers from it endure in silence. Is it any wonder that many of these lads become in time domestic despots, selfish husbands, grinding masters, showing one element of character in all their dealings with their fellow-creatures, namely, dominancy, wherever and whenever it is possible to assert it?

Meanwhile the sisters of these students go to no public place without protection. They are chaperoned to balls and concerts; they never walk out at dusk alone; and even at midday, if young ladies of twenty-five or thirty choose to attend a garden promenade, propriety approves the lad of seventeen escorting them as guardian ! This gives *him* a delicious sense of power, and observe, the dangers he protects them from are purely *ideal*, whilst the inflation of his egoism is very *real*.

Again, the ball-room, whether public or private, is always a field for masculine supremacy. A young girl, unless she is willing to be thought loud and rampant, can never venture to choose a partner for the dance. The young men are masters of the position, and girls must sit and wait, very much as in Mr. Long's Babylonish marriage market ! Indeed, one is forcibly reminded of the picture; for what guides selection with the men is personal appearance (with good dancing). If girls are plain-looking, they may very probably sit still all evening, or hang upon the hostess' hands like a drug in the market.

Skating is, for girls at least, a more recent amusement, and *because modern*, it is not ruled as dancing is by tyrannic custom. To skate in pairs, seems as pleasurable as to dance in pairs, but girls never stand waiting on the ice for men's selection. They pair off with one another, and the glow of healthy exercise and happy life is on the cheek, although no flirting in the case is possible.

The mixture of the sexes, both in study and in all innocent amusements, I look upon as *eminently desirable;* but in a country where we have *all but* a million of surplus women, and in every district except the great business centres, at least six marriageable girls to one marriageable man, sex equality *numerically* is quite out of the question.

In many places the giving of a ball depends upon facilities for importing men from a distance, and therefore opportunities for dancing are to the natives a rare occurrence, confined to once or perhaps twice in the year. We seem to me to be everywhere, except in large towns, in positive danger of dropping out of our social life one of the greatest, most healthful enjoyments of youth, and that simply because we are slaves to conventional custom. To move in rhythmic measure, to skim the ground, and float along in waves of delicious music, is as natural to some young people as play to little children ; and the keen sense of pleasure, the delicate exhilaration caused to the whole nervous system by dancing, is a precious boon to be guarded by society, and to be freely and impartially distributed as a means of innocent happiness during the period of youth.

But in the present day, not only does society fail to fulfil this mission, but it fails more thoroughly than it did some thirty years ago.

In a small University city, where there has always been a select and cultured society, a lady called my attention to this fact. "When I came here," she said, "twenty-five years ago, the social visiting was easier and far pleasanter for the young. Parents and old people were invited to dinner at five or half-past five o'clock, and when they left the table, the young people had arrived and were assembled in the drawing-room ready for music, games, or, what they liked best of all, a carpet dance. The elders looked on, or had their whist-table in the corner, and after a slight refreshment, all would go home together, and that before midnight. Now everything is different. The dinner-parties are very late, and the old and young seem separated. The latter, too, are no longer unsophisticated. They want elaborate preparations for a dance, an elegant supper, champagne, a floor with no carpet, an instrumental band. Dances are therefore rare, and I am not sorry; for I look upon them now as too *exciting*, a sort of midnight dissipation."

This sophistication of the young is a peculiarly melancholy feature of modern society, and yet how natural it is! In proportion to the tender susceptibility of the young human plant, is the readiness with which it imbibes the mental and moral hue of the surrounding social atmosphere; and if we want simplicity to be the genius of character in the young, we *must* make the conditions of their life, simple, easy, and natural.

Amongst the wealthy classes in this country, however, there is now *no* simplicity in social life, and very little in domestic life. Lavish expenditure has brought about luxury and conventionalism, and these have eaten out the heart of social intercourse and left it hollow. They have utterly destroyed the ease, the charm, the tender grace, the genial warmth of English society.

The remedy lies in a public, a universal awakening to some of the important facts of human life, and in adjusting public and private conduct to these facts.

One of these facts is that wealth does not always minister to happiness, but on the contrary, it very frequently obstructs and injures it, and at the present moment it is largely obstructing the happiness of youth, and frightfully injuring its natural simplicity and innocence. A perfectly free and spontaneous intercourse between the sexes is necessary to happiness in the period of youth; and it is possible even now under wise regulations to obtain that innocent freedom, in each and every social class. The stiffness of the ball-room, the late hours, the conventional rules, the stimulants, and aids to artificial excitement, however, *do not* promote that end; and dancing should be rescued and set free from the moral degradation and formal conventionalism that surround it, at a public or private *ball*.

My reader at this point may plausibly suspect my logical consistency, for whilst I have assumed that too much freedom is permitted to young men, I here assert that what we want is *greater* social freedom! I grant the inconsistency, but it belongs not to my logic but to the action of society. Young men should not be free to cause discomfort or loss to fellow-citizens. They should not be free to indulge in irreverent discourtesy to teachers or their elders. They should not be free to create social disorder; and clearly civic authority ought firmly to suppress such ebullitions. Again, young men should not be free to seek female companionship in *any* class without regard to social equality; and to approach with insolent familiarity girls whom they would not choose their sisters to associate with. But here public opinion is the only controlling force that can legitimately be exercised; and with right training in regard to sex relations, and perfect liberty of social intercourse within their own class, no other force would be required.

I think my reader will agree with me, that a young man of ordinary education and average good taste, would never prefer to flirt with barmaids and form friendships with young ladies of cigar-shops, if he felt free to do the same with girls of his own social grade, at his and their sweet will. And *this* is the freedom that society ought to accord, but with a perverse inconsistency society checks his impulses, and trammels him where he ought to be free; and on the other hand gives freedom where its exercise brings degradation to his character, and causes misery in a social class in which his presence is nothing less than an impertinent intrusion.

There is no middle course possible to us here. Respect for womanhood is what society must aim at impressing on her masculine youth, and that has two implications. It implies, first: not a superficial acquaintance, but a thorough knowledge of the womanhood within his own class, and more than this, freedom and power to enter into all the varied ties that are unhurtful to society and mutually agreeable. Second: it implies a courteous distance from the womanhood of a different social class ; a distance free from all personal contempt, and simply resting upon a necessity of nature, viz. the need of sympathy in tastes and habits, in order to form *true* sex relations.

I of course admit that social inequalities and class distinctions all tend to disappear with the steady growth and spread of rational education ; but in the meanwhile social justice must wear this twofold aspect, for without some repression, womanhood within the classes that are socially weak is literally preyed upon by men belonging to the classes that are socially stronger.

Now, to parents who have never carefully examined (as I have done) the evils of the present system, and whose minds are thoroughly occupied with other, and I venture to say, much less important subjects, my proposition—that young ladies and young gentlemen should enjoy intimate, free, and frequent social intercourse—will appear startling and perhaps strongly objectionable. Visions will rise up before them of grossly imprudent marriages, of friendships that *look* improper, of love affairs of boys and girls in their teens !

I am not myself a parent, but with all parents I have the deepest sympathy. I am not ignorant of the many difficulties of their position, of the grave sense of responsibility that weighs upon them, and of their natural and strong desire to keep their daughters morally pure and safe. Nevertheless my slowly formed but now calm conviction is, that with the best intentions, the highest aims and purposes, the generality of thoughtful parents in all classes are at the present day acting *irrationally* towards their children, and of their own free will are subjecting them to habits and customs that will inevitably degrade them, and in the end will thwart and defeat these higher aims and purposes. I say *higher*, for no doubt there are moments when parents feel that if their sons are outwardly respectable, and daughters well married in the mere worldly sense, success has been or will have been achieved. But there are other moments when the mind is in an elevated and abstract rather than practical mood, when what *might* be, rather than what seems *likely to be*, is thought of ; and at these moments parents reject, as quite *inadequate*, the outward comfort and respectability, unless adorned with inward purity, and that sympathy which in their secret hearts they know to be the very wine and elixir of life.

Now, if I seek the confidence of an affectionate mother, and ask her privately—" Is the moral purity of your sons less dear to you than that of your daughters?" she answers, "*No.*" " Is it less important in respect to the happiness of married life?" she answers, "No." " Or less important as regards parentage? I mean less likely to affect for good or evil the health and dispositions of your future grandchildren?" and again she answers, "No." If I demand a deeper confidence still, and trusting to the nobility of her mind, which will enable her to see through the apparent indelicacy to the true motive of the question, ask—" What has marriage been to you?" she replies at once—" A disappointment! But," she explains, "I was a dreamer in my youth, a pure idealist. I married without any *real* knowledge of what life is, or is meant to be. I throw no blame upon my husband. He is practical, a man of the world; not always, nay seldom *sympathetic*, but kind to me, and outwardly we are a united couple. In the inner depths of my own consciousness lies the fact that marriage has been a bitter disappointment; but I am wholly ignorant of whether the *cause of this* lies in my objective experience, or my romantic expectations. It may be, I freely admit, that I have looked for *too much*, rather than that the relation has yielded me *too little.*" The personal question *she* would gladly solve, I must not dwell upon.

The point of grave importance for us is, that under the fair surface of decorous married life, there exists in this country in hundreds and thousands of cases a vacuum, I mean a space, empty and void of real, substantial happiness: and although I have represented it from one point of view only, my reader, if he or she has any experience of life, will fill up the picture, and recognize that men suffer from uncongenial partnership as *frequently*, as *truly*, and as *deeply*, as women suffer.

Congenial marriage is, and perhaps always has been, extremely rare; but if society ever approaches the goal of general happiness, it will be by a path strewn with this flower of life, a path in which mock unions will give place to those that rest on the *only firm* foundation—deep and true sympathy of nature. This *sympathy of nature* in adult life is what we have to preserve, where it exists, and to produce where it does not exist.

A community which has outlived its infancy and childhood, and has ceased therefore to act from either blind instinct or archaic authority, but has entered upon a stage of rational and self-conscious existence, will see clearly that pure and innocent happiness is absolutely dependent upon broad sympathy between the sexes; and it will take the forces of evolution into its hands, and apply them to *produce* that sympathy.

In my last chapter I have shown how forces of evolution are perpetually at work within the sphere of human emotions, and how steady growth in all the noble sentiments that elevate mankind above the brutes is attainable both in general society and in individual social units. Sympathy in the young, as compared to sympathy in adult life, is very much like pap or baby-food, as compared to all the rich and varied diet of the strong man's table. A sympathetic friendship between an innocent but shallow-minded Fanny Dover and her male prototype is music of a weak and trivial order; nevertheless, it is by practice of that music, by interchange of these poor harmonies, such as they are, that the young instruments gain in capacity, that all the notes of tender feeling deepen and grow clearer, all the chords of human sympathy grow strong, rich, and full, and at last the music that each or both are capable of bringing into life, is like the glorious harmonies, the combinations of delicious sound, conceived and executed by a great master such as Handel or Mozart.

But here let us suppose that instead of general harmony and individual sympathy between the sexes, what society requires is the *very opposite of this :*—society desires men and women to be discordant, disunited, to have no earnest thoughts, aims, and purposes in common, no community of enthusiasms, no similarity of habits, tastes, and dispositions, but, in one word, to be as thoroughly *unlike* as nature will permit. Then, undoubtedly, no better system than the present could be devised; for observe, just at the period when the character is growing, and habits, tastes, and dispositions forming, we carefully separate the sexes. We hedge our girls round securely, we tether them with strict conventional rules; we keep them ignorant of real life, and feed them on romance and fairy-like illusions; we lift them above all *necessary* work, and bid them amuse themselves and ornament society. We hint at marriage and a possible future of grave and important duties, but as no certainty of this in their own case exists, we *only hint.* We give no positive instruction in reference to wifehood or motherhood, but leave all vague, and vagueness within a girl's mind we call innocence, and treat it as a charm.

The general result is that our girls are discontented, often miserable. The bolder natures rebel, and take a path their parents can scarcely approve; they make excitement their one pursuit, and follow it to the loss of all womanly dignity, until "the dyed and painted, loud and rampant modernization" appears upon the stage of social life. The gentler natures inwardly repine, they suffer both in mind and body; hysteria takes possession of their frames, and lowers every pulse of life within them, till they become timid, shrinking, narrow in intelligence and judgment, but intense in purpose, fastidious in taste, and morbidly emotional.

Meanwhile, our boys are maturing in a wholly different school—a school as much too wide as that of girls is too narrow, and one in which the refining influence of mother and sisters is often wholly absent. They "see life," and that technical phrase implies something which refined women never see. And the general result is that our lads are full of worldly wisdom, which means gross selfishness; they are wide awake, and broad, and somewhat coarse, tolerant to every masculine vice, but strongly, fastidiously proper on the one point—female conduct within *their own class;* and to the girls within that class they show themselves conceited, presumptuous, and tyrannical.

Now, dropping the fiction that we want discord and antagonism between the sexes, we face the fact, that here are the elements from which there must be organized the music of the future—the harmonies of character—the quick response to thought and feeling; the delicate play of sympathetic emotions that make the real charm of life, and on which depends the happiness of the race, when we ourselves shall have passed into "the silent land."

All these young people will form relations of some sort with one another, and weave the web of social life destined to adorn or to disfigure England in the twentieth century. Perhaps five out of every six of the lads, and four out of every six of the girls, will form the closest of all ties, a relation so near, and requiring such intimate and tender sympathy, that failure to apprehend a momentary thought, a chance impulse, is capable of causing a jar;

> "A little rift within the lute
> That by-and-by will make the music mute,
> And ever widening slowly silence all." *

---

* Tennyson.

But alas! that power to apprehend a husband's thought, that delicate susceptibility to a wife's (perhaps morbid) feeling —dare we of the nineteenth century expect it? Have we prepared for it by giving them in youth the opportunity of intimate knowledge, and of growing each to each in fitness for the needs of life? Have we not deliberately caused them to become as unlike as nature will permit? And is it not true, that judging from our action in the present day, what we may *rationally* anticipate in the future is a society more miserable, more hollow and empty of real joy, and above all, more feeble and diseased than even the society which surrounds us now? Unless we adopt a new policy, our girls are certain to be weakly, delicate, and fretful wives and mothers; and the very higher education with which we try to make them happier is straining the nervous system, and bidding fair to make matters by-and-by considerably worse.

There are comparatively few girls with strongly intellectual proclivities; and when these proclivities do not exist, to force upon them an education beyond their natural assimilating powers is to inflict an injury in addition to the negative result of doing them no good. The average British lad, although more rational than the average girl, is by no means intellectual, and even amongst men of highest education and attainment, it is rare to meet with one who values culture in a woman, and will deliberately seek to give and take the intellectual intercourse with her, which might add so immeasurably to the happiness of life.

When, in consequence of this, I say to parents, do not let us think too much of this higher education, and aim at cramming girls with Greek and Latin, I am not traitor to my principle, that woman exists, not for man, but, as man exists, for the development of her varied faculties, for the achievement of the highest work within her power, and the attainment of the greatest goodness and happiness possible. Goodness and happiness will not be promoted by artificial education, any more than by artificial restraints upon her freedom, that is certain, and yet we are in danger in the present day of looking on this higher education as a panacea for all the evils that I have pointed out. The movement, with its *admirable results* in offering facilities to woman for the development of all her varied faculties, is (if not abused) a glorious step of social progress, and gives abundant promise of a happier life to come.

But to parents I respectfully address a word of caution.

Your girls are for the *most part* of the Fanny Dover type. Whilst full of buoyant life their mental calibre is commonplace; and Cambridge scholarship or Girton Greek and Latin may add no *real* dignity to their womanhood, but only prove veneer that hides the homely grace, the solid merit of the human nature that lies below.

In my young days all parents in my social class strove to give their daughters what was called " polite accomplishments," and many who could ill afford it spent money freely on music lessons for girls who had neither ear nor any aptitude for music. Piano-strumming was the order of the day, and oh! the weary hours of wasted youth spent in producing agile fingers, capable of sending forth a rapid sequence of hard sounds with mechanical precision—the whole performance as different from true music as the jingle of a baby-rattle from the delightful language of Herr Joachim's violin. A smattering of music is a poor attainment, and a smattering of anything, whether Greek and Latin, Mathematics and Algebra, or even Moral Philosophy, is not one whit superior. It may give vanity, but it will add neither *worth* nor interest to the social unit.

Where we err is in our lack of *reverence for human nature*. We must remember that no two human beings born into the world can ever be alike. Nature intends no dreary sameness; each human plant has individual instincts, aptitudes, and tastes, and the surroundings of each should permit of free development according to its own inborn pattern. No parents need regret that not very many of their daughters are of an intellectual pattern. Amongst the other sex, as I have said, there is but little of that pattern too; and what we really want for goodness and happiness are unanimity of aims and purposes, community of interests, affinity in tastes, harmony of habits, and sympathy of sentiment.

If girls of commonplace intellect are guarded from the vulgarity of false assumptions and affectations, if they are simple and natural, if they are instructed in all the ordinary duties of life, and practised in the use of their reasoning powers within the sphere of these duties, if their dispositions are trained, and their emotions regulated to social life, and if they emerge from a happy youth, *not delicate*, and *not morbid*, but healthy in mind and body, then, even although destitute of polite accomplishments and high culture, they will form a solid foundation for a superstructure of social life, a *century after this*, vastly superior to what we have now—a social life in which men and women, of stronger mental calibre than their parents or grandparents, will unitedly enjoy high intellectual intercourse.

Meanwhile, there are exceptional girls, with latent powers for full reception and enjoyment of the highest culture, and these in every case should be enabled to drink deeply at the purest fountains of knowledge. No smatterings of science or of art should be deemed sufficient for the lady-student, and no public or private countenance should be given to the false, corrupting theory, that it is right and proper for the female sex to make a toy, a plaything, or at the best an ostentatious personal ornament of culture, whilst men must use it for the public good.

Pursuit of knowledge in every direction should by a civilized community be regarded as *sacred;* and whenever trifling takes the place of earnest study, it ought to be condemned as vicious and degrading. For this reason, it betrays an immoral attitude of mind towards *truth*, and hinders the development of the highest, most important modern moral sentiment, viz. *reverence for truth.*

But in reference to *intellectual* girls, there are some points we must consider. A strong head and a tender heart are not incompatible. On the contrary, the girl with large brain is endowed with emotional as well as mental capacity, above average. If her intelligence is fed, and well digested knowledge causes her mind to expand, the development will be according to the inborn pattern, and she will become a woman of unusual powers of loving her fellow-creatures, as well as of unusual powers of apprehending and enjoying truth. Now, in our present social state these are not altogether favourable to happiness; for with strong impulses towards " a life filled with emotion," she will find it difficult to form close ties. In respect to marriage her requirements will be greater than those of ordinary women, and her chance of marrying *much less.* She would not be happy to decline upon a lower level than her own, and be mated with a clown like the heroine of Locksley Hall; and whilst men of intellectual tastes in England are comparatively few in number, a considerable portion of that small number are unsympathetic and even antagonistic towards women of intellect. The vulgar prejudice against the "blue stocking" has its faint reflection in every circle, and although the best minds see clearly the justice of sex equality, hearts have not yet embodied the truth, and there is no warm welcome frankly offered to the female " blue."

"It has been said that the man of the nineteenth century insists upon having for a wife a woman of the seventeenth century. It is, perhaps, nearer the truth that he demands the spirit of the two centuries combined in one woman ; the activity and liberality of thought which characterizes the present era, with the *intellectual* submission to authority which belonged to the past." *

Parents should be open-eyed to the disabilities that culture entails, and deal candidly with their daughters in reference to marriage. In our present position, when so many men emigrate, and the sexes are wholly disproportionate in numbers, marriage is to all girls only a probable contingency ; but to intellectual girls it is *not* a very probable contingency, it is a *possible* but *improbable* contingency, and life without it should be faced and carefully prepared for. They should be strengthened to withstand the temptation of forming false ties, and to pursue, if need be, a lonely path without anti-social bitterness. But above all they should be trained to make their culture useful to their fellow-creatures ; and in the noble effort to benefit humanity and to accomplish the highest work within her power, the largest-hearted woman will find some satisfaction in her life.

Society has in female culture a social force to utilize, which has as yet greatly run to waste. Some openings have been made for it in educational and municipal responsibilities ; but until every profession is open to woman, and every public duty devolves upon the social units (able to perform it), irrespective of sex, there can be no certainty that the force I speak of, is not lost to public service and creating private misery.

My reader has no doubt observed that my whole treatment of the subject of the *Period of Youth* is built upon this argument. After a community is established in peaceful life, its warlike stage passed through, and it has become secure in the mere *necessaries* of existence, without which life of any kind is impossible, the thing of greatest importance is human *sympathy*. In every relation of life, whether near or distant, sympathy, that is kindliness of feeling arising from affinity of sentiment and disposition, is what makes the wheels of the vast machinery of social life move easily and pleasurably. Naturally, then, in judging of a community we look first to the

* "The Relations of the Sexes," *Westminster Review*, April, 1879.

closest of all relations, the tie which certainly tends most to make or mar the happiness of individual life, in order to see if this sympathy, this great factor of human happiness, exists strongly *there*. And what we find is, that in the British community, behind the fair outward surface of decorous married life, there is very little of this intimate and all-important sympathy. The causes of this we trace without difficulty to the surrounding conditions of the two sexes at the *plastic* period of human life; and we maintain, that by the application of the mature reasoning powers of the general community, we are able to take into our hands the forces of evolution, and so alter the environment of youth in our own generation as to evolve or develop sympathy to an extent hitherto unknown, and lift the next generation to a level vastly superior to ours, both in regard to happiness and goodness.

But now it is always in reference to goodness that earnest, thoughtful minds deprecate reform of system, and shrink from change. These minds conceive that whatever may be the faults of our present system, and however great its failure in producing happiness, it at least secures *social order and domestic purity;* and these two elements are of such transcendent importance, that we may well be willing to pay for them a costly price. I grant the importance of these elements, but I firmly deny that our present system secures them; and in proof of this I turn to our historian of European morals, and quote his facts. Of the most degrading and corrupting of sex relations, prostitution, he says, " However persistently society may ignore this form of vice, it exists, nevertheless, and on the most gigantic scale, and that evil rarely assumes such inveterate and perverting forms as when it is shrouded in obscurity, and veiled by a hypocritical appearance of unconsciousness. The existence in England of unhappy women, sunk in the very lowest depths of vice and misery, and numbering certainly not less than fifty thousand,* shows sufficiently what an appalling amount of moral evil is festering uncontrolled, undiscussed, and unalleviated, under the fair surface of *decorous* society." †

My fellow-compatriots, let us not deceive ourselves. The price we are paying for the purity of daughters is the moral degradation of sons, and the *necessity* for reform lies in this, that our apparent goodness is not real. It is only as the

---

* Some writers estimate that in London alone there are eighty thousand prostitutes.
† " History of European Morals," W. E. H. Lecky, vol. ii. p. 301.

sounding brass and tinkling cymbal. If we fear to break the
"fair surface of our decorous society," lest it should cave in,
the system is not worth preserving. To break it up with conscious mind, and strive to make it pure and sweet *throughout*,
would be no revolution of blind force, but a *re*-formation—the
rational action of a free and civilized race, a race resolved that
the surroundings of its social life shall be such as will present
no vicious temptations to young men, no constant strain upon
their moral instincts, tending to reduce the innocent, impulsive,
unreflectiveness of youth to mean and despicable hypocrisy.
A fair and decorous surface is no compensation for an inwardly diseased organism, and this great social evil festering
in our midst is like a canker that saps and undermines our
truthfulness, and utterly destroys our *moral* growth.

The changes that are necessary are numerous and complex
in detail. But they admit of being generalized in very simple
form, for one and all of these changes will be movements on
the lines of—greater reverence for human nature, greater simplicity in social life, equality and unity of the sexes, and enlightened training of the young to social sympathy, and to the
steady repression of every anti-social feeling.

CHAPTER XI.

THE SUBJECTIVE REQUIREMENTS OF SOCIAL LIFE.

"In any case, let us know the facts about human nature, the pathological facts no less than the others; these are the first thing, and the second, and the third also."—JOHN MORLEY.

"Passion may burn like a devouring flame; and in a few moments, like flame, may bring down a temple to dust and ashes; but it is earnest as flame, and essentially pure."—*Autobiography of Mark Rutherford.*

IN the year 1789 all British subjects sufficiently developed to be altruistic, *i.e.* individuals possessing over and above the self-regarding qualities, intelligence to apprehend and sympathy to feel the struggles of Humanity, engaged in a great effort to break down the tyranny of circumstance in search of happiness, had their whole attention fixed upon events in France. To such individuals it was well known that the miserable condition of the down-trodden French people made revolution justifiable, and the great idea which lay at the heart of the revolution, viz. simplification of life—a return to nature, accompanied by the shaking off of all the artificial, conventional intricacies of an old and corrupt civilization—was in itself a noble ideal that kindled the enthusiasm of distant spectators as well as of the actors in these memorable scenes.

In proportion, however, to the expansion of feeling experienced by distant onlookers (in proportion to the strength of their generous desires and ardent hopes for the French people) was their disappointment when the collapse came. The Great Revolution culminated in a paroxysm of savage violence, a debauch of every barbarous and brutal passion that humanity is capable of; and sympathy for the sufferings of the lower classes at the moment was quenched in *horror* by the spectacle of their bloodthirsty and insatiable revenge. But we

who look back upon the awful tragedy and calmly survey
events from a distance in time as well as space—we at least
perceive that it left valuable fruits behind. With these fruits
we are not, however, here concerned, but rather with its errors
and the causes which conspired to make the judgment of a
thoughtful, *not unsympathetic*, British mind find expression in
these words : " *The French Revolution disgraced reason.*" * One
great factor of the phenomena was the spirit and teaching of a
single man. Rousseau had, as Mr. Morley tells us, "hold over
a generation that was lost amid the broken maze of fallen
systems." Now, Rousseau died in 1778 ; but what Mr. Morley
means is, that in the minds of the leaders of the Revolution
Rousseau's ideas held dominion, and, unfortunately, his *central*
idea was "disparagement of the reasoning faculty."† It was
a true case of the blind leading the blind. Rousseau was a
man of unusual sensitiveness, and his whole surroundings in
childhood and later life tended to make his character morbid
or ill-balanced. His self-regarding feelings became excessive
and exaggerated, his intelligence was untrained, his social
sympathies were repressed, and because the artificial system
of Parisian life was wholly uncongenial to him, he developed a
marked *unsociability*, and gave himself up to a brooding imagination, a sensuous expansion over a purely ideal "natural
life," which had no real basis in the facts of human nature.
But all reform, to be true and lasting, *must* be based upon the
facts of human nature, and the false notes in Rousseau's teaching
were bound by law (as sure and natural as the law of gravitation) to produce discord and confusion. Rousseau was not
responsible for the defects of his training. Before he was
seven years old he had learned from his father to indulge a
passion for romance. His father, Isaac Rousseau, and the boy
of six would together spend whole nights in reading romances
to one another in turn till the morning note of the birds recalled
them to actual life, and the elder would cry out that he was
the more childish of the two ! In reference to this practice,
Rousseau himself says, "It gave me bizarre, romantic ideas of
human life, of which neither reflection nor experience has ever
been able wholly to cure me." ‡ Every wide and powerful
stream has its many tributaries, and each tributary has its tiny,
thread-like sources, could we but trace them out. Now here,
in this fact of life, this young imagination feeding day by day

* George Combe.   † " Rousseau," by John Morley, vol. ii. p. 206.
‡ Ibid., vol. i. p. 11.

on fiction, we come upon a source, a tiny spring of cosmic force, destined to converge and to contribute to the great outbreak wherein "a generation was lost amid the broken maze of fallen systems." But we see in it more than this—a partial explanation of one of many causes why the great French Revolution was *not* a rational reform throughout, but *objectively* a wild saturnalia, and subjectively a phantasmagoria of ideas, false and true, inextricably mixed. Rousseau's psychology was defective, and the modern moral sentiment of universal justice was in him little developed. True, he desired equality of class, but of sex equality he had simply *no* conception. He revolted against all tyranny of aristocracy over plutocracy and of the rich over the poor, but of private or domestic tyranny, such as the despotic rule of husband over wife, he took no notice whatever. His ideal of womanhood was not even up to the civilization of his own period and race; it was *Oriental*, and his conception of the natural state meant social freedom for man, but for woman entire subjection! In his great work, "Emilius," of which Mr. Morley writes, "It is one of the seminal books in the history of literature. . . . It touched the deeper things of character. It was the veritable charter of youthful deliverance," manhood is extolled. The hero, Emilius, was to be first and above all a *man*. Not so with the heroine, Sophie. Her individual dignity is nowhere recognized. The duties taught to women are to please men, to be useful to men, to console them, to make their lives agreeable and sweet to them. Woman's orders are caresses, her threats are tears. Now, even in the seventeenth century, says Mr. Morley, French history had shown a type of womanhood in which "devotion went with force . . . divine candour and transparent innocence coexisted with energetic loyalty and intellectual uprightness and a firmly set will." * Clearly, then, on *this* point Rousseau was behind his age, and his doctrine was a false one. But, again, his conception even of manhood was incomplete. Because he saw the goodness of humanity under its coarsest outside he scorned contemptuously all literary culture, social position, and social accomplishments. He taught the supremacy of emotion over reason, and never analyzed emotion or pointed out that there are certain phases of human emotion which, if indulged in, will degrade man below the brute. In the training of the young he imagined that self-love is the one quality of embryo character on which to work, and did not recognize that

---

* "Rousseau," vol. ii. p. 247.

"sympathy appears in good natures extremely early, and is susceptible of rapid cultivation from the very first." *

That the great French Revolution was not productive of unmixed good, that it disgraced reason, and that a generation was "lost amid the broken maze of fallen systems" creates now no surprise in minds accustomed to look to antecedents and see throughout phenomena the course of natural sequence. But it stands for us a terrible warning—a warning of how false teaching plays a part in bringing into action forces that are overwhelming in their destructive power, incapable of all control, and how in all forward movement intelligence *must* be the guide, the pioneer, the supreme ruler of human destiny. For social units to act aright spontaneously as Rousseau desired is, no doubt, a beautiful dream; but before that dream can become a reality, habits that harmonize with general well-being must be formed in the social units, and under the directing intelligence of an adult generation each young generation must be trained into fitness for social life *far more thoroughly* than has ever as yet been accomplished or attempted.

The known facts about human nature are quite indispensable to a theorizer upon social systems, and some comprehension of these facts must be established in the general mind ere we can hope to see a widespread, individual, and yet general preparation for a better system. The facts are, of course, both physiological and psychological, and these terms embrace the emotional nature of human beings as well as their intellectual nature and bodily structure. The purely animal appetites and propensities are of transcendent importance, because unless these are exercised there could be no life at all, and unless exercised in proper measure, there is no healthy foundation on which to raise a superstructure of elevated, widely intelligent life.

The importance of eating and drinking has never been wholly misunderstood, and the British race has long since passed out of the stage of mental ignorance in which fasting and inanition could be regarded as more holy than paying due attention to the laws of nature. Whilst, however, the necessity for sufficient eating and drinking is recognized, and we are in no danger of loss of life through wilful abstinence, the pleasurable sensation given by exercise of the function is often unduly esteemed and overstrained, and the epicure is as much the victim of this error as the glutton or the drunkard, although

* "Rousseau," vol. ii. p. 217.

in a somewhat different form. In reference to this appetite, what we want is the rational treatment of it, and rational training to the proper exercise of it in childhood and youth. And it belongs to the *conscious epoch* that we are entering upon, to think out and devise the method we shall adopt.

There must be thorough, *adequate* teaching of physiology in schools, that children may have correct ideas regarding the structure and functions of the body as soon as the mind is capable of understanding and assimilating the facts. This teaching should be entirely oral (no text-books given to children), but accompanied by the use of prints, models, and all appliances that can aid a child to receive knowledge through the eye and ear without strain upon the imagination. But here I am compelled to pause and guard my reader against the supposition that I would add to the already far too burdensome daily lessons of our children. Were it so, I should deserve to be condemned as the wilful placer of the straw that will break the camel's back. No! What I desire to see is a lightening and simplifying of the daily tasks, a sweeping away and discarding of all lessons that are uninteresting and useless, and substituting in their place only what is useful and can be made thoroughly interesting to a child. Amongst the upper classes, at least, books in the present day take far too prominent a place in a child's daily life, and the evils resulting from this are so great that I think it matter for grave consideration whether we should not banish all lesson-books from primary schools, and only permit reading to be *publicly* taught from the age of eleven or twelve. I suspect that what we should gain in health of body and in the natural development of general intelligence would more than compensate for our loss in mechanical proficiency in reading.

In the *Century Magazine* for November, 1882, Miss Mayne says, " I can tell you of a little girl, six years old, who cannot read a word, but who knows many flowers, and how they grow . . . who knows some of the planets and constellations, and where to look for them, who delights in watching the ants, bees, and birds, and in hearing stories about them, and who expresses her ideas with ease and accuracy ; and all this has been accomplished without perceptible effort on the part of the child." However, apart from this there are lessons in history and in book-taught geography * that may well be

* " I remember," says Mr. Henry George, "a little girl, pretty well along in her school geography and astronomy, who was much astonished to find that the ground in her mother's back yard was really the surface of the earth."—" Progress and Poverty," book vi. ch. ii.

banished to make room for the oral teaching of physiology. We have only to recall the long lists of names we painfully committed to memory in our own youth—names of rivers with their tributaries (tiny streams we have never heard of since), or of towns noted for this or that production, and our after disappointment to find that no shawls are made at Paisley; that but few of the carpets we meet with are manufactured at Kidderminster; that Coventry is by no means famous for ribbons —to perceive how useless this kind of teaching is. Could anything be more irrational than to impress upon the plastic brain of childhood fixed ideas about conditions that are continually altering?

And worse than this is done in reference to history. We feed a child's imagination upon all the bloody deeds, the cruel carnage, the hideous warfare that was necessary as our race gradually took possession of and settled itself upon that portion of the earth's surface which we call our native land, and in the process we foster every barbarous, cruel instinct which the child has inherited from warlike ancestors, and think that this will fit him for peaceful, gentle, professional or commercial and domestic life, and a social system in which tyranny and masterful ways are simply *intolerable*. Again, we load his memory with dates which, unless he is to be an historian, he will never require, and, if he is, he could easily find them stored in books (their proper place), instead of uselessly engraved on the delicate, precious tissue of his brain. We require him to remember the names of Henry VIII.'s six unhappy wives, and which the royal monster divorced, and why—facts which we ought to bury in oblivion rather than plant them in the pure garden of a child's innocent thoughts.

Now, if instead of (not over and above, but instead of) all this, we have our children taught the functions of the heart and lungs within themselves, the structure of the eye and ear and brain, the active process started in various glands when they have swallowed food and caused digestion to begin, and so on, we exercise the mind and draw out intelligence without effort; and we give them some simple facts of nature that will never change, and are certain to prove useful in their own individual lives. More than this, we steer clear of two dangers: first, that of fostering anti-social feeling; second, that of warping the moral nature, by bringing before the infant mind human conduct which requires explanation or comment, but the true analysis of which is quite beyond the infant comprehension.

After this long but necessary digression, let us return to our main subject—the purely animal propensities or appetites. Besides the direct teaching of physiology and training the young mind to reverence for all laws of nature, and to personal obedience to some of the simple laws of health (as, for instance, the avoidance of over-eating, or too frequent eating, because of the evils that must follow if the organs of digestion have not sufficient rest, etc.), we have further to *beware* of indirectly causing the young to overvalue the pleasures of eating and drinking, and this is not a trifling or unimportant matter.

The kindest parents thoughtlessly draw their children's attention to this pleasurable sensation by using it *as a reward*. "If you are good I will give you sweets," a mother will say; or, "If you are at the top of your class this week you shall have whichever pudding you like best on Sunday," and so on. At that early age the habit of thinking of the stomach and acting with a keen eye to its gratification is soon acquired, and slowly but surely another habit grows as well—I mean the tendency to discriminate and over-analyze the sense of taste, and to reject wholesome food, simply because it does not give the *greatest* pleasurable sensation. Fastidious children are far too common; and in my own social circle I know many young men and women, apparently in good health, who are thorough epicures. I do not mean that they always desire luxurious dishes, but that they are incapable of taking without comment, and disposing of with healthy appetite, whatever good food is placed before them. Now, this adds immensely to the trouble and cares of housekeeping. I do not hesitate to say that an enormous change in favour of health, comfort, and happiness might be secured by attention paid to this minor point. I mean that parents, and all who have the guidance of the young, should carefully avoid what may tend to exalt the pleasures of the table in their minds, and train them from the first to keep this appetite in its truly natural position, which is an entirely *subordinate* one.

Next in order to the animal instinct or appetite of eating and drinking comes that of the sex instinct or feeling. Of equal importance to life—that is the life of the *race*—this latter differs in its essential nature from the former. Mr. G. H. Lewes, in his important work, "The Problems of Life and Mind," makes clear to us, that whilst the individual functions of man (alimentation being one of these functions) arise in relation to the cosmos, his general functions, including sex-appetite, arise in relation to the social medium; and "animal impulses," he says, "become blended with human emotions," till "in the process of evolution, starting from the merely animal appetite of sexuality, we arrive at the purest and most far-reaching tenderness." The social instincts, which he calls analogues of the individual instincts, tend more and more to make "Sociality dominate Animality, and thus subordinate Personality to Humanity." *

This appetite, then, holds in reality a higher position than the appetite we have just been considering. It is not purely egoistic, it has a wider range; and when exercised under moral conditions it calls into play emotions that are of the highest, most purifying order. The indulgence of it in measure purely normal and healthful has always been encumbered with difficulties arising from the law of population, that is, the tendency of man to increase more rapidly than the means of his subsistence can increase; and the outward consequences are seen in the various forms of social life, the promiscuous intercourse, the polyandry accompanied by destruction of female infants, adopted under barbarous conditions, and when the tribal solidarity had not yet been secured. In days when a considerable development of tender human feeling and the dawn of moral sentiment added new factors to the phenomena, the *difficulties continuing just the same*, false theories regarding this appetite quite unavoidably sprung up. The instinct was regarded as in itself impure, and the self-control that was capable of suppressing it and living a life of celibacy (or entire abstinence) was exalted as the highest form of human dignity, regardless of the injury to health and happiness that such a life entailed.

Another outcome of the difficulties, and also of the peculiar conditions that surround maternity, appears in this: that in the present day, the strength of this inherited instinct differs in the sexes, although the appetite is common alike to both. The difference varies of course in individual cases; but as a rule, the instinct is in all men keen and strong, whereas in many women who have tenderness and all the sympathetic qualities largely developed, the appetite from which these latter have arisen is in itself extremely weak, and sociality in them entirely dominates animality. In Mr. W. R. Greg's essay, called "Malthus Notwithstanding," he thus refers to the strength of human passion. "If Mr. Malthus' doctrine be

* "Problems of Life and Mind," vol. i. p. 159.

correct, the great majority of men and women must not only keep within moderate bounds the strongest propensity of their nature, but must suppress and deny it altogether—always for long and craving years, often, and in the case of numbers, for the whole of life. Observe, too, that the desire in question is the especial one of all our animal wants which is redeemed from animalism by being blended with our strongest and least selfish affections, which is ennobled by its associations in a way in which the appetites of eating and drinking and sleeping can never be ennobled—in a degree to which the pleasures of the eye and ear can be ennobled only by assiduous and lofty culture. . . . Is there any other instance," he goes on, " in which nature says in the most distinct and imperious language, ' Thou shalt do this'? and also in language equally imperious, if not equally distinct, 'If thou dost thou shalt be punished'?"*

There is in the general or popular mind of the present day a tolerably correct estimate of the importance of this appetite, and a recognition (although not publicly expressed) of the fact, that wherever it exists nature requires its due and healthful gratification; and by a few thoughtful individual minds one of the *greatest* and most *pressing* problems is plainly seen to be —How shall this function be exercised during the whole period of virile life, without injury to society and with all due regard to personal dignity and purity? To those reformers who perceive that human intelligence has found a means by which to subjugate the law of population and regulate the birth-rate, an enormous obstacle is, in their minds, swept away. Humanity needs no longer to tread upon its own heels and pour in countless multitudes, further human beings, upon a stage already *overcrowded*, and where the actual generation gets crushed out ere it has had its fair share of comfort and happiness. By limiting the family it is possible to find a solution of this problem in *early marriages;* but the inevitable corollary to this proposition is, that our marriage laws must adjust themselves to entirely new conditions; and this corollary would make imperative a thorough search for truth in reference to marriage—a sifting of the whole question to ascertain what are the proper objects, the true purposes of married life.

This subject, however, we shall not here consider. At present I must say a word regarding artificial checks to reproduction. There are many parents in the middle classes, who, aware that they have brought up their children luxuriously, and that their

---

* "Enigmas of Life," W. R. Greg, p. 71.

personal requirements are therefore great, dread early marriage for them on account of the expense and cares that an establishment and a young family would necessarily bring. On the other hand, they plainly see the dangers to sons and daughters of a dissatisfied and restless youth, and they know that in many cases, although not in all, the true means of giving equipoise, *i.e.* a healthy balance of mind and body, would be by entrance into married life. To such parents the thought of obtaining for their children the advantages, without the drawbacks, of early marriage would be hailed as an inestimable boon but for ideal scruples which confuse the judgment and perplex the mind. It has been said that artificial checks do not suffice, and that if they did, to use them is improper and contrary to nature. Now, my reader will easily perceive, that in a work of this kind it is the abstract aspect of this question *only* with which I have to deal; but when we look at France, a nation quite as civilized as our own, and see that her population is all but stationary, and that the small birth-rate is not confined to cities (where licentiousness might be suspected as a partial factor), but extends to country districts, where the industrious peasantry lead a pure, domestic life, it becomes evident that there is *no* practical and insurmountable obstacle to be overcome.* Again, the second difficulty is one of sentiment, and the principle of reverence for the vital nature of humanity is precisely what ought to lead us in the path I indicate. To reverence human nature is to give it freedom for the exercise of every pleasurable function, whether physical, mental, or emotional, which is unhurtful to the individual and to society. When the interests of these two conflict, individual interests are subordinate, and rightfully must suffer; but the moment that the two

---

* The following paragraph appeared in the newspapers early in January of the present year (1885):—

"POVERTY IN PARIS.—Pauperism in Paris is not increasing, but slightly diminishing. The triennial census just taken by the authorities shows 123,324 indigent persons, being 411 less than three years ago, the households they compose being, however, 47,627, an increase of 812. The families are therefore becoming smaller. The percentage of pauperism to the population has sunk from 6·22 to 5·43, but in two arrondissements it is over 12. Provincials and foreigners form the majority of the indigent, and of every 1000 foreigners in receipt of relief, 407 are Germans, 356 Belgians, 72 Dutch, 51 Italians, and 10 English."

The permitted influx of foreign labourers into France keeps wages low and partly defeats the prudence of her people. According to Mr. Gosselin there were in France, in 1851, 378,000 foreigners; since then they have increased to more than 1,000,000 in 1881.

can harmonize, the aspect of the question alters. If we *then* refuse to take advantage of a rational discovery, and still permit individual suffering, when the general welfare no longer makes it unavoidable, we are convicted of a *double* irreverence. We show ourselves irreverent of human happiness, and irreverent towards the highest, noblest of humanity's endowments— I mean the reasoning powers of man, by which he may control the forces that are adverse to his happiness. To feel *thus* is to be akin to Rousseau, who exalted one part of human nature, whilst his central idea was "disparagement of the reasoning faculty." But, as we know, the intellect may see truths which the heart will not embody, and the sentiment of impropriety is apt to linger and cause pain even after the principle that condemns the sentiment is well established in the general mind. What I have said on the evolution of sentiment in Chapter IX. I need not here repeat. There was an epoch when to think the earth moved round the sun was impious and perverse; there was an epoch when to talk of substituting steam for honest ships'-sails and the wind of heaven on Neptune's broad bosom, was felt by many hearts to be a desecration; there was an epoch when to allay maternal pangs by chloroform was to evade a providential discipline; and there was an epoch when a loving father said to his son: "It is not *right*—it is not *becoming*—it is not *proper* to play upon the violin, or even to laugh"! When sentiments, then, are out of harmony with the principles that intelligence accepts, what we must do is strive to alter *sentiments*, or wait till slowly but surely the master-hand of time alters them without our willing aid.

And now one other point. Our greatest living systematic thinker holds the theory, that as civilization advances, all the higher pleasures that arise from culture of the intellect, the feelings, and the æsthetic taste, will be preferred to those of *sense alone*. My method does not oppose this theory—in effect it would hasten progress in this true direction. "There is nothing degrading," says Mr. G. A. Gaskell, "in the enjoyment of a good dinner, and the interest taken in it; but as it is despicable for a man to live only in order to eat, so it is *despicable* (because merely selfish and narrow to human nature) to make animalism in any sense his supreme good." Where animalism is strong and instinctive, and love of offspring deficient, the tender joys, the anxious cares of parentage are not desired, and will be, by means of artificial checks, easily and intentionally avoided. Not animalism alone, but animalism combined with philoprogenitiveness will be the complex force to bring into the world the coming generations, and two results will follow. The inestimable blessing of parental love and parental responsibility, *assumed by choice*, will be the outward heritage of all, and inwardly, the new race will by inheritance partake of natures that are broader and fuller than the type of man in whom the pure and tender love of children has no existence, and animality dominates sociability.

This hopeful forecast is no wild dream. It has a scientific basis in the organic laws of human nature. Meanwhile, the daily record of our police courts furnish melancholy proof that we are surrounded still by human beings in whom animality, and not sociality, is the predominating force, and in whom, when passion is fairly roused, there is no strong enough tender feeling or moral sentiment to check or control the brutal instincts. This is a painful fact; but we must act in accordance with my motto: "In any case let us know the facts, the pathological facts no less than the others," otherwise we cannot *rightly* deal with questions of reform. "Sensuality," Mr. Lecky tells us, "is the vice of young men and of old nations." *
The latter half of this proposition would lead us to consider the many outward temptations presented by a corrupt civilization and an artificial social life. But these questions are apart from the subject of this chapter, whilst the question, how are young men trained and guided as regards this vice to which they have a strong natural tendency? bears directly upon my subject. The answer is, that in the present day there is *no* training or guidance systematically given, but *all is left to chance.*

A wise and thoughtful writer of the male sex gives evidence on this point. He says, "So far as a knowledge of the duties of sex is concerned, an Englishman is born and brought up in Egyptian darkness.

"I cannot help thinking that in all stages of human society, the community at large ought to know all that can admit of any practical application in the physiology of sex. It seems clear that the confused ignorance in which Englishmen grow up is unnatural and vicious, and that this fundamental falsehood is one chief root of the social evils we deplore. The ancient lamentation of the prophet holds good here: 'My people is destroyed for lack of knowledge.'

"We do, however, in an indirect way, give our boys

* "History of European Morals," vol. i. p. 152.

lessons on the subject which we know must powerfully awaken their natural curiosity. We set them to study a literature, which however invaluable or indispensable as an instrument of education, has this most serious drawback, that almost all of it is more or less permeated with a spirit of falsehood and licentiousness, such as would never be tolerated in the literature of the day, but which is glorified and excused because it is classical. There would be no harm in this if boys were trained in the knowledge which would enable them to detect the falsehood and to despise the licentiousness. But here the schoolmaster is silent.

"We must not, however, imagine because parents and schoolmasters say nothing, that no teaching goes on. . . . Curiosity is one of the most irrepressible tendencies of human nature, and . . . at the time of life when the animal instincts begin to develop into conscious activity, curiosity in regard to all matters pertaining to that instinct is sure to be specially strong. The only remedy possible is to destroy curiosity by giving such full information as will leave nothing to be inquired about. That this information will very soon be obtained in one way or another is a matter of absolute certainty. . . . We have no choice . . . except between true and wholesome information, given by the parent . . . and the one-sided, false, and sensual teaching which boys are certain to derive from each other. Let no one imagine that this curiosity can or ought to be suppressed and stamped out by any measures of supervision and restraint. It is indeed God's voice within the boy crying out for light ; and if we refuse to answer that dumb, inarticulate cry, or endeavour only to stifle it, because it troubles us, we are fighting against God, and the guilt of whatever consequences may ensue will rest upon our heads."

For the method to adopt in training to a correct knowledge of the physiology of sex I must refer my reader to the important article from which I have been quoting. It is in the *Westminster Review* for July, 1879, and the title of it is, "An Unrecognized Element in our Educational Systems."

Mr. Morley, in his interesting and otherwise highly sympathetic "Life," makes one severe remark on poor, simple, untrained Jean Jacques Rousseau. "Rousseau," he says, "is the only person that ever lived who proclaimed to the whole world . . . the ignoble circumstances of the birth of sensuality in boyhood." * In the conscious epoch of humanity it will be

* "Rousseau," vol. i. p. 15.

held a grave reflection on the *adult* generation, a shameful disregard of social responsibility, if any youth has his boyhood stained, his pure mind sullied, by passing into manhood in surroundings that can fittingly be spoken of or thought of as "ignoble circumstances."

But now we must beware of thinking for a moment that a knowledge of the physiology of sex would be an all-sufficient equipment of the young for the duties and responsibilities of social life. There are many relations between the sexes with which sensuality has nothing whatever to do, and in which *not* physiology of sex, but psychology, and a knowledge of the emotional nature of human beings, is the necessary guide to noble, true, and upright conduct. The differences of sex do not belong to bodily structure alone, they extend to the intellectual sphere, to all the varied faculties of thinking and feeling; and throughout their whole range, these modifications present a foundation for sympathy without uniformity, for a harmony wider and fuller than unison, which ought to give a special charm and interest to the multitudinous relations of our complex life. This subject is only now beginning to be considered.*

During the epoch of the entire subjection of women, female education was completely neglected; since that epoch it has been too superficial, and even now is not sufficiently solid to make apparent what are the real and fundamental sex distinctions in intellect; and besides this, comparisons have been little brought out, because men are not yet free from the ideas of the past. They ignore the possibility of close mental companionship with women, and look for the attractions of the sex still in "tender little ways and bashful modesties;" or build their hopes upon a fancy picture such as Lydgate's, whose dreamland contained "a perfect piece of womanhood, who would reverence her husband's mind after the fashion of an accomplished mermaid, using her comb and looking-glass and singing her song for the relaxation of *his* adored wisdom alone"!

So far as we at present know, one broad divergence seems this: the sensual and intellectual faculties predominate in man, whilst in woman it is rather the emotional and moral faculties that predominate. I have spoken in my last chapter of the extremely emotional nature of some girls, and of the

* Miss Sara Hennell treats of it in her works, and Mr. Higginson in his "Common Sense about Women."

misery they endure from the simple fact that there is no wide field in which these emotions can disport themselves in joyous freedom, ready to seek, take, and respond to every impulse that is innocent and pure and true. The false notion that every girl instinctively desires a husband, and the gross, vulgar ideas regarding sex relations that permeate society disgust and shock her nature; and the more susceptible she is to all the finer issues of a life of feeling, the more she is compelled to self-repression, and to wear an outward, artificial garb that falsifies her nature. George Eliot's Dorothea is a true and noble picture of the type of womanhood to which I here refer. "No life would have been possible to Dorothea which was not filled with emotion," and when Lydgate (the man whose ideal woman was of the mermaid order) was brought into closer contact with her, and perceived the tenderly sympathetic nature of her intellect, it filled him with astonishment, and he mused over it thus: "She seems to have what I never saw in any women before—a fountain of friendship towards men; a man can make a friend of her!"*

In such women (and there are many such) sociability dominates animality, and provided they are trained in, or have themselves attained to, correct ideas regarding human nature and wherein its dignity consists, to set them free from all conventional repression would be a gain, a wholesome, pure advantage to society in general. Affection, not passion, is the mainspring of their being. They may be safely trusted to form ties with men that will not degrade, but rather elevate the character and sweeten life, and if they marry, it will be because some human being has approached them through the higher channels of their nature, and not through that of animal instinct or appetite.

But my reader may *object* in this wise—have I not spoken of morbid emotions in women, and the female tendency to dwell too much upon the feelings? yet I seem to advocate a policy that might foster and increase that tendency. My reply is this: morbidity is an unhealthful condition caused by repression. Where emotional pleasures are proportionable to an individual's emotional instincts, physiological benefits will arise, and every power possessed, whether bodily or mental, will be strengthened and increased. To be morbid is to suffer from the craving which accompanies the under-activity of an

* "Middlemarch."

organ; what I propose will give the comfort and happiness that always accompany the normal activity of an organ.*

But I do not deny that in this sphere of human nature as in every other, *instruction* and *training* are required, and during the period of my own youth it was to me a marvel, a source of great surprise, that my elders never deemed it necessary to aid me in the understanding and regulating of a highly emotional nature. The suffering caused by mere bewilderment and ignorance was very great, and much of vital force or energy was literally wasted in combating imaginary fears, that ought to have been spent in brightening life around me.

If we are to inculcate and train the young to reverence for human nature, we must *not* leave out of count the master passion, the ruling force of human nature, but we should rather seek to give direct and forcible lessons upon the *dignity of love*. The term love is so indefinite that my reader may very possibly consider this suggestion vague and unpractical. Love in some minds means simply sexual feeling, in others it implies all the tender ties of sympathy that bind us to our fellows; and again, in other minds it points to that wide charity which covers up a neighbour's sin and mingles deep compassion with its sympathy. From pulpits and in sabbath schools there has been direct teaching upon the latter, but upon the first and second there has been little or none; and on the first I have myself heard *indirect* teaching given in schools, which would be certain to produce injurious effects upon the children's minds; and I do not here refer to classical stories, the evils of which have been already pointed out.

But let me illustrate by what I heard on one occasion. A history lesson was given to a mixed class of boys and girls of ages ranging from eight to twelve. The school was not a second-rate one. The children were of the upper middle class and the teacher was outwardly a gentleman. The siege of Quebec was the subject, and to give added interest to the lesson, reference was made to a somewhat similar siege of Edinburgh Castle, as the children were well acquainted with the latter. The difficulties presented by the impregnable position of the castle, the despair of the besiegers, etc., were graphically described—the attention of the children was rivetted. Then came the telling incident. A soldier from the fortress betrays his countrymen. He offers to lead the enemy up the most precipitous side of the rock, which is therefore the side

---

* *Vide* Herbert Spencer's "Data of Ethics," p. 91.

least defended, and so to give them entrance to the castle. The English leader asks for what purpose he had adventured his own life in descending such a precipice? "Now, children, can *you* guess?" says the teacher. "No? well, that man came down the steep rock in the darkness every week to *visit his sweetheart.'* Don't you think he must have been awfully fond of her to risk his neck just to see the girl he loved?" and the children broke into laughter, and the teacher smiled, if he did not laugh. And there was not a young mind there that did not momentarily think *that* soldier was a fool, nor a young heart there that did not expand with a sense of its own infant superiority!

Such impressions are not easily effaced, and the impression undoubtedly was, that of *irreverence* towards one form of human love. And yet the teacher was not, I am convinced, consciously culpable. He was intent on making the history lesson interesting; his business was to fasten facts within the children's heads, and he was doing that extremely well. Parents and the public have not yet demanded more from teachers, and until they do, we must not blame the teacher who, in fastening facts within the head, may chance to warp the feelings, and to demoralize and hurt the young emotional nature. I have said that gross, vulgar ideas regarding sex relations permeate society, and make it impossible for women of refinement to be as frank and free in intercourse with men as would be true and natural to their womanhood.

Now, *here* is *one* source and origin of the evil, and if gentlemanly teachers are careless how they misrepresent the emotion of love, and cast the shadow of frailty or indignity upon it, what dare we to expect of servants with whom our children (under present social arrangements) are in daily, hourly contact? That class has not as yet the germinal ideas of a sound philosophy of sex. What wonder is there that, no sooner does some affinity of nature begin to attract and draw to one another a girl and boy, than the feelings aroused are—here is something to be *ashamed of*, something to *fear*, or something to *conceal*.

But the clear and definite teaching regarding this emotion should be, that love is essentially noble, having nothing to conceal, nothing to fear, nothing to be ashamed of, and it is *only* when mixed with some anti-social feeling, such as jealousy or tyranny (that would coerce another to respond to its advances, and would inflict thraldom upon a free fellow-creature) that there *is* something to excite in us a sense of shame. In such cases, however, the so-called love is spurious. It is not the true, genuine emotion; it is a homelier instinct, viz. egoism! A selfish, despotic passion, wearing the garb of a tender, delicately minded, sympathetic emotion. True love repudiates tyranny and masterhood. It would not, if it could, control another human being's feelings—it glories in its freedom, and it rejects as worthless all response to its advances that is not equally spontaneous and free.

But are such lessons required? I think I hear my reader ask, and is it not to take the bloom, the modesty, the delicate refinement off a maiden's mind to speak to her of love at all? I reply—it is possible to treat the whole subject with refinement; and if this is not done, the vulgar aspects of the question are certain to be forced upon minds unfurnished with true ideas to counteract the false. The merest children are talked to and joked with upon the subject of marriage. A year ago my niece. a child of ten, said playfully, " I'll do so and so when I am married." " Perhaps you'll be an old maid," said her smaller brother; to which the child with confidence replied. " No! Mr. W—— said when he was cutting my hair that I'm a pretty little girl and sure to get married, for young men fall in love with a pretty face."

Nor is it only what is *said* to girls that may mislead them. If Rousseau suffered all his life from "bizarre, fantastic notions," derived from romances, assuredly our girls are exposed to the same danger. The frivolous, idle life so many of them lead, just at the period when the organism is specially impressible and emotional, causes novels to be devoured, and it would be difficult to exaggerate the confusion of ideas, the conflicting principles, the illogical, contradictory states of mind and feeling produced by this promiscuous and undirected novel reading.

Sometimes by chance a happy lesson is taught, as in this instance. A girl I knew began to think of love as early as fourteen. She formed opinions on the subject, and one opinion was, that for a girl to fall in love was quite disgraceful. A woman should be mistress of her feelings, and only let affection go out towards the man who is in love with her! At sixteen the girl left school and entered on a new experience of life. A brother's friend, a lad of seventeen, became an intimate companion, whose society she thoroughly enjoyed; when suddenly her theory alarmed her. Could it be possible that *she* was guilty of that heinous sin against the modesty, the dignity of womanhood? Had she *disgraced* herself by *falling in love?* Closely and anxiously she examined her heart; but truly she did not know! for where was her standard of measurement? She liked the lad extremely, but whether she was in love with him or not was a mystery she could not solve. Meanwhile she had no self-deception as to his feelings. She did not highly estimate her own attractions, and was sure that he was *not* in love with her. But all this thinking on the matter, this self-examination and morbid fear of doing wrong, created a new symptom! Self-consciousness was painfully aroused, and every time the lad addressed her she would blush, to her own inward confusion and disgust. New fears assailed her. This tell-tale symptom was a false witness. What if it told the lad she loved him, when in reality it only meant that she had been thinking much about him? She determined to avoid him altogether; but behold, he seemed to guess her purpose and boldly to counteract it. He sought her openly, drew forth her blushes, and attracted notice to them. At first this puzzled the girl, for she knew she had become awkward, stupid, unentertaining in his presence, and yet he seemed to like her company *better* than before! But also she perceived that what he liked was not her company so much as the painful state of shyness and distress into which he alone was able to throw her! It gratified his *sense of power*, and slowly the truth dawned on her. True love, she thought, is sympathetic, kind, considerate, and *he* was showing himself selfish and tyrannic. Her hero fell in her esteem, but *that* did not give her back her self-respect. For months she was a melancholy misanthrope, all relish for outward life was gone; and what she cared for only was feeding her emotions on religious books and novels.

Now, amongst the latter she chanced on one that threw a new light upon her trouble, and literally restored her self-respect. The heroine of the tale, a girl called Maud, was something like herself—not prepossessing in appearance but full of inward life. She was a lonely creature, until a bright, frank, and free young man became a boarder in her father's house. The opposite natures attracted one another, and Maud expanded and blossomed into gaiety and exuberant happiness in this new companionship. After a time, however, a temporary separation takes place, and Maud makes the discovery that she has given her whole heart to the young man. Nevertheless, no humiliation overwhelms her. She accepts the fact calmly, and has every reason to suppose that he is equally attached to her. Meanwhile a pretty sister, just returned from a boarding-school, has taken her place at home, and when Maud reappears there, the young man is in the full swing of a violent flirtation with Cecilia! Maud receives a shock, and naturally draws in her tentacles, and lives beside them " shut up in egoistic sorrow." She makes no attempt to recover her faithless lover, who sees her now gloomy and morose, as compared to her more lively sister. The flirtation reaches a climax, and when Maud is suddenly made aware of the fact that she must regard as her future brother the man whom she had looked upon as her lover, a passionate burst of emotion sweeps all before it. Reserve is gone, and to him she speaks out the simple truth. She had given him all her love—she could not help it, she was not ashamed of it, why should she be? But now, now that his best love was not for her but for her sister, he must be kind, and aid her to avoid his presence until her peace of mind has been restored, and she can meet him with feelings rightly fitted to their new relations.

Now, in this picture in fiction, the girl in actual life saw the *dignity of love*. Maud had to bear the shadow of grief and disappointment for a time; but her fine character was not degraded or injured, but rather ennobled and enriched, by having had called out in her and exercised, the latent faculty of unselfish love; and even the volatile lover respected her not less, but more, after she had spoken out and met him freely in the strength of artless, unaffected truth. The girl discarded her false theory of love; but how much happier for her and those around her, if at fourteen, when she began to ponder on the subject, her parents or teachers had known what was in her mind, and had prepared her for social life by giving her the true ideas. Besides, as I have said, novels are no safe guides, and it seems perfect folly to permit the young of either sex to pick out at random and store up from novels alone all their ideas regarding the emotional relations of life.

Girls are perhaps the greatest novel-readers; but young men read them also, and many a student, who is outside the range of home influence, spends his Sunday in a perhaps solitary lodging over a pipe and a novel.

To the question then—is teaching upon this subject of love required? I answer—yes. Nor do I think *that* teaching sufficient which would deal only with the abstract quality of love. We must associate it with the human being, and strive to fasten in the minds of the young of each sex, a worthy, elevated ideal of the opposite sex, so that they may be attracted to one another by what is admirable in character, as well as by the outward charms of beauty and of manner. If instead of the ancient classics, where the predominant attractions of womanhood are naturally beauty of form and feature, we educated young men, far more than we at present do, in English classics, and from the works of genius of Thackeray, of Robert Browning, and, above all, of George Eliot, gave them pictures of modern social life, inculcating at the same time the moral principles that should direct and control their conduct in social life, we should have no difficulty in familiarizing their minds with ideals of womanhood that would preserve them from the errors in which so many now make shipwreck of their happiness.

In "Middlemarch" the noble nature of Dorothea is so minutely described, that the study of it would prove a valuable lesson in psychology. Young men would learn to see that girls are not for the most part like waxen dolls with pretty faces—that they are not in the least perfect like angels, nor are they soulless as the mermaids! but flesh and blood human beings, *nearer* to themselves in thought and feeling than they had dreamed. With natures somewhat gentler perhaps, and with an inward nerve equilibrium very easily overbalanced, the best of them are seeking and striving like themselves to understand the great mystery of life, and to lay hold upon the true, the beautiful, the good. And then, again, in Lydgate's personal history, the miserable consequences flowing naturally from his false ideal of womanhood are all so true to nature and so graphically told, that unless our lads were dolts or dunderheads they would be sure to carry the picture in their minds, and find it as a beacon or signal to preserve them from similar dangers.

But in our present educational system we give the first place not to the best light literature of our own age, but to that of a past age; a literature which carries with it moral ideas completely out of date, and sensual pictures that unmistakably tend to corrupt and degrade the mind. Sensuality, says Mr. Lecky, is the vice of young men, and it seems an undoubted fact that not emotional sympathy but sexual feeling is the predominating force, at the present epoch, in *early manhood*. In an old country like ours, where a high and increasing birth-rate causes an intense struggle for existence in the lower classes, and causes also social pressure in the upper classes, mercenary marriage unions or gross licentiousness result. The sensuality of youth perpetually tends to social vice and individual degradation. But if we can succeed in lowering the birth-rate, and relieving both the struggle for existence and the social pressure, and thus altering the surrounding conditions of youth, on the one hand; and if on the other hand we can succeed in impressing upon the mind in boyhood a noble, yet true ideal of womanhood, we may anticipate that the whole *strength of sensuality*—this predominating force of early manhood—will set in the direction of virtue, and tend to elevate and ennoble character.

For observe, its gratification will depend on new conditions. Purchase alone will not suffice. Brutality, tyranny, masterhood in any form will prove an obstacle, and it will only be by kindness, gentleness, sympathy (all which are social virtues), that love will be responded to, and passion be gratified. There was wisdom in Mr. Garth's judgment of Fred Vincey's case. "The lad loves Mary, and a true love for a good woman is a great thing, Susan. It shapes many a rough fellow." And we may depend upon it that when false love will no longer pay, true love will be rapidly developed.*

But again, as regards training, or the subjective requirements of social life, something more is necessary than even the right directing of emotional forces. The young ought to be enlightened on the subject of how to regulate and modify emotional by intellectual forces; and for two reasons, one of these individual and the other social, such lessons are more important in relation to girls than boys. Girls have fewer outward distractions than young men, and have, besides, an inward tendency to dwell upon their feelings. George Eliot points this out in "Middlemarch." "Rosamond," she says, "had no pathological studies (as her husband had) to divert her mind from that inward repetition of looks, words, and phrases, which makes a large part in the lives of most girls." Now, in view of the fact that in this country the female sex is in a large majority, we must not shirk this other fact, that girls very frequently suffer from disappointment in love; and if greater social freedom is accorded to them, these disappointments are at first likely to be more rather than less frequent.

Ought this to deter us from adopting a freer social system?

* Miss Martineau, in her "Morals and Manners," remarks: "If women were not helpless, men would find it far less easy to be vicious."

I answer—no. But it should make us doubly anxious to prepare our girls by *training* for the experiences which we know must come to them. Nature has already done its part, for, as I have pointed out, they are capable of forming other pure and happy relations with men, besides that of marriage, and their lives may be made full and satisfactory, although there be truth in Mrs. Cadwallader's shrewd observation: " That is the nonsense you wise men talk! How can she choose if she has no variety to choose from? A woman's choice usually means taking the only man she can get." Mrs. Cadwallader was, of course, looking at society under the old *régime* of man's supremacy and woman's subjection; but even under the *régime* of equality of sex, where no girl would need to check her spontaneous impulses or repress her kindly feelings, many an one will yet find herself (like the Maud whose position we described) compelled to alter the attitude of her mind towards the man who has won her warmest love. Now, in such a position, the first impulse is always towards concealment and isolation. Like Desdemona the sick heart says : " Do not talk to me, Emilia ; I cannot weep ; nor answer I have none." The wounded deer hides from the herd to suffer alone, and although we seek help from one another under physical hurts, when our hurt is emotional, we act similarly to the wounded deer. It may be pride that causes us to do this, but (again I quote George Eliot) " pride is not a bad thing when it only urges us to hide our hurts—not to hurt others." And it may even be wise policy; for human sympathy has not as yet the delicate touch it will no doubt in time acquire. It is apt to be a little rough in its handling although it means well.

But seeing, then, that there is this tendency to concealment, the hour of trial is *not* the time for elders to intrude advice. Such lessons *must* be given in *advance*, and girls should have the knowledge *how*, when a crisis of personal grief overtakes them, to set about at once acting nobly and bearing wisely. The possibility of giving the lessons I do not for one moment doubt, and I can conceive of no more attractive branch of education to a girl of fourteen, than that which deals with the principles of action that should guide her conduct on her entrance into responsible social life, provided these lessons are given by a thoughtful, tender woman, who freely admits that the principles are not of arbitrary authority, but are to be tested by the girl's own experience in the years to come.

I well remember the effect upon my mind when an old lady of eighty gave me touchingly the history of a chapter in her life. "I had," she said, "a grief that almost drove me mad, and for a time it ruined my health. Day after day I rose with effort from my bed, and in a donkey-chair drove in the Devonshire lanes, with book and vasculum and lens, to study mosses, never returning till sunset and darkness compelled me to desist. Fresh air and all the healing influences of outward nature restored my bodily health, and as my knowledge of these wonderful, though tiny plants, in all their beauty and variety, grew each day greater, my interest increased, and to my own profound surprise and inward thankfulness, the sorrow I had suffered passed away, and my mental peace and happiness were *completely* restored." Now, this lady discovered for herself that just as pure air and sunshine oxygenate and regenerate the blood and restore physical health, so fresh ideas and mental exercise in a *new world of thought* will revive the drooping spirits and regenerate the sorrow-stricken heart. But there is surely no reason why this should be an individual discovery, made as it were afresh by girls who *might* and *ought* to benefit by the experience of their elders.

My reader may feel inclined here to remind me that I have said few girls are strongly intellectual, and it is possible that such a science as botany might fail to interest a Fanny Dover. Just so; but occupation of *some* kind suited to the capacity may be within the reach, and let me quote what Mrs. Ellis tells us in her "Daughters of England": "I fear it is a very unromantic conclusion to come to, but my firm conviction is, that half the miseries of young women and half their ill-tempers might be avoided by domestic activity, *because* there is no sensation more cheering and delightful than the conviction of having been useful, and I have generally found young people particularly susceptible of this pleasure." Activity (bodily and mental) is the panacea for the emotional pains of youth, and it would be easy to show girls, how, when grief assails them, true courage and true wisdom consist not in sitting still to bear—not in resignation and endurance only, but in a brave setting of the face to some intellectual pursuit —if not science, perhaps language, literature, poetry, music, according to the individual taste, and a patient persistence in the effort to acquire, until the reward which they may certainly anticipate breaks in upon them.

Now, if individuals are furnished with this knowledge, and if they have the moral sense—an instinct already largely developed in the average young of our race—we need have very little fear concerning love-disappointments in early life. It is the period of frequent change of feeling, and if useful activities take the place of lassitude and idleness in suffering, each trial of the kind will lift the individual in the scale of being, and enlarge and widen character; whilst we shall cease, in great measure at all events, to be surrounded by hysterical girls and reckless lads.

There exists, however, an altruistic as well as an egoistic reason for *regulating* human emotions. If friendships between men and women are to be freely formed, it is clear that these relations will cross and intersect the closer ties, and that the latter must be carefully respected. An ardent, warm-hearted girl enters into friendship with a married man who reciprocates her affection. Here, besides respect for social order and for mutual gratification, the happiness of a third individual is an element in the position, that makes right conduct depend upon intelligence and a sympathetic understanding of the feelings of others. I have already pointed out that the great movement towards equality of sex will advance rapidly or slowly according to our success or non-success in divesting ourselves of the anti-social passion of jealousy. But jealousy is a word of wide import, and I do not mean only the fierce passion that raged in the bosom of Othello, but all the petty jealousies, the mean, silly grudgings of another's superiority or happiness, that simmer in the breast of a narrow-minded, perhaps malicious, woman.

In proportion to the repression and gradual disappearance of the anti-social feelings small and great, including petty jealousies of every kind, will general happiness be built upon the giving of greater and greater freedom to the expression of every social feeling that humanity possesses or will in time develop. And if right training regarding these emotions is carefully bestowed upon the young, it would, I am convinced, give an enormous impetus to progress by teaching them to *literally kill* the germs from which the anti-social passions spring, grow up, and are ultimately *reproduced*. Parents forget their own long past experience, and perhaps few are aware at how early an age a child may feel and keenly suffer from jealousy. Mrs. Jamieson, the well-known writer, gives in her " Commonplace Book " some particulars of her childhood. In reference to this subject she says: " I was not more than six years old when I suffered, from the fear of not being loved

where I had attached myself, and from the idea that another
was preferred before me, such anguish as had nearly killed
me! Whether those around me regarded it as a fit of ill-
temper or a fit of illness I do not know. I could not then
have given a name to the pang that fevered me. I knew not
the cause, but never forgot the suffering. It left a deeper
impression than childish passions usually do, and the recol-
lection was so far salutary that in after life I guarded myself
against the approaches of that hateful, deformed, agonizing thing
which men call jealousy, as I would from an attack of cramp or
cholera. If such self-knowledge has not saved me from the
pain, at least it has saved me from the demoralizing effects of
the passion, by a wholesome terror and even a sort of disgust."

No doubt this self-protection was to Mrs. Jamieson a great
gain; but I hold that even a wholesome terror or disgust of
the passion may cause a girl to make mistakes in reference to
jealousy, unless her mind has clear ideas on the subject. And
here I will quote a case in point exactly as I know it to have
occurred. Two girls became strongly attached to one another.
They worked together, studied together, and for a time the
friendship made the very sunshine of their lives. After some
years one married, and, of course, *outwardly*, conditions somewhat altered. But *inwardly* they altered too, and without the
same necessity, as it afterwards appeared. The girl who did
not marry felt herself forsaken. She suffered from jealousy,
and imagining that if she kept her hold upon her friend the
husband would suffer *similarly!* she set herself to loosen the
bond between them, for this she deemed her duty. A husband's right was in her view a monopoly of a wife's affection,
and for an old friend to make a claim was possibly to cause a
breach in married happiness! She strove to cool her own
affection, and she hid her misery from her friend, lest it should
dash *her* newly found joy. The sentiments were generous and
the action was brave, but from the want of an intellectual
understanding of the true position the whole thing was a mistake. A few short years of married life and then the wife spoke
from her death-bed to her former friend: "*Why did* you separate
yourself from me? What made you think my love for you
would change? I have been happy in my husband and my
child, but love never *narrowed* me; it *widened* all my nature,
and there was room enough for friendship too. I have sorely
missed my friend, and felt that her loss has thrown a shadow
over my married life."

Now, clearly in this instance wrong action simply arose from ignorance, and not from any want of earnest desire to do right. The girl was thoughtful and meditative, and the only direct lessons she had ever received upon personal conduct were from religious teachers. The general drift of these lessons had been that self-sacrifice is always right, and that in every crisis, where how to act seems to present a puzzle, the *safest* rule is to choose the most difficult path, and look upon *it* as the path of duty. This rule the girl had acted on, at great self-sacrifice, and the result was, that besides the immediate pain to herself, she had hurt the happiness of the very friend she would have died to serve, and brought upon her own future a cloud of mingled self-reproach and grief that shrouded her life for years. This fact alone speaks volumes as to the necessity and propriety of giving lessons to the young upon the guidance of their conduct in reference to the emotional part of human nature from a purely secular standpoint. It is a significant fact, that one of the professorships of Harvard University in America, was founded by a woman, who proposed to call it "A professorship of the heart;" nevertheless, at this moment the chair goes by the name of a "professorship of Christian morals."* The truth is *this*—the *scientific* aspect of ethics is waiting still for man's consideration, and the field of education in this direction has never as yet been entered. Mrs. Charles Bray, however, has written a small educational work upon the true lines;† and when a rational public opinion calls for the effort, there are plenty of minds capable of producing such manuals as may be required in the practical work of instructing children.

The French Revolution was, as we have seen, a cataclysm, a pulling down of system before intellectual ideas were ripe for building up a nobler, simpler, purer form of social life upon the right foundation, viz. knowledge of human nature. The British race avoids French error, and that, partly at least, through absence of French enthusiasm. Both in political and educational reform the work *we* do is never radical. As new light comes to us, we alter or add a little here and there, but as a rule we do not pull down and sweep away. The bulk or structure of our system remains as before. As a consequence of this our politicians are overweighted and frightfully encumbered by the load of lumber from the past, which they carry with

---

* "Common Sense about Women," Thomas Higginson.
† "Elements of Morality."

them in every progressive step; and our children are burdened by an education which prematurely exhausts the brain, and leaves them, after all, without the really necessary *subjective requirements of social life.*

These requirements embrace some knowledge of bodily structure that is correct and solid, not superficial—an understanding of the dignity of the animal appetites and instincts, but at the same time the necessity for wholly subjecting these to the still higher requirements of social man; an apprehension of the delicacy and beauty of the emotional part of human nature, and the *extreme* importance of *regulating* it by development and exercise of the intellectual faculties, and a definite conception of moral principles that will guide to right action in social life, without reference to disputed theories of the universe, and subjects which young minds have not the power to judge of, or comprehend.

CHAPTER XII.

INDIVIDUAL RIGHTS.

"Eccentricity has always abounded where strength of character has abounded. That so few now dare to be eccentric marks the chief danger of the time."—JOHN STUART MILL.

"Do you think it worth while for people to make themselves disagreeable by resenting every trifling aggression?" an American asked Mr. Herbert Spencer, adding: "We Americans think it involves too much loss of time and temper, and doesn't pay." The tenor of Mr. Spencer's reply was, that the political corruption, the decay of the free institutions of America, springs from the habit in individuals of "easy-going readiness to permit small trespasses, because it would be troublesome, profitless, or unpopular, to oppose them;" and he quoted to the questioner words spoken by one of their own early statesmen: "The price of liberty is *eternal vigilance*." Free institutions, said Mr. Spencer, can only be properly worked by men, each of whom is jealous of his own rights, and sympathetically jealous of the rights of others—who will neither himself aggress on his neighbour in small things or great, nor tolerate aggression on them by others.

I have hitherto condemned jealousy, but my reader will readily perceive that what Mr. Spencer here indicates is an emotion of a different nature, spoken of under the same term. Evolution in thought goes on so rapidly that evolution in language does not keep pace with it. The jealousy I have condemned is anti-social, an emotion in entire antagonism and opposition to general happiness; whilst the sympathetic jealousy which Mr. Spencer commends and approves is in perfect conformity with general happiness.

The first has been already described and exposed as a selfish passion, spontaneous, tyrannic, capable of producing brutal cruelty, or mean, malicious action. The second is intellectual in its genesis, and beneficial in its results. It arises from respect for individuality—an intelligent apprehension of the fact, that general happiness is absolutely dependent upon freedom (to a certain extent) being accorded to each individual. With intellect thus advanced, the greater our sympathetic desire for the happiness of *all* becomes, so much the greater also will become our *jealous regard* for the individual rights of *each*.

This state of mind, which the poverty of our language forces us to call a sympathetically jealous one, is social in the widest sense. It is possible only to a race which is so advanced in civilization as to repudiate tyranny and oppression, and form correct conceptions of social justice. To the children and children's children of that race, however, this sympathetic jealousy may become an unconscious, spontaneous instinct, a habit of the mind, requiring no justification or enforcement, and causing no jar or discord when humanity is in its highest, noblest states of feeling. The sentiment is in fact founded on that delicate sensitiveness to quantitative relations which we call justice, and which, as we have already stated, is certain to increase with the advance of civilization. It may be, and assuredly will be, in time, exercised without anger or cruelty, but with calmness and personal gentleness, without attributing blame to ignorant trespassers or aggressors, and yet with all the unswerving firmness that belongs to each force which rests upon a rational or scientific basis. To understand this state of mind and to aim at producing it in ourselves and others, is of vast importance to human progress; therefore we are bound to inquire, what are the rights of man? why are they rights? and what will be the result to society in general, if the faithful guarding of these rights is a duty universally discharged?

I chanced to read lately a short newspaper article upon Mr. H. George's work, "Progress and Poverty," in which occurred the following statements: "There *are no laws of nature*, and there are no *rights of man*. We are simply using meaningless phrases when we speak of either. Law is command, control; but nature is instinctive force, and can therefore neither give nor receive laws. Man has no natural rights any more than the wolf or the bear. All rights are *conventional*." Now, in these propositions there lies a mixture of truth and error dangerously misleading to ignorant minds. The word law is at all times apt to perplex the reader who knows nothing of science, for it certainly carries with it the notion of arbitrary authority, and that simply because its *first* use was to indicate human regulations or systems of control. Of these law was the symbol. But when thinkers adopted and adapted it to the scientific conception of nature, the reference to authority was put aside. It then became a term to signify a certain order in nature—the sure and constant co-existences and sequences in phenomena, which man can ascertain and guide his conduct by. To say then, with the writer of the newspaper article, that there are *no* laws of nature, is simply to reject the term fixed on by men of thought and culture as the best expression within their reach to represent a concept or idea, which we shall no more in this nineteenth century get rid of than we shall see the moon depart.

The laws of health, for instance, are perfectly invariable, as are indeed all laws of nature; but they do not *necessarily dominate*, since other and opposing laws of disorganization may at any moment get the upper hand. Man is competent to disregard the laws of life. If he so acts, another course of *natural order* is initiated; the man becomes subject to pathological laws, which conduct him steadily to the grave.

Necessity then, without command, leaves us no choice. If we would gain the end that all humanity desires, that is, happiness, we must conform to all the *laws of nature* that favour that end; and in man's intellect which is slowly developing, aided by the scientific method now discovered, we are on the road to control the forces of nature, by paying constant regard to nature's *unchanging laws*. Within, man has the laws of his organization; without, he has the laws of circumstance or his environment. By intellect, by science, by union or co-operation, he has the means to take a firm advantage of these laws, avoid the pitfalls of the past, and move with energy and joyous freedom along the path of progress.

But now we turn to the *natural rights* of man, which have no existence, says our newspaper philosopher, any more than natural rights of a wolf or a bear. If he had said a cat or dog, a donkey or a horse, we should have replied: "But these dumb creatures too have rights, and now civilized man protects them in their rights." The wolf and bear, we concede to him, have none; but we deny that man's case is analogous to theirs. For observe, these brutes have no control over their breeding instincts. Their numbers are held in check by forces external to themselves, and no law is possible with lower animals, save that of force.

The African elephant, the Siberian bear, the wild horse, in all its freedom, are alike under mastery. The forces of carnivora or other destructive creatures, of starvation, of disease, etc., control them, and the fierce struggle for existence has its outcome in "survival of the fittest." When the horse, dog, cat, pass from the mastery of carnivora and other external forces to that of man, *he* controls their breeding powers, and by a gentle force, renders their existence harmonious with the rights of Humanity. A happy life is possible to horse and dog, without infringing on the liberty of man. Fear gives way to love. The canine tendencies to growl and fight become transformed, and visible in playful gambols; and the law of gentleness is henceforth supreme—a law which recognizes the individual rights of every sentient creature, unhurtful to the higher life around it. Society, supremely powerful in its humane emotions, dominates all domesticated animals, and makes them subserve general happiness; whilst admitting the obligation to give to each a happy, though useful life, and a painless death.

Vivisection is in accordance with this law. The practice is justifiable only in so far as it subserves general happiness; and society rightly claims to guard it carefully from all abuse.

Now, the law of *universal sympathy* is only slowly rising in the broad domain of human thought and feeling, like the morning star appearing in the east—a star of hope, low as yet on the horizon, but giving promise of a glorious day, filled with the sunshine of an all-embracing happiness. Are rights conventional? Undoubtedly they are, in the highest, truest sense; for conventional means, by *tacit agreement;* and we have now a noble band of social units whose development, intellectual, emotional, and moral, raises them above the law of force, and ranges them in tacit agreement with the higher law of love. So long as the natural man has self-regarding qualities only, individual rights have no existence. But when evolution has brought the natural man intelligence to grasp the fact, that general happiness depends on individual interests, and a moral sense, with the delicate perception of quantitative relations which we call justice, then, but not till then, individual rights exist. They are claimed by the reason, and yielded by the sympathy of the civilized man; and they belong to every social unit, from the prince to the pauper, from the refined and cultured sage to the ignorant street Arab, and down to the babe in the cradle, the old maid's gentle cat, and the sportsman's ill-used pigeon.

What, then, are the individual rights that a high morality will sanction? I find the answer to this question in a natural principle or law of happiness applicable to perfect man, which has been formulated thus: "Every man has freedom to do all that he wills, provided he infringes not the equal freedom of every other man." This is not an arbitrary law of man's devising. It is the order of nature which will hold good in the social relations of a perfectly civilized race; or, in other words, the condition of social life by which that race will attain to happiness. And humanity knows this, not by authority or arbitrary command, but by the study of the intrinsic qualities of human nature, and the extrinsic elements essential to happiness. Meanwhile, we are a very imperfect race, with but few social units as yet prepared for this delightful freedom. The liberty of semi-barbarous man is often as inimical to general happiness as the liberty of the lower animals; and outward restraints *must be maintained*, until an inward force of self-control is everywhere efficient in restraining evil, regulating society, and promoting order, peace, and concord.

By virtue of this force—interior, not external—each separate unit will be controlled, held, and balanced in its true social position, without being pressed upon, and without pressing on its fellow-units. But what is this force? It has a twofold aspect, and may be described in two ways. We may call it a man's sentiment of reverence for humanity in his own person, and his reverence for humanity in his fellow-creatures; or we may call it the individual's self-respect and instinct for personal rights, and his sympathetic jealousy for the rights of others. The instinct of personal rights is already *firmly* implanted in our race. If my reader doubts this, let him pass a day in any well-filled British nursery, and closely observe what goes on there

Baby and little Jessie, a child of four, are on the floor. Jessie plays with her doll, whilst baby creeps around her on the carpet. An impulse seizes baby to clutch the doll which Jessie holds firmly. Baby screams, and nurse turns round and raises him. "See, Jessie," she says, " he wants your doll, and " (with a chiding tone) " you should be kind to baby-brother." She takes the doll from Jessie, and gives it to the infant. Jessie throws herself upon the ground and kicks and screams. A paroxysm of emotion sweeps over her, and until the wave has spent itself, tranquillity in nerve or muscle is simply impossible. But now the nurse angrily commands her to cease crying, although for the moment the child is utterly bereft of any power of self-control ! As the screaming still continues, she scolds her for disobedience and bad temper, and finally thrusts her into the corner, to sob and weep till pure exhaustion brings passivity, dejection, and a sadly abject tone of mind, which the ignorant nurse calls being good again, and stamps with her approval by a kiss !

Now, the feeling which little Jessie experienced, and which the nurse stupidly (because unnecessarily) called out, and then did her best to crush and injure, is a *valuable element* of human nature, an instinct of the civilized being, which ought to be guarded, preserved, and trained, or directed, but never crushed, and never trampled on, as in this case. To take her doll, was to trespass on Jessie's innocent freedom (for she was not injuring baby) ; it was to hurt her proper pride, her infant sense of personal dignity, and to bruise the precious seeds of justice germinating in her mind. The paroxysm that ensued was no indication of a barbarous nature ; it was a virtuous emotion, a just, rational indignation, and proved the child to be the offspring of a self-dependent, self-protecting race, and not an embryo slave or serf. It was, in fact, a wave of the very force of which we have spoken, as certain in the future to regulate, restrain, and hold together social beings in the only relations conducive to peace and concord.

Later in the day, poor Jessie has her doll restored, and with it in her arms she feels happy again. Baby plays with his rattle on nurse's knee. Thinks little Jessie, " My dolly is a baby too, and wants a rattle." She takes the rattle out of baby's hands to give to dolly ! Baby screams, and nurse is furious. She snatches back the rattle, slaps Jessie, and calls her " a naughty, naughty child." This time Jessie does not even cry, although the tender little fingers ache from the rude blow they have received. No hot indignation fires her blood ; her pulse is lowered, and she shrinks away silently. Why is this ? Has nurse's violence frightened the child ? No, not at all. But within that infant bosom, there is a vague, indefinite feeling that, this time, she herself was in the wrong. She was a trespasser on baby's rights. Now, if the child is of a reflective nature, the little brain begins to work. " Why may baby take my doll if I may never take his rattle?" Ah, why indeed! The gropings of the young intellect come to no light on this problem, and by-and-by, poor Jessie's ideas of right and wrong depend entirely upon the frowns and smiles, the slaps and kisses, of an ignorant woman, who rules the nursery with barbaric authority, and kills, or at least warps, each budding conscience that comes within her sway.

At this point, I suspect my reader's sympathies will all be roused in favour of the nurse whom I appear to be belabouring with blows as arbitrary as hers on little Jessie! It may be said: "Nurse did her best to show the principle on which she acted. Her words, 'You should be kind to baby-brother,' at least inculcated sympathy, and have you not harped upon the string that sympathy is precisely what we want most developed in our children?" I reply—Yes! But sympathy is *never* developed by *command*, and to order a child to be kind, at the very moment when an aggression has been made upon his or her individual rights, is as insane a proceeding as to command a steam-engine to move forward without turning on the steam! Moreover, baby received a lesson, young as he was—a lesson that was sure to start him on a vicious course. He learned that if he cried, he probably would get all that he wanted, and depend upon it, he would unconsciously but vigorously act out that principle.

But tell us then, my reader says, what should the nurse have done? She should have instantly removed the baby, saying gently, "Children must never take things from one another. Not even baby can be permitted to do that; we must teach him better. But see, he is so young, he does not even know yet that the doll is yours, not his. Would you like to lend it to him for a little? No? Ah, well! he cannot have it, then; but come and help me to amuse him, that he may forget the doll." The little maid puts down her treasure to kiss and fondle her baby-brother; and *ten chances to one*, that by-and-by her sympathy, called out naturally and not by authority, carries her a step further, and she says, "Nursy! baby may hold my dolly for a little now." Then, later, when the brilliant idea occurs, that dolly would like the rattle, Jessie knows that she must not trespass on baby's rights, and she restrains her impulse, and thereby exercises the noblest faculty of her nature, viz. self-control, under the dominion of the rudimentary moral sense of *justice* or *equivalence of rights*.

Is, then, the charge I brought against our typical nurse a

true one? My charge is both negative and positive. On the negative side, she misses the precious opportunities of developing sympathy, and training conscience in the children; and on the positive side, she injures them, by confusing the infant intellect, crushing or trampling out the instincts that are valuable, and shaping them so far as she can do it, to abject, craven servility, under ignorant, barbaric authority.

But would it be in accordance with my principles to bring an indictment, simply for the purpose of condemning individuals? Certainly not; and let us pause and think for a moment of the antecedents of our ordinary nurses. The crowded homes in, perhaps, the back slums of some factory town, the loveless childhood, the many slaps and buffetings of their own youth, the school-board school, where scarcely anything is said to them on such subjects as how to treat a little child, or how to curb their own barbarous instincts, and check their tendencies to tyrannic despotism! The whole tone of this work is misunderstood, if any one supposes that my purpose is to blame, for ignorance, individuals who have never enjoyed the true conditions of enlightenment. What my argument really leads up to, is the condemnation of our system. When the *whole* of my premises are taken into account, the logical deduction is, that hitherto the vast importance and the many difficulties of training *during infancy*, have not been recognized, but that in the future the nursery *must* become a field, where intellect and culture *rule supreme;* in order that the young generations may enter upon responsible social life with all the subjective endowments that an advanced civilization is capable of bestowing.

The instinct of personal rights which, I again repeat, we must look upon as a *master power*—the great force, in short, that will rule and regulate the social order of the future—is in many of our units ill-balanced; that is to say, it is developed on *one* side only, viz. the egoistic, whilst the altruistic side of sympathetic jealousy for the rights of others, is latent only, not active. Now, this is somewhat like the position of a planet, pulled from its course by an overdue proportion of centripetal to its centrifugal force—equivalence of each being the necessary condition to its retaining the true orbit. And the problem is, how to increase the force on the one side, without in the very slightest degree injuring it on the other. The practical work of accomplishing this, in the case of each little individual organism, in view of all the grand *possibilities* of life, is, I maintain, a labour of such extreme delicacy and interest, as to be quite worthy of all the attention and intelligence that the most cultured women can bestow.

A gardener thinks but little of the lower order of plants, that are simple, not complex. He roots them in his garden, under natural conditions of space, air, and sunshine, and leaves them alone. But with plants of a more composite, a higher order, he acts differently. *They* have to unfold a greater variety of delicate, sensitive, beautiful organs. He watches them closely day by day, in order to place around them the conditions necessary for bringing into functional activity *all* these delicate organs; and by his fostering care, he secures the healthy development of each, through his knowledge of, and attention to, the laws of their being, in spite of the fact, that (in another sense) he has no power to make even a blade of grass to grow.

In the nurseries of the past, the primitive arrangements were suited to a very low order of human beings, and no doubt the little inmates were comparatively simple or uncomplex. But now, our nurseries are filled by the children of an advanced civilization—the offspring of a highly organized, a widely complex humanity; and can we in the name of common sense keep round them a social medium of archaic authority and barbaric force? The primitive nursery, it is true, has in some degree altered; for, within the last few years, the natural cosmic conditions of fresh air and sunshine have been more considered; but the true and natural *social* conditions for the development or unfolding of the varied mental, moral, and emotional faculties within the nurseries, are at this moment in complete and utter abeyance or neglect.

But I must pass from the nursery to the school. In the play-ground, quarrels amongst boys arise, not so much from purely barbaric impulses, such as simple cruelty, anger, hatred, etc., as from a wholly different cause—an efficient sense of personal rights, accompanied by a deficient sympathetic jealousy, or sense of the personal rights of others. "It isn't!" or "It wasn't fair!" is a phrase very frequently upon a school-boy's lips; and it is quite remarkable with what a manly courage and dignity a little urchin of ten or twelve will criticize his master's treatment of him, and tell the man of fifty to his face that he put him down in class when "it wasn't fair." Now, if each boy in the school were as eager that all the other inmates, masters included, should be treated fairly, and get *their* dues, as that he himself should get *his*, all that would be wanted for regulation, would be intelligence to discriminate in each case of quarrel or misdemeanour where lay the right, and where the wrong. As matters stand, however, both sympathy and intelligence are deficient, and the former is lacking especially towards *masters;* for in a measure sympathy springs up naturally amongst the boys, through community of interests, and takes the form of an *esprit de corps*, leading them to espouse a companion's cause in antagonism to the cause of a master, without much regard to whether the boy be right or wrong. In such cases sympathy actually becomes a force, *deflecting* the whole school from the path of rectitude, and under these conditions deceit and cunning flourish to an alarming degree. To have successfully hoodwinked a master, in order to save a companion from some deserved reprimand or punishment, becomes a schoolboy's virtue!

Seeing that intelligence and sympathy are the two things urgently required, a system of pure authority which begets antagonism is unsuited to British young people in the present day; and this has, throughout the whole nation, been dimly felt, although not as yet frankly considered or openly discussed. The feeling, however, has wrought out certain results within almost every school. It has caused arbitrary authority to relax, whilst no other form of strict discipline has taken its place, and there follows as a very natural consequence, all that I have already described as frequently occurring in university towns—the public suffer acutely from the self-assertion and ill-regulated conduct of young men.

A very pressing and important want of the present day, is an *efficient discipline* of the young; but whatever system may be adopted, it must be free from external despotism. It must differ from the old form of arbitrary authority upheld by rewards and punishments; and it must appeal to the instinctive sense of justice and of personal rights which is strong in the bosom of every juvenile John Bull. The reform instituted by Dr. Arnold at Rugby seems to have been on right lines, *so far as it went;* but it was deficient, and its fruits have not been in any great degree a marked national success. The fact remains, that whilst during the last forty years the authoritative control of the young (especially of the male sex) has relaxed in an astonishing degree, there has been very little, almost *no* progress made in the development of the inward regulative force which must ultimately take the place of outward discipline.

At the beginning of this century, the wise, philanthropic

Emmanuel Von Fellenberg, had upon his estate of Hofwyl, two leagues from Berne, a self-governing college, which for a considerable number of years was, in reference to three points —first, the happiness of the students; second, the order and discipline of the school; third, the development of self-control in the individual members—a *triumphant* success. This college had a constitution drafted by a select committee of students, adopted by an almost unanimous vote of the whole body, and approved by Mr. Fellenberg's signature. The professors had no authority except within their class-rooms. The laws, whether "relating to household affairs, hours of retiring, and the like, or for the maintenance of morality, good order, cleanliness, and health, were stringent, but all were strictly self-imposed."* A breach of the laws was an offence against the union, and in reference to such the students alone had sole jurisdiction.

Mr. Fellenberg meanwhile kept watch over the doings of the students, but he appears to have seldom or never interfered. Punishment by college authorities had literally no place in the system, and neither was any outward stimulus of reward, or even of class rank, admitted. Emulation was in no way artificially excited. There were no prizes, and there was not even the excitement of public examinations. As regards the tone of feeling amongst the students, one who was at this college for some years records that there was no bitterness or ill-will there, no coarse incentives to right action, no mean submissiveness, no selfish jealousies. "There was pride," he says, "but it grew chiefly out of a sense that we were equal members of an independent, self-governing community, calling no man master or lord." The system nurtured a conscious independence that nevertheless submitted with alacrity to what it knew to be the will of the whole. It gave birth to public spirit, and to social and civic virtues.

The students, about a hundred in number, were divided into six circles, and each circle elected a councillor, whose jurisdiction extended to the social life and moral deportment of each member of his circle. Each circle was a band of friends, and the chief, chosen by itself, was the friend most valued and loved amongst them. The circle had its weekly meetings, held usually in summer in a grove near by. The councillors formed a grand jury, holding meetings occasionally,

* This account is taken from "Threading my Way," by R. D. Owen, p. 124.

and having the right of presentment when an offence occurred. The judiciary consisted of three judges, whose sessions were held with due formality in the college hall. The punishments within its power to inflict were, a vote of censure, fines, which went to a fund devoted to relief of the poor, deprivation of the right of suffrage, and degradation from office. But trials at any time were rare.

Now to us, with our British experience of the irrepressible nature of the student *vis vitæ*, the organization of this college appears almost Utopian, and when told of its success, our impulse is towards incredulity! Nevertheless, it is a fact, that order and discipline were maintained, and that the method of control sufficed even to overawe such lads as from time to time transgressed the rules. An instance of this I may give ; but first let me mention that, strange to say, one rule of this student-made constitution was, "The use of tobacco is forbidden."

Two German princes, sons of a wealthy nobleman, were more liberally supplied with pocket-money than the other students ; and as there was no outlet for the legitimate expenditure of so much at Hofwyl, it only proved a snare to them. Now and then they would get up at night, and go to the village of Buchsee, and spend an hour or two in the tavern, smoking, and drinking lager-beer. This irregularity became known to one of the other students, who communicated it to the councillor of his circle. He called together his colleagues in office, and they, having satisfied themselves as to the facts, presented (as the term was) the two princes, Max and Fritz, for breach of law : and the brothers were officially notified that on the second day thereafter, their case would be brought up before the Tribunal of Justice, and they would be heard in defence. To face this Tribunal, composed of students, was an ordeal the princes were unable to bear ! They decamped secretly, and returned to their parents. Mr. Fellenberg took no action in the matter ; he allowed the fugitives to tell their own story at home in their own way. In a few weeks the father came, bringing the runaways, and asking as a favour that they might be again received on probation. This was agreed to, and no unkindness was shown—no allusion to their offence was ever made by the other students. Max and Fritz remained in the college several years, and proved perfectly quiet and law-abiding members of the union.

Since Mr. Fellenberg's death the college has declined, and

lost all its distinctive characteristics; but the years of its success suffice to prove, to my mind at least, that a method may be found by which to reconcile the sturdy independence of the young generation—an independence which we dare not, *must not crush*—with a controlling power, a stringent discipline, properly adapted to youth in the present stage of our civilization. As soon as girls and boys are old enough to unite for self-regulative purposes, by placing upon their shoulders the responsibility of their own and their companions' social conduct, we may *exorcise* the very spirit of rebellion against authority, and at the same time immensely exercise intelligence and develop sympathy. Do I mean that masters and professors should have no authority? They would formally abdicate in favour of an authority, which in relation to the young would be both educational and disciplinary. They would remain the teachers and guides in every position of perplexity—a court of arbitration or appeal—but never rulers in any sense that would appear arbitrary or antagonistic to the young. When bad conduct came before them they would decline to punish, and hand over the delinquent to the judgment of his equals; only requiring that they should carefully consider the case from their own standpoint of personal dignity and individual freedom, and judge it on behalf of social order, in accordance with the principles of *equity* taught them by their professors.

But it may be said: "Would not this system cause great waste of precious time that ought to be devoted to book-learning? for we may assume that misdemeanours, which a master of experience would summarily dispose of, would as certainly create great discussion in a court of juveniles." Possibly that might be so; but the effort to judge justly, the sifting of evidence, the forming logical conclusions from the evidence that all could accept, would call out intelligence in many cases far more effectively than book-learning, and would in fact form a training of the *greatest value*. After all, whilst we no doubt desire our children to become cultured, accomplished men and women, what we absolutely require for happy life is *practical, moral* Humanity—creatures of noble character, full of dignity and self-control, yet brimming over with generous, spontaneous sympathy for others.

One certain element of good in adopting this method, would lie in directing special attention to personal conduct, and its bearing upon general happiness. To be critical of his own individual conduct is foreign to the nature of the average

British youth, but self-criticism would become a mental habit, where duty required the carrying out of a self-made disciplinary system. Again, the faculty of placing one's self for a moment ideally in the position of another would also grow; for effort would be made to feel as James or Robert felt when they were tempted to break a rule, and committed the trespass which brought them up for judgment before their fellows; and the stimulus given to this faculty would assuredly promote a rapid development of sympathy.

It is a curious fact of evolution, that in our transitional state there are many human beings, especially men, with a very large development of benevolence, who are yet destitute of real sympathy. They cannot image to themselves a fellow-creature's outward position and inward condition, so as to discover his or her desires, however much the individual may be loved by them. The representative faculty had no exercise of this kind during the period of life when the brain was essentially plastic, and in adult life, rigidity of the mental structures renders the acquisition impossible. But there is much in this fact that they themselves deplore. Under such limitations, kindness often misses its aim, and with all the will in the world to make some deeply loved one happy, alas! there lacks that kinship of spirit which is the only key to all the delicate, tender, secret emotions of the human soul.

The present position, then, is this. From the nursery, where authority is purely barbaric, and the budding conscience is hurt, our children pass into schools, where they are drilled and moulded into book-learning machines, but no attempt is made to train the emotional nature; from these again to high schools or colleges, where, except in class, they are emancipated from control, and as a natural result of his want of training the British student's future is gravely uncertain. He presents in many ways an element of social danger, simply because his bumptious independence and sense of personal rights is not duly balanced by an intellectual understanding of, and a sympathetic feeling for, the equal rights of others. Authority I have said has unconsciously relaxed and fallen back before the bumptious freeborn British schoolboy. Dare we expect that he will ever be submissive to a self-made system of control?

I think we may. The study of boy-nature shows me *imitativeness* largely developed. What one boy does another will do, no matter at what cost! The customs and fashions of equals are adopted and followed with slavish, punctilious obedience. If a collar or necktie, a cane or pipe, is not exactly what companions use, a young man will discard it in defiance of a parent's wish, and that, not from an impulse of rebellion, but simply from inability to bear the pain of being unlike his fellows! Now, this imitativeness will prove an immense factor in self-regulation. When the British nursery is under educated supervision and control, and the higher social qualities of our children are from the first carefully guarded and developed, we need have no fear that the schoolmaster will fail in organizing unions for self-government. In most English schools combination already exists for purposes of regulation in reference to sports and pastimes; but a good cricketer may be a public nuisance in many ways, without the members of his cricket-club deeming it their duty to call attention to his breach of social morals, and requiring him to reform, under penalty of the censure of his comrades. What we stand in need of is, *conduct-clubs*, and I am convinced that in the generous ardour of youth, an equal vigour and energy to that shown in the foot-ball and cricket competitions (which take place over the whole country) would be called out, in a nobler emulation, to develop social virtues and repress all anti-social and barbarous vices.

What human nature has already accomplished, on however small a scale, it would be presumption to deny, may on a much larger scale be by-and-by worked out by a Humanity improvable and improving; and in Mr. Nordhoff's account of the Oneida Communistic Society, I find an institution of criticism described, and pronounced effective in maintaining order within the group, and restraining every vice, whether petty or great. These people, says Mr. Nordhoff, "depend upon criticism to cure whatever they regard as faults in the character of a member; for instance, idleness, disorderly habits, impoliteness, selfishness, a love of novel-reading, 'selfish love,' conceit, pride, stubbornness, a grumbling spirit."

Now, these are precisely the faults that we want our young people to correct, and although in our huge, ill-regulated community, we do not as yet recognize the fact, we are undoubtedly in the exact position of these Communists thus far, that we have *no* force to depend upon *but* criticism to effect what we wish. We differ from them, however, in this—whereas their criticism is systematic, and in a great degree effective, ours is a matter of pure haphazard. It is carried on in slapdash style, in the public prints (often, however, anonymously), and in private circles it takes the form of backbiting, even slander, and the result, in respect to what we want of criticism, is an utter, unmitigated failure.

Its working at Oneida, I give in Mr. Nordhoff's words: "On Sunday afternoon, by the kindness of a young man who had offered himself for criticism, I was permitted to be present. Fifteen persons besides myself, about half women, and about half young people under thirty, were seated in a room, mostly on benches." Mr. Noyes (the head of the community) in a rocking-chair. Charles, the young man to be criticized, "sat inconspicuously in the midst of the company." He was asked if he desired to say anything. He replied (whilst retaining his seat), "that he had suffered for some time past from . . . intellectual doubts—a leaning especially towards positivism," and so on.

Hereupon, a man being called on to speak by the leader, who was *not* Mr. Noyes, remarked, "that he thought Charles had been somewhat hardened by too great good fortune; that his success in certain enterprises had somewhat spoiled him; if he had not succeeded so well, he would have been a better man; that he was somewhat wise in his own esteem, etc.

"One or two other men . . . had noticed these faults in Charles, and that they made him disagreeable; and gave examples to show his faults. Another concurred in the general testimony, but added that he thought Charles had lately made efforts to correct some of his faults, though there was still much room for improvement.

"A young woman next remarked that Charles was haughty and supercilious, . . . that he was needlessly curt sometimes to those with whom he had to speak.

"Another woman said Charles was often careless in his language; sometimes used slang words. . . . Also that he did not always conduct himself at table . . . with careful politeness and good manners.

"A man concurred in this, and remarked that he had heard Charles condemn the beefsteak on a certain occasion as tough; and had made other unnecessary remarks about the food on the table while he was eating.

"Amid all this very plain speaking which I have considerably condensed, giving only the general charges, Charles sat speechless, looking before him; but as the accusations

P

multiplied, his face grew paler, and drops of perspiration began to stand on his forehead.

"Finally, two or three remarked that he had been in a certain transaction insincere towards another young man, saying one thing to his face, and another to others; and in this one or two women concurred.

"The remarks took up about half an hour, and now each one in the circle having spoken, Mr. Noyes summed up.

"He said that Charles had some serious faults; that he had watched him with some care; and that he thought the young man was earnestly trying to cure himself. He spoke in general praise of his ability, his good character. . . ." He remarked that, on one occasion, Charles had come to him for advice, and had acted upon the advice given, at considerable self-sacrifice. Taking all in view, "he thought Charles was in a fair way to become a better man, and had manifested a sincere desire to improve, and to rid himself of all selfish faults."

Thereupon, the meeting was dismissed. In reference to it Mr. Nordhoff says, whilst Charles "might be benefited by the 'criticism,' those who spoke of him would perhaps also be the better for their speech; for if there had been bitterness in their hearts before, this was likely to be dissipated by the free utterance." *

Now, it is not as a model for us to imitate that I speak of this Oneida Community, but simply to corroborate my proposition—that human nature may by *free-will* adopt methods that outward authority could never successfully *compel* it to adopt. Throughout our whole social body, there is much less of docile subjection and instinctive veneration towards superiors than formerly, and also much more of union and co-operation (conscious and unconscious) amongst equals. The latter is a force that at the present moment tends frequently to personal degradation, and always to class antagonism; nevertheless, it is capable under enlightened guidance of becoming the direct agent of social morality and general happiness.

I have shown how an *esprit de corps* amongst boys turns them from the path of rectitude, and in kindness to one another they prove false to masters and teachers. Let us see how the very same spirit of union and brotherly fellowship affects the morals of the working class. I give a homely

* "The Communistic Societies of the United States," by Charles Nordhoff, pp. 290-293.

illustration. In a late snowstorm the roofs of houses were endangered, and a widow lady sent for her plumber to remove the snow from her roof. The plumber was unusually busy, and to overtake all his engagements he employed unskilled labour, with this result. The snow, in this case, was not thrown off the roof, but only piled up on one side, whence it fell in masses and completely unroofed a scullery which projected at the back of the house! Two lawyers told the lady that the plumber was legally liable for the £4 expended on repairs; nevertheless, the man repudiated the claim and scoffed at the idea that she could compel him by law.* On consultation with legal friends, the advice given was: "Threaten as much as you like, but do not take legal action! That would entail further expense, and probably you would obtain no redress! Tradesmen are banded together in such fellowship that the chances are, no plumber will give evidence against his brother in the trade."

Behold, then, clannishness amongst workmen actually renders the civil law in such cases *nil;* and it does this unconsciously, for in trades unions there is no rule compelling members to refuse to give evidence against a comrade. Personal conscientiousness, the instinctive love of truth and justice, is in our labouring classes subordinate to clannishness or fellowship amongst themselves; but we should expect that amongst the highly educated members of our social body at least, conscientiousness would rule supreme, and at all times subordinate class feeling. Let us judge if this be so.

Some of my readers may remember the case of Thomas Pooley, the poor Cornish well-sinker, who, in 1857, was tried for blasphemy, and because he had written upon a gate some foolish words concerning Jesus Christ, was convicted, and condemned to imprisonment for a year and nine months. The man was very poor, but very industrious and perfectly harmless. He was weak in the intellect, and when cast into prison he lost his reason entirely, and had to be conveyed to an asylum for

---

* The following, from a newspaper of January 11th, 1883, will show how the law stands in such cases :—

"PLUMBERS' WORK.—An action was brought by a plumber named Dee, in the Croydon County Court, on Monday, against a civil engineer named Dalgairns, for upwards of £30 for the erection of a lavatory. Defendant made a counter claim of £120, on the ground that, the work being improperly done, sewer air escaped into the house, and caused the illness of six members of the household, and the death of his son. He therefore claimed the doctor's bill and other expenses. The judge struck out the plaintiff's claim, and gave judgment for the defendant."

the insane. The whole proceeding was an offence against humanity, and in opposition to the spirit of the age. Our blasphemy laws had for years been practically obsolete. The brain of our social organism, that is, the thinkers amongst us, had perceived that vital truth needs no protection from law or any outward authority. It prevails in virtue of its own vitality; and its discovery and progress are best aided by the natural condition of *perfect freedom* in the expression of all speculative opinion. Persecution for any doctrine, not seditious, or not libellous on individuals, is detrimental to truth, and as general happiness depends on *truth in all things*, to punish any man for merely saying in his own way, what he thinks true, is contrary to the general well-being.

Enlightened men of that day—John Stuart Mill, and many others—came forward to protect poor Pooley, and thus showed their intellectual standpoint, their moral rectitude, their sympathetic jealousy for the rights of other men, embracing even the poor, the imbecile, the lowly. The man was restored to his family, and again the blasphemy laws, which are barbarous, belonging to a primitive age, fell into proper abeyance. Strange to say, however, they have reappeared recently upon the stage of public action, and barbarous tyranny has once more striven to overcome freedom, and suppress by violence the purely speculative opinions held by individuals who are in a minority, therefore socially weak.

Since the year 1857, a great advance has undoubtedly been made in intellectual freedom throughout the whole nation, and also an advance in knowledge of the moral principles which *justify* toleration. It would therefore seem natural to expect that all individuals who are in the flow of intellectual progress, would spontaneously combine on the first symptoms of an outbreak of the Old World spirit, and vigorously call to order (that is to adaptation with our own age) those that are tending to relapse. No such action, however, was taken, and we are bound to ascertain the cause.

The three men who suffered in 1883 from this new outbreak of persecution do not hold a high social position, and the penny journal which was prosecuted belongs to a class of literature thought vulgar. It is destitute of high culture and refined taste, and it circulates only amongst the comparatively uneducated of the lower middle class. Nevertheless, the editor in his defence showed great erudition, and a clear understanding of the principles by which all men are in duty bound to maintain and defend the liberty of the press. He pointed out that the doctrines pronounced libellous in the *Freethinker*, have been over and over again expressed in decorous language by our most prominent leaders of thought; and since these men have never been placed in the dock as criminals, he inferred that the social force which carried him there, was no earnest, bigoted, mistaken love of truth, but a sentiment of vindictive prejudice and dislike towards a special class of individuals. "You ought to show," said Mr. Foote, in his address to the jury, "that the liberties of those who seem friendless and poor shall not be rashly imperilled in the interests of classes; but that every man, whether he addresses his fellow-men through the medium of a penny paper or a twelve-shilling book, has precisely the same rights." It seems clear enough that the master of literature who spoke of the Trinity as three Lord Shaftesburys, is as guilty of blasphemy as Mr. Foote; and if public opinion took no offence at the latter expression, it was bound to put down this persecution, and rebuke the men who set it in motion.

A curious contrast was presented by this case and Pooley's. Pooley was prostrate in face of the law. He could neither speak in his own defence nor pay others to speak for him; whereas Mr. Foote and his companions boldly and ably defended themselves. The law crushed Pooley, but before the bar of public opinion his weakness and utter helplessness proved favourable; for benevolence and compassion were thoroughly aroused, and as thousands of individuals are swayed by these sentiments alone, apart from any rational principle bearing upon the special subject before them, the social force was broad and strong—it swept aside the ancient bulwark, and promptly rescued Pooley. The law had not an easy task in crushing Messrs. Foote, Ramsay, and Kemp. They showed vigorous fight, and loudly appealed to public opinion for aid; but benevolence and compassion were by no means excited in *their* favour. What they demanded was simple justice, not justice allied with patronage.

Now, although the social force of rational, moral principle in favour of free-thought is immensely stronger in the present day than in Pooley's day, that force has not overcome prejudice. The *Freethinker* is a vulgar publication, and its writers support their doctrines in a combative spirit which is wholly repulsive to men of refinement, who, thinking very much as they do, are nevertheless able to express their own thoughts in good taste, and with philosophic calmness. These men shrink from the suspicion of alliance or fellowship with Messrs. Foote, Ramsay, and Kemp; and whereas Mr. Buckle and John Stuart Mill led public opinion in 1857, to the rescue of Pooley, the men whom we look up to now for intellectual strength and culture held back, and silently, we may say, basely, permitted smaller men to suffer persecution for merely exercising a freedom of expression which they themselves enjoy. Am I not justified in saying that the sentiment that attracts like to like, in other words, our class feeling, clannishness, or *esprit de corps*, controls our morality, throughout the whole social system, in the highest circles, as well as the lowest; and unless we lay hold of this social force, and use it to develop conscientiousness and support character, it will continue to subordinate reason and equity, and tend to the greater degradation of humanity.

I have already alluded to the threefold regulative system of control which has outwardly restrained our anti-social tendencies, and shaped our life to some degree of order, during our long progression from savagery to our present mixed, semi-barbarous, semi-civilized condition.

Of this threefold system, two-thirds—I mean Law and the Church—have greatly lost in outward authority and influence, whilst the other third, viz. custom or fashion, has (through growth of social feeling) rather gained than lost in its power of *arbitrary control*. Liberty in some directions is far from implying liberty in all. Our great labouring class, for instance, is shaking off its legal disabilities, and wrestling with the ruling class for liberty; but within his own class the working man is all obedience. Trades Unions rule despotically, and each individual member imitates his brethren, and submits to the rule with touching, childlike docility.

Speculative minds, again, have freed themselves from archaic dogmas. Truth is the divinity they worship, but not as a fetish to serve blindly. Open-eyed, they range through the universe of thought, seeking truth everywhere and revelling in freedom. Nevertheless, in daily life and conduct they show themselves not free men, but slavish conformists. They follow custom, however irrational, without protest! *They*, too, simply *imitate* their brethren. Their love of truth does not control and shape their social relations and their daily life, because in these familiar domains they are not mentally free. The pressure of opinion outside, combining with a spirit of fellowship, a tendency to union within themselves, form a stream upon which they drift along in the current of every senseless fashion of the day—examples of a kindly but, alas! a weak, frivolous conformity. "That so few now dare to be eccentric, marks the chief danger of the time," is a reflection as true now, as in the days when Mr. Mill wrote his Essay on Liberty.

Outwardly we are a free nation, taking off our hats less than any other European nation, which is a very significant symbol, as Mr. Spencer shows.* In religion, we are thorough non-conformists—Protestants protesting against popish or clerical supremacy—but inwardly we are prostrate, letting the tyranny of that monstrous pope, *conventional custom*, crush out of us the pith of individual character, and the sweetest honey of our social life. "Send no poet to London!" cried the sweet singer, Heine; "that colossal uniformity, that mechanical motion, that irksomeness of joy itself, that inexorable London stifles phantasy and rends the heart." This tendency to uniformity is adverse to progress and to happiness.

The path of evolution is one that leads from the *homogeneous* to the *heterogeneous*, but our impulse to conformity turns us the other way. It gives supremacy to the mediocre natures that always preponderate in number; it suppresses genius, which is ever exceptional, and grinds our common humanity down to the universal level of the commonplace. "Is there no remedy for this?" Yes, there is a remedy within our reach, so soon as leading minds awake to full consciousness of the danger of our present position. On examining the question of our material welfare, in a former chapter, we found that Mr. Matthew Arnold enlightened us as to the great centre of evil from which there radiate—poverty, crime, disease, and exhorted us to let the light of reason play freely upon the population question, whereupon it became clear to us that intellect is capable of controlling adverse forces, and working out human happiness, in spite of the instinctive tendency which he showed as the great obstacle to our material progress.

Now, similarly here, Mr. J. Stuart Mill points to a central evil—to the special obstacle, in short, which obstructs our mental and moral progress; and as in the former case, so here also we find that intellect can guide us. It must teach us how to take the *force of union* that is diffused amongst us, the gentle spirit of conformity that is at this moment working our

* In his Essay on Manners and Fashion.

ruin, into our own hands (as it were), and forge out of it a strong chain, to hold back the tyranny of social conventionalism, and place a barrier around each and every individual, whether high or low, rich or poor, strong or weak, to guard his rights, and keep sacred his liberty of thought, word, and action, within the whole sphere of his personal or individual life.

For political liberty, for religious liberty, for class liberty, we have long been, and at this moment are still bravely fighting; but for social liberty—that is, individual freedom from the *subtle tyranny of conventionalism*—we have not yet begun to fight; and until we do, we are not upon the evolution path to natural social heterogeneity, the only true condition to bring about rapid mental and moral development. But my reader may say, "Why fight at all? What hinders individuals from preserving their own freedom in all social matters? Surely a man needs no aid to eat, to sleep, to dress, to see his friends, in short, to live out his personal life, according to his own ideas, and precisely as he chooses?" Strange as it may seem, he does require this aid. As a matter of fact, the tyranny of custom and fashion so press upon him at every point, that alone, he cannot maintain freedom without a fight, and if he fights unaided one primary result is injury to character.

Now, it so happens that development and preservation of character are precisely what we desire, and if we mar or hurt it in the individual (by thrusting upon him a struggle for personal liberty), we are simply frustrating our direct aim and object.

And here we have a point in reference to human nature which we must carefully observe. If an individual has worked out intellectual freedom for himself, and is compelled to live a life of nonconformity in matters that he deems essential, he is certain to yearn tenderly and deeply for conformity with his fellows on every other side; and in all that he thinks less important, he will avoid eccentricity. He will eat, dress, live, like others if he can, although in doing so he suffers greatly. Individual strength of intellect does not suffice to oppose the social strength of intellectual weakness and folly; the latter is much the *more powerful;* and what we have to do is to subdue it by another *social* force that will dominate the tyranny of numbers, and protect the individual in his freedom.

The only force available for this purpose is a well-directed public opinion. Society in the aggregate is bound to protect individuals from tyranny, and the peculiarity in this case is that the tyranny is its own. Society, then, has to renounce tyranny; it has to change its own mental attitude, to cease from meddling interference, from petty exactions, from despotism in trifles, from all desire and demand for *uniformity*. It must set itself to desire variety, to show reverence for personal freedom, and sympathetic jealousy for individual rights, and, above all, it must brace itself to an energetic guarding of the liberty of each and every social unit with an eternal vigilance.

By union and co-operation we may (if we will) bring about general enlightenment—the opening of all eyes to the true nature of this despotism which enthrals our minds. Conventional custom means the rule of the unintelligent. We ought consciously and intelligently to reject its authority, and adopt that of principle or law, but law not of man's devising, but of his discovery only—the social law of civilized human nature, that "Every man has freedom to do all that he wills, provided he infringes not the equal freedom of any other man."

In some few directions the pressure of tyrannic conventionalism has led to union in opposition. For instance, a Rational Dress Association now exists which thus sets forth its purpose: "Female dress has reached an extreme pitch of extravagance both in form and expense, and even rich and fashionable ladies moan over the slavery to which they are subjected. The objects of this society are to promote the adoption according to individual taste and convenience, of a style of dress based upon considerations of health, comfort, and beauty, and to deprecate constant changes of fashion, which cannot be recommended on any of these grounds." This movement, then, appeared progressive. It was calculated to lead public opinion in a matter important to health and happiness, to ventilate the whole subject of dress, and gain freedom for individuals. "Women's dress," said one of the promoters, "ought to facilitate freedom of movement." But behold, upon the 8th of February, 1883, at a meeting held in London, a lady speaker stated that the society had arrived at three conclusions, and one of these was, that "fluctuating fashion must be crystallized into some unchanging shape of dress"! Alas for freedom! We are to be as individuals physically free within our clothes, but we are not to be mentally free to choose our own clothes! Fashion is to rule us still; only, instead of a changeful dame who keeps us for ever on the move, she is to become rigid, and settle once for all what we poor women are to wear down the dark ages to the crack of doom! Let us beware. The law of social freedom is not to be obeyed by shaping new forms for tyranny to assume, and even union and co-operation may prove obstructive unless *intelligence* and *knowledge* guide them.

At the beginning of this chapter I quoted Mr. Spencer's words to an American ; and now in ending my chapter I will remind my reader of his wise comments on the social state of the Americans. The American people, he said, retain the form of freedom, but there has been a considerable loss of the substance. The sovereign people are fast becoming a puppet which moves and springs as the wire-pullers determine. The remedy is a question of character. There is a lack of moral sentiments. . . . In short, the Americans and we sail in similar boats. Both nations are outwardly free, inwardly slavish. Both require conscientiousness ; that is, love of truth and justice —a sense of individual rights, a sympathetic jealousy over the rights of others, and willingness to pay the price of liberty in an *eternal vigilance.*

CHAPTER XIII.

OUR UNORGANIZED SOCIAL LIFE.

"Institutions left us from the past are no more diabolical than divine, being the fruit of necessary development."—PROFESSOR SEELEY.

"It will be our own fault if in our own land society is not organized upon a new foundation."—MISS SEDGWICK.

IN this country, as we have seen, parental and other despotisms have marvellously relaxed. Arbitrary, visible authority has given way to respect for the individual; but a subtle, invisible authority—the rule of Mrs. Grundy—has, on the contrary, vastly increased. This rule of Mrs. Grundy is a social force of entirely different nature from the inward, invisible force, which ultimately will regulate a perfect society, holding its units together in such relations as cannot fail to produce pure and innocent happiness. The latter is centred in each individual—a power of self-control making pressure from without, to keep him in his true position, unnecessary, and removing from him all impulse to press upon or tyrannize over others. The former is a tyranny as complete as any system of outward control; only it is a tyranny of the mind rather than of the person, ruling through fear, and not through love as the future force will rule. It tends to crush individuality and repress development of those qualities that beautify and ennoble human character. Whereas, on the one hand, the regulative force of self-control causes each social unit to resemble a majestic planet, revolving upon its own axis, and freely radiating forth in all directions its little portion of the cosmic light, this other force makes of the individual a piece of mechanism to be wound up and moved hither and thither, as Mrs. Grundy wills, and often put by her to base and ignoble uses.

"*But*," says a fashionable lady of my acquaintance, "you make a mountain of a molehill. It pleases you to inveigh against conventions—it pleases me to follow them! I want, for instance, to entertain my friends. I find it easier far to do it just as others do, than to invent a method of my own. I ask my hired cook what dishes are in fashion. She tells me, and I get them. I bid the waiters do precisely as they did at Mrs. D——'s. I find from the fruit merchant, to be sure, that this season no table is complete without a pine-apple, and one costs thirty shillings—too dear by half! But still my husband likes to have all things *comme il faut*, and we must keep quiet now for a time. I have had three dinner-parties in a week, and three are quite enough for a year. In our set more are seldom given." What can one say to this? That lady's entertainments are pure machinery—the springs or catches that hold her securely in her own set; and since she wants nothing more let us leave her in peace.

But society is not wholly composed of Mrs. Grundy's puppets. We have flesh and blood human beings, minds thirsting for social intercourse, hearts palpitating with social desires; and how do *these* fare in this luxurious age, when nothing is too costly to grace our dinner-tables? Mr. Herbert Spencer says: "Not a few men, and not the least sensible men either, give up in disgust this going out to stately dinners and stiff evening parties; and instead, seek society in clubs, and cigar divans, and taverns. 'I'm sick of this standing about in drawing-rooms, talking nonsense, and trying to look happy,' will answer one of them when taxed with desertion. 'Once I was ready enough to rush home from the office to dress; though I found each night pass stupidly, I always hoped the next would make amends. But I'm undeceived. Cab-hire and kid gloves cost more than any evening party pays for. No, no; I'll no more of it. Why should I pay five shillings a time for the privilege of being bored?' If, now," continues Mr. Spencer, "we consider that this very common mood tends towards billiard-rooms, towards long sittings over cigars and brandy-and-water—towards every place where amusement may be had—it becomes a question whether these precise observances which hamper our set meetings, have not to answer for much of the prevalent dissoluteness. . . . Men must have excitements of some kind or other; and if debarred from the higher ones will fall back upon the lower." This passage, we must remember, is not the utterance of a superficial thinker or a commonplace mind. It is the thoughtful, deliberate conviction of a great philosopher, who has studied human nature, and knows both men and manners—who knows the world in which we live, and the beings to whose happiness that world should minister.

We have already found that our Fanny Dovers pant for innocent excitement, and complain that as society is constituted they cannot get it without "being bad;" and now we find another class, called by one of their own sex, "sensible men," in a precisely similar position. They want excitement too, and when they cannot get it *with* propriety they take it *without!* On most occasions, to be sure, they pay outward homage to propriety, and carefully conceal what Mrs. Grundy might object to, or deplore; but on the Epsom holiday there is open revolt! Mrs. Grundy may frown as she likes. The English gentleman, for one day, soars mentally above her strict, despotic rule, and (lest we should doubt it) gives *public demonstration* of the truth of Mr. Spencer's allegation.

It is not to abuse or condemn, however, that Mr. Spencer chronicles the fact that sensible men desire excitement. He goes on to say: "It is not that those who thus take to irregular habits are essentially those of low tastes. Often it is quite the reverse. Amongst half a dozen intimate friends, abandoning formalities, and sitting at ease round the fire, none will enter with greater enjoyment into the highest kind of social intercourse—the genuine communion of thought and feeling; and if the circle includes women of intelligence and refinement, so much the greater is their pleasure."

I pause here to state that I know many women of intelligence and refinement to whom the easy intercourse above described would be an inestimable boon—women shut out of all conventional social life (because of limited incomes), and, as a rule, indifferent to *that* effect of poverty—but who, in their solitary lives, would find infinite refreshment in mental contact with the other sex. As society is constituted, unless they have the courage to be eccentric, they cannot seek men's society, and men do not seek theirs; and in their minds has grown the somewhat sad conviction, which I have elsewhere expressed, that men have no desire as yet for communion of thought and feeling upon a platform of mental *equality* with women, but care only to exercise these faculties in intercourse with their own sex.

Now, it is only from this intimate communion of thought

and feeling that true friendship can evolve, and friendship is the social relation which, *above all others*, tends to enlighten and elevate human character, and is least liable to any base alloy. But the averment made is that men have no capacity for that kind of friendship which, in addition to the pure enjoyment that results from intimate intercourse of sympathetic minds, would add the distinctive charm of sex differences in thought and feeling.

This opinion has again been privately expressed to me quite lately, by a woman of intellect and culture, and how deeply she regrets this state of things is evident from her published works. Here, for instance, is *her* view of friendship: "It is curious that so little enthusiasm is ever stirred by descriptions of friendship, whereas this order of affection is the real salt of human life. In it there are no heart-burnings, or jealousies, or darkening of the understanding by the fumes of passion or the mists of despondency; calm, clear, and harmonious, it strengthens, elevates, and satisfies; it gives cheerfulness and beauty to the most rugged and monotonous career; while the want of it is desolation."*

This *desolation* is the lot of thousands of my sex. There are tired governesses who have outlived the spring of youth, and lost the personal attractions that many men most prize; but whose life, whilst making them unfit to sympathize strongly with the purely domestic pleasures and cares of their own sex, has eminently fitted them for intellectual interests and pursuits, and to whom the bracing influence of contact with a masculine mind, and the soothing pleasure of a strong man's friendship, would prove a luxury indeed.

There are widow ladies who, unable to give stately dinners and stiff evening parties, find themselves thrust out of the social circle of which they previously (as wives) had formed a part, and without intimate companionship, except that of little children. "I have trustees, and I have male acquaintances," a widow will say, but "not one single *masculine friend* to whom I could with perfect freedom and confidence resort, when some new development of character takes place in my young sons, and I feel sadly ignorant and perplexed. I would not cast on others a responsibility which must be borne by me alone; but what I long for is, the moral strength that comes through frank, intelligent discussion of a subject with minds that are mature, and have a wide experience of life." The

---

* "The Admiral's Ward," Mrs. Alexander, vol. ii. p. 219.

very joys of motherhood are overborne by anxious cares, when there is no husband to share these cares; and in this country, where widows are shut up to the privacy of domestic life with children only, and there is absolutely *no* easy, intimate social intercourse with men, and nothing to divert the mind from care, the position is a cruel one, and *desolate* in the extreme.

Again, we have thousands of women in this crowded land more desolate still, because, without a mother's ties, cares, and occupations—single women, with means enough to live the idle life approved for the sex as *genteel*, and no call to usefulness, save that of the Church—a little patronage of the poor, which they perceive is, at the best, a very partial and unsatisfactory good. They do not give stately dinners, because there is no host to fill the usual place. Their entertainments are the old fashioned tea-parties, and not a man is found to fit in suitably with these simplicities. All men dine late, and only women go out to tea. Society, then, to them means a few intimate acquaintances of their own sex, who are similarly situated to themselves. Can any one deny that such a life is cramping to the nature and inadequate for happiness? These ladies in their youth were members, perhaps, of a large family circle, which has diminished one by one, to the saddening of their hearts. Some brothers may survive, but they are busy men. They have an active public life, a private domestic life, and take their part as well in all the stately ceremonious conventions of their circle. No wonder they have neither time nor inclination for women's tea-parties.

Miss Martineau has written, and I think her words are strikingly true: "The fact, the painful fact in the history of human affections, is that of all natural relations the least satisfactory is the fraternal. Brothers are to sisters what sisters can never be to brothers—objects of engrossing and devoted affection." When all the lines of individual life have gradually diverged, and middle-aged sisters only vegetate, brothers may be kind and often helpful to their sisters; but between them there *is not* the genuine communion of thought and feeling which Mr. Spencer rightly calls "the highest kind of social intercourse."

When I say, however, that *desolation* is the term by which to characterize the lives these ladies lead, I do not mean that they themselves are always miserable, or that they feel society has wronged them, and show it by envious thoughts or spiteful words. *Far from it.* The picture familiar to us all from books, of a cross-grained, gossiping, malicious old maid is, in my opinion, a mere survival of the past. The general evolution of humanity has carried us beyond the point when such developments were common, and now a specimen of the genus is extremely rare. But here is the point. When any human being, male or female, finds the conditions of life narrowed, and his or her faculties are in excess of those conditions, a state of restlessness, a dissatisfaction which has been well called "negative pain," sets in. *Disuse*, however, in time kills the faculties that are in excess of the conditions of life, and gradually contentment takes the place of restlessness, and the organism itself becomes adapted to the gray life and still monotony of a sunless, cheerless existence. The pussy cat that purrs upon an old maid's hearth is not more inoffensive, sweet, and gentle than her mistress, who sits knitting by that hearth alone. Yet when we think of that mistress's bright dreams in early youth, of all the strong capacity she once possessed for fulness of life, and when we contemplate the ideal of a rational and social life which science and philosophy alike pronounce possible, we turn sadly from this peaceful picture of the British domestic hearth and call it nothing *less* than *desolation*.

I do not expect this sentiment will be re-echoed in the minds of my readers of the masculine sex, for the average man, so far as my experience goes, has quite a marvellous conception of the adult feminine nature, its homogeneous feebleness, and the primitive simplicity of its requirements. A middle-aged, unmarried woman of my acquaintance tells how she at first rebelled against her fate, and desperately shrank from a lonely life in lodgings; and how she told her feelings to a brother who, she believed, loved her, and whose own domestic life was rich and full, with wife and children, and every luxury that money could bestow. But her confession that she wanted more than lodgings—that she desired a *home*, and power to have friends, and a social life around her—was received with grave surprise and not one spark of sympathy. The words spoken to her were these : " I cannot conceive that any unmarried woman really requires more than a couple of apartments comfortably furnished and means enough to ask a friend to tea now and then." This view of feminine wants is not at all exceptional, and the gentleman, so far as I understand, showed *ordinary* benevolence and brotherly spirit towards his sister.

But now the conventional social life which, in innumerable

instances, brings to the one sex desolation, and in the case of
the other leads to irregularities of life, is surely quite unworthy
of our nineteenth-century civilization. In effect, all thoughtful
far-seeing minds amongst us have long perceived that some
great and radical changes must take place. But *when* these
changes shall be initiated must necessarily depend upon progressive movement within the individual social units. My
personal conviction is that the first stage of preparation for a
better social system has already been passed through, and that
throughout the whole community there are, here and there
and everywhere, isolated individuals whose inward condition
of thought and feeling eminently fits them to *initiate* change,
and introduce the *order of the future*. To my mind, what
Mr. Spencer says of his own sex is joyfully indicative and
reassuring: "Men as well as women long for substantial mental
sympathy, and, abandoning formalities, delight in easy social
intercourse; and if the circle includes women of intelligence
and refinement, *so much the greater is their pleasure."* If this
be true, it gives, I think, a hopeful promise for the *immediate*
future.

"It is," Mr. Spencer tells us, "because men will no longer
be choked with the mere dry husks of conversation which
society offers them, that they fly its assemblies and seek those
with whom they may have discourse that is at least real, though
of inferior quality. The men who thus long for substantial
mental sympathy, and will go where they can get it, are often,
indeed, much better at the core than the men who are content
with the inanities of gloved and scented party-goers—men who
feel no need to come morally nearer to their fellow-creatures
than they can come, while standing tea-cup in hand, answering
trifles with trifles; and who, by feeling no such need, prove
themselves shallow-thoughted and cold-hearted." * What Mr.
Spencer here asserts in reference to *his* sex, I calmly and
deliberately now assert in reference to mine. It is outside the
ring of strictly conventional society, in solitary nooks and
hidden corners, that individuals dwell, who, "better at the
core" than gloved and scented party-goers, are longing day
by day to come morally nearer to their fellow-creatures than
superficial flippant talk could ever bring them, even if they
attended every stately dinner that takes place throughout the
kingdom.

But in facing any social problem, we are bound to look at

* "Manners and Fashion," *Westminster Review*, Jan. and April, 1854.

it from every point of view, and to remember that when misery arises from non-adaptation of organism to environment, there are two ways in which adjustment may be effected. The organism may alter and become adapted to the environment, which *remains the same;* or the environment may change to suit the organism.

Stately dinners, crowded *soirées*, brilliant balls, and a wealth of luxury, show, and glitter form an artificial environment to humanity—the organism—and from the interaction of the two, there results a large amount of misery and desolation, with a small amount of general happiness. But from these premises alone we are not logically entitled to infer that change of system *must take place.* To some minds it will appear more rational to expect that humanity itself will alter, and find means to bring to this artificial environment a charm hitherto unknown.

I listened, many years ago, to an earnest sermon upon this subject, viz. our conventional social life. Its emptiness of joy was distinctly pointed out, and human beings at stately dinners and crowded *soirées* were pronounced to be as destitute of liveliness and grace as the soulless furniture, the chairs and tables, that at each entertainment played an almost equally effective part! The preacher's view of the position, however, differed from Mr. Spencer's. He thought the organism, not the environment, was in fault; that each individual bidden to a feast was bound to bring with him a portion of the *elixir* of social life, like bottled champagne,* and not to draw the cork until the company could share the effervescence. Instead of that, individuals, for the *most* part, with the scent and gloves and company attire, assumed a stately artificial manner, and, except that the outward appearance was often extremely pleasurable to the eye, *their* contribution to social intercourse was of the flat, stale, unprofitable order, without the quality of ministering to "the feast of reason and the flow of soul." His remedy was more of duty-spirit within the individual. He eloquently exhorted his hearers to exercise force of moral constraint within themselves, in order to cause to bubble up and overflow a stream of joyful, exuberant life in every social circle of which they formed a part.

Shortly after this discourse, the clergyman himself gave an evening entertainment, at which I was a guest, and somewhat closely observed him. As *host*, he had "the value of position,"

* I am not quoting the sermon, only recalling its general tenor.

and I am sure he strained every nerve to carry out his theory; but I am also sure, that the theory *broke down!* There was no "feast of reason and flow of soul," no mental friction of a sparkling kind, no genuine warmth of fellowship, no spiritual abandonment of all formalities and breaking forth of keen enjoyment; and, on the whole, time passed on that occasion as heavily and stupidly as at any ordinary evening party, and gave the usual sense of disappointment.

Nevertheless, for many years that sermon influenced my mind. It never once occurred to me to doubt the doctrine, and I continued blaming my fellow-creatures and myself at every fresh experience of the blank, hollow nature of our conventional festivities. As time passed on, I ceased to wish to go to any; and, perhaps, had I been "a sensible man," might have deemed myself free to act as Mr. Spencer describes—to abandon formalities in favour of an unpolished society and irregular life.

The *true* nature of the problem stands before me now in an entirely different aspect. The good clergyman's argument is, to my mind, a *fallacy*, and *his panacea* an utter *delusion*. What is the use of carrying with you to a stately dinner your bottled champagne, if the companion allotted to you, and with whom you ought to share it, is a teetotaller? Or, to drop metaphor, suppose a man is full of wit and humour, and of the strong resolve to use these faculties, these bright endowments of his nature, for the enlivening of his fellow-creatures; and suppose his two next neighbours at a dinner-party of twenty, where no general conversation is possible, are, on the one side, a lady, whose real interests centre in domestic management (to which occupation her life is principally devoted), and whose knowledge of the literature of the day is strictly confined to *Chambers's Journal* and *Good Words*, and, on the other side, a lovely girl of Fanny Dover type, who devours Miss Braddon's and Ouida's novels in her boudoir, but thinks Thackeray's works intensely stupid, who very naturally finds dinner-parties tame, and hates them, and whose mind is at the moment vividly recalling her previous night's flirtation with a brainless youth, whose waltzing was, to use her own words, "simply delicious;"—in *this* environment, how *can* the man of wit or humour kindle Promethean fire? and what *will duty-spirit*, or a strong resolve to master circumstance, do for him? He has the flint; but no amount of force would bring an answering spark. Is he not right to husband his resources, to spare his moral energy, to simper weakly and talk nothings gravely, like his fellows?

But if, by any chance, the being seated at his side is of a mental calibre similar to his own, the duty-spirit or moral force is quite superfluous. The intellectual life will bubble forth and sparkle in the light of mutual sympathy; and if the sympathy extends to feeling as well as thought, the genuine communion will follow, and the delicious sense of moral fellowship, or nearness of spirit, be enjoyed, without the aid of any inward or outward compulsion. It is true, that the set formalities of the surroundings may, for the moment, blight a delicate organization, and congenial spirits in close proximity may fail to strike the vein of happiness. But if that be so, inward *stiffness* of *resolve*, the *duty-spirit*, would only hamper further, and prove obstructive to the birth of what, to be of real value, *must* be spontaneous.

What, then, are the *chances* for a man of noble nature, cultivated mind, and wide capacity for brilliant conversation at a stately dinner-party? What are his chances, I say, for enjoyment of a *higher kind* than the eating of a well-cooked dinner? Are they equal to *one* chance in a hundred? Perhaps not *even this low calculation!* For consider, first, congenial intercourse is not the primary aim and object of these entertainments. The primary object is a commercial one, viz. to repay in kind—that is, in substantial food and unsubstantial show and glitter—the debts incurred by the host and hostess, and to prove to demonstration that their command over available commodities—that is, the wealth of the country—is *quite equal to that* of their neighbours. Second, the hostess knows nothing of her guests' particular social powers. She has no intimate knowledge of any man, save her husband; therefore, even if she wished to select his partner with some view to his intellectual pleasure, the result would equally be one of *chance*. The host should know his guest well, for he is a club acquaintance; but even in clubs, acquaintance may be merely superficial. "In the clubs I saw," says Mr. Nadal, American Secretary of Legation, "there rarely seemed to be any *abandon* or heartiness."* And Mr. T. H. S. Escott remarks: "Of society in the sense of fellowship, a club does not necessarily give anything; indeed, the genius of modern club life may be almost described as that of isolation. A new-comer into the community

---

* "Impressions of London Social Life," E. S. Nadal, p. 15.

will probably find that he is not the less completely alone because he happens to be in the company of some score of his fellow-creatures."* In any case, our host would not dream of arranging partners at a dinner-party—the function belongs by prescriptive right to his wife ; and a woman shut out from public duties is apt to be tenacious of her private despotic rule. In the matter of companionship, then, our brilliant social unit must take his chance.

As he enters his host's drawing-room, he sees in a corner a girl, with whom he once spent half an hour in intimate, genuine communion. How long ago? Ah me! he thinks, two years, at least. And yet the flavour of that talk is with him still, and he has never had one glimpse of the girl since. What would he not give to go straight to her now, and saying, "You and I are congenial spirits; let us for this one evening keep together," boldly conduct her to the dinner-table? Alas ! he dares not for his very life. A sin against conventional etiquette so great would certainly be punished by social ostracism. And, still more sad, the girl might, he thinks, mistake his motives, and attribute too much significance to his attentions. So he lets slip the opportunity, and I confess my own opinion is that he could not do otherwise. Obedient to authority, he gives his arm to one of the bright leaders in the world of fashion, whose ornaments are all external, and whose mind, compared with his, is as a vacant chamber empty of ideas to one furnished to profusion. Meanwhile, the girl he would have chosen for companion is as much bored as he. Her allotment in this game of chance is a young squire full of enthusiasm for horses and dogs, but whose interest in life does not extend to higher phenomena than fox-hunting, cub-hunting, and horse-racing. The girl has never seen a race or hunt, and has no desire to do so. She thinks of these pursuits as barbarous, and already she is conscious of a strong antipathy to *her* partner ! so, after a few remarks upon the weather, she allows silence to reign between them, whilst her thoughts are busy elsewhere.

Now, it is possible, that at this moment my reader also may feel bored and somewhat impatient. It is always painful to have our attention called to evils which we see no means of remedying, and I am dwelling upon the disappointments that a stately dinner-party brings to select individuals, and yet I admit that an exceptional individual here and there cannot act

* "England, its People, Polity, and Pursuits," vol. ii. p. 60.

differently from his neighbours. More than this, the *law of freedom*, which I adhere to and uphold as our only right principle of action, does not permit me to desire that any arbitrary external authority should exist, by which to put down stately dinners or crowded *soirées*, and I would not, even if I could, compel a fashionable lady against her own desires to desist from giving three stately dinners in one week. "What is your object, then?" my reader asks; and I reply, "A simple one." I address myself only to individuals who dislike the stiff and formal mode of entertaining, who feel repressed and miserable at stately dinners, and who at intervals are conscious of inward disgust in the very midst of brilliant outward surroundings; and my object is to bring such minds to the *logical conviction* that their human nature is worthy of a very different environment—that it is one of the very highest and noblest duties of life, to act in accordance with their *own instincts* in this matter, and instead of constraining themselves to bear with gentle patience the tedium of an artificial, conventional society, to put it for themselves firmly and resolutely aside, and seek a new form of social intercourse.

Am I, then, advocating the irregular life which Mr. Spencer tells us some sensible men have adopted? Certainly not. I cannot advocate that of which I know nothing, and I have no right to judge it, either with approval or contempt. But what I say is *this*. There are now amongst us a sufficient number (not, perhaps, a majority, but a large minority) of individuals who are by development thrown out of any *possible* adjustment with the prevailing forms of social intercourse, and who, although they can do nothing singly, are yet capable by union and co-operation of evolving for themselves an environment at once nobler and simpler. I mean a social life, not rigid and uniform or ruled by tyrannic custom, but shaping itself from inward forces according to the infinite variety in human nature, and everywhere giving freedom to kindred spirits to come together in close communion, to the enjoyment of mental and moral fellowship without "let or hindrance." "The price of liberty is *eternal vigilance*," said the American statesman; and the free institutions of America, says Mr. Spencer, must be *guarded* in this spirit of eternal vigilance. In the ordering of our life, however, with the view of producing genuine social intercourse, we have as yet to strike our *first blow for liberty*, and to establish the institutions which will *secure individual freedom*. Now, this we are bound to accomplish without belligerent attitude and without aggression.

The problem stands thus. How shall we subdue the subtle tyranny of Mrs. Grundy by a force of union in opposition to Mrs. Grundy? How shall we be able to put down the domineering, interfering spirit which is the originator and upholder of strict conventionalism, whilst we also *raise* individuals *above* the slavish and imitative spirit which supports and fosters strict conventionalism? If by union we only aim at causing fashion to crystallize in some new form, as, for instance, the Rational Dress Society proposes to do, then, no matter how superior the new form may be, *we shall fail* in aiding progress. To supersede an old tyranny by a new tyranny is a very poor performance, and when we consider the heterogeneous elements of a society, the infinite variety of its units, and the complexity of their social feelings and desires, we may surely see that *rigidity* or *uniformity* would be certain to cause repression somewhere, and interfere with happiness. To kill or exorcise the very *spirit of fashion* is what we must aim at. We want to make our social organism more *organic*, and to have done with this "colossal uniformity, this mechanical motion, this irksomeness of joy itself, that stifles phantasy and rends the heart."

"You are very illogical," is the taunt with which some bright young friends of mine meet my remonstrances on their pursuit of fashion; "you like variety, yet abuse us for love of change! We are glad that fashion constantly changes; we hate to have our skirts trimmed, and sleeves cut in the same form, as three months ago. We prefer freedom to change." "Freedom!" I cry; "pray which of you would venture to design a form for herself, without scrupulous regard to what other girls wear, or are likely to wear? And as for variety! why, at this moment I am surrounded by uniformity! Your figures are all alike—small waists and high shoulders. Your heads are alike—fringed on the forehead. Your feet are alike—high-heeled. Your dresses are alike—ornate and tight-sleeved; and on the breast of each of you I perceive a flower fastened on precisely the same spot—one inch from the centre! No, no, my young friends, there is nothing admirable, nothing to defend, in this mental slavery to fashion. Viewed personally, it is contemptible; viewed socially, it is hateful; for it causes you to tyrannize over the old. I perpetually see and feel that the aged, for peace' sake, yield in this matter to the imperious will of the young, and pay a tribute to fashion which causes them bodily discomfort and mental worry."

There is also another moral aspect to this question. A passage in a journal which I chanced to meet with lately has taken hold of my mind. It is oddly expressed, but contains a truth we should all do well, I think, to ponder. I give the passage as it stands. "In the present stage of human progress, the æsthetic and the moral are conterminate at neither end. Æsthetic emotion may be roused in us by that which is ethically odious, and moral feeling may be called up by that which is artistically ugly."* When a girl compresses her waist, she commits an immoral action. It tends to evil, and the evil may be propagated and injuriously affect a portion of the coming race. To think small waists beautiful, then, is to have æsthetic emotion running counter to morality, and this presents a very *serious danger* to human happiness.

Again, my young friends constantly assert with confidence that this is *right* and that is *wrong*, in reference to dress. And when I ask, "Do you mean artistically or morally right or wrong? What is your standard?" they recognize at once that the form which they at the moment approve might be condemned under the artistic standard; so they defend it upon moral grounds, falling back upon the theory that it is *right* and *proper* to do as others do, and carefully to avoid making one's self conspicuous. In regard to smoking and drinking, however, girls judge differently. They have no toleration for the lads who cannot resist drinking as much beer as their neighbours, without any reference to their own individual capacity; and they scoff openly at the boy who, in spite of painful sickness, perseveres in learning to smoke, and thinks it *his duty* to be seen with a pipe like other boys, and to avoid making himself conspicuous. The present fashion of tight trousers amongst young men is a subject of derision amongst young ladies; and no doubt the tight sleeves, tight waists, high heels, etc., of the latter, are equally derided by the former! And, in effect, the similar condition of thought and feeling in the two sexes begets *no* mutual sympathy, but, on the contrary, pushes them apart. Now, why is this? Simply because the subjective state is *in itself immoral.* The subtle tyranny of fashion, the slavish imitation of our fellow-creatures, is utterly degrading to humanity. The mind becomes bowed to the yoke. The higher faculties, capacities, and susceptibilities are

* D., in *National Reformer.*

crushed out and an unintellectual, unsympathetic nature results —a nature so dulled and blunted that it neither sees things as they really are, nor feels so strongly as to *care* to look below the surface.

In speaking of sex differences, I alleged that moral instincts appear to me more firmly implanted and slightly more developed in woman than in man, and that therefore a far closer union between the sexes throughout the whole social body, than at present exists, would prove beneficial to our highest interests. But in this tendency of woman to be thoroughly conventional, this deflection of her moral nature, which causes her to think, that what the majority do in small matters is generally right, and that in any case it is admirable to acquiesce, there lies what I cannot conceal from myself is a very serious drawback to my theory, viz. a force antagonistic to general welfare and progress. Moreover, I am compelled to admit that with advance of sex equality this injurious, immoral force may temporarily increase. If freedom from man's rule is followed by a willing, docile obedience to Mrs. Grundy's rule, and if liberty is exercised merely in a mechanical although kaleidoscopic dressing, a luxurious high-living, and what has been well called "conventional foolery," progress will undoubtedly be retarded.

On the other hand, this danger may, and I trust will, be overcome by the *higher education of women*. But I must again guard my reader from supposing that I indicate by that term what is commonly understood by the name. A knowledge of Greek and Latin, or even what in one word may be styled culture, is not to be depended upon as a force likely to counteract social evils and guide us to right social conduct. My meaning is, that the growth of general intelligence, the development of common sense, the cultivation of the logical faculty in woman, the systematic training that will harmonize her emotional with her moral nature,—these tend to true elevation of character and correct guidance of private and social life. That false logic which sees in the changes of fashion the spontaneous variety that healthy human nature requires will become impossible to the female mind; and the female heart will cease to expand with æsthetic enjoyment when brought into contact with what is ethically odious.

Education must embrace direct teaching upon the moral aspect of æsthetic emotion, and Mr. Mill has furnished us with an excellent text for such lessons in these words: "It is not by wearing down into uniformity all that is individual in themselves, but by cultivating it and calling it forth within the limits imposed by the rights and interests of others, that human beings become a noble and beautiful object of contemplation; and as the works partake of the character of those who do them, by the same process human life also becomes rich, diversified, and animating, furnishing more abundant aliment to high thoughts and elevating feelings, and strengthening the tie which binds every individual to the race by making the race infinitely better worth belonging to." An instructor's skilful handling of this text would speedily convince girls that fashion, even in the matter of dress, is no unimportant question, and that to resist the tyranny of fashion at every point, and aid others to do the same, is to accomplish real work in the moral elevation of humanity.

"It is ideals," says Mr. John Morley, "that inspire conduct;" and a worthy ideal of *social life* has never as yet been thought out and placed before girls in graphic words calculated to lay hold of the imagination. They catch glimpses of such an ideal here and there in works of fiction, no doubt; but that is insufficient, and the actual social life around them is often basely artificial, only fitted to inspire conduct in a wrong direction. To quote Mr. Spencer again, "Our social intercourse as commonly managed is a mere semblance of the reality sought. What is it that we want? Some sympathetic converse with our fellow-creatures; some converse that shall not be mere dead words, but the vehicle of living thoughts and feelings; converse in which the eyes and face shall speak, and the tones of the voice be full of meaning; converse which shall make us feel no longer alone, but shall draw us closer to one another and double our own emotions by adding another's to them." Lord Bacon upon society says, "A crowd is not company, and faces are but a gallery of pictures, and talk but a tinkling cymbal, where there is no love." These passages (and many others from our best authors) contain a lesson concerning social life which, if impressed by elders upon the enthusiastic mind of youth, would make its mark upon a generation. It would turn into a nobler channel the exuberant life, the ardent activity which at the present moment is carrying us in a vicious course; it would inspire the young to wrest conventional society from its downward tendency, and raise it in the scale of civilization by giving to it heart and soul.

It is now thirty years since Mr. Spencer wrote in the

Westminster Review, "The time is approaching when our system of social observances must pass through some crisis, out of which it will come purified and comparatively simple." Thirty years is a long period in the life of a generation. That period, however, has passed away and brought no *crisis*, the harbinger of relief. Social intercourse was never more mechanical, fashion never more imperious than now. There has been *change*. Dinners are later, invitations are more formal, *menus* are more extended and more varied. Balls are more magnificent, *soirées* more crowded. Fashion is more extravagant, and etiquette more rigid. And as a consequence of this, there are more human beings than there ever were before thrown out of the *unsocial* social ring, leading desolate lives, with heart and mind starved for lack of the bread and wine of human fellowship, of intimate, warm, genial converse with their fellow-creatures. Surely in this epoch of *awakening consciousness*, of passing out of childhood into the strong manhood of our race, we ought to do something more than stand and wait, longing for a crisis that never comes.

And again, thirty years ago Mr. Spencer wrote, "To reform our system of etiquette and fashion, is an aim yielding to few in urgency." We may at the present moment urge, I think, upon the young that *no reform* in which they can take part is more absolutely *necessary* than reform of etiquette and fashion, and that to aid in this great work is an aim worthy of their highest endeavours and warmest enthusiasm. We, who are old, are little fitted to turn the stream of our existence into any new course. *Inertia* has in a *large measure* laid hold upon the most of us. We have borne the burden and heat of the day; we are somewhat weary with striving, and we have more of the longing for quiet rest than readiness to embark upon any new adventure. But we may be generous to the heart's core. We may lay before the young the facts of our experience with a *noble candour*. We may tell how the organized social intercourse of our day has failed to make us happy, how fashion, yielded to, has tyrannized over our minds and hurt our sentient or emotional nature, until we are painfully conscious of being *more* the puppets of society, ruled by *it*, than spontaneous efficient social units ruling society and using it to evolve the human happiness, which is the great end and object of life. We may point out the truth of Mr. James' proposition, that "the general efficiency and spontaneity of a people is the very essence of civilization." And we may exhort them to learn wisdom by our experience, and build up a noble social structure upon the ruins of our failure.

Can we guide them in the *actual* work? That is not for *me* to say. There is no one method of social life superior to all others. The fitness of any method to a group of human beings (and *fitness* means its success in producing happiness without social disorder or any hurt to society in general) must be its entire justification. And in our present semi-barbarous condition—a society composed of individuals in every stage of civilization, from the simple primitive savage to the highly complex human being, the inheritor of both refinement and culture—the outward life must display an infinite variety of forms, if it is to be the spontaneous outcome of the varied inward condition. "To give any fair play to the nature of each, it is essential that different persons should be allowed to lead different lives. In proportion as this latitude has been exercised in any age has that age been noteworthy to posterity." *

We want, at the present epoch, *experiments in living* under the law of equal freedom and in the spirit of brotherly union—a spirit which, as I have shown, is already so widely diffused amongst us that it is capable now of becoming, *with direction*, a thorough and efficient disciplinary force. Experiments in living of any kind, whether public or private, have been in this country as yet *extremely rare;* and as those on a large scale, such as Robert Owen's experiments, the Agapemone founded by the Rev. Henry Prince (on a religious, not a purely socialistic, basis), and some others proved failures, they, no doubt, have tended to check within the cautious British mind all impulse to further experiment on the same lines.

A thorough understanding, however, of Mr. Owen's schemes, and of the adverse conditions under which they nevertheless secured a *partial success*, would bring out the fact that, from a *rational* point of view, they ought to be regarded as giving distinct encouragement to further effort and ground for hope that in the present day such efforts would be crowned with success. If my reader will carefully peruse the little work called "Cottagers of Glenburnie," by Elizabeth Hamilton, he will perceive what were the habits of the Scotch peasantry at the beginning of this century. The confusion and dirt inside the poor houses of one apartment, the dunghill close to the door outside, the idle, gossiping, slatternly women, whose usual reply to any remonstrance was, "I canna be fashed,"

* Essay on "Liberty," J. S. Mill, chap. iii.

presents a perfectly *true* picture of the peasant homes of the period; and the mill-workers at New Lanark, Scotland, were in no way superior to their class, when Mr. Owen brought to bear upon them his gentle rule and his methods of organization.

One of his first steps was to give lectures to his workers on order and cleanliness, and then from among themselves a committee was formed to promote these social virtues by *simple criticism*. The duty of the committee was to visit each family weekly, and report upon the condition of the house. At first, a storm of rage and opposition was roused amongst the women. A majority locked their doors, or met the inquisitors with abuse. "They had paid their rent and did no harm to the house, and it was nobody's business but their own whether it was clean or dirty." The committee was encouraged by Mr. Owen to persevere, but also instructed to act with perfect gentleness, and only to request admittance as a favour; whilst the minority, who welcomed the visits, were distinguished by having a few plants sent to them from the master's greenhouse. These conciliations worked wonders, and the weekly reports of the committee gradually became full and favourable.* A few short years sufficed to lift the New Lanark mill-workers from the state of moral and social degradation, in which Mr. Owen found them, to one so vastly superior as to attract the attention of all benevolent minds interested in the welfare of the lower classes. Between the years 1815 and 1825, no fewer than 19,500 persons recorded their names in the visitors' book at New Lanark. There came there, not only nobility and gentry from every part of Great Britain, but also many foreigners of rank from the Continent, and many individuals of the highly educated classes heartily approved of Mr. Owen's experiment, and publicly witnessed to its success.

Mr. Abram Combe, a man of judgment and ability, deemed that success so great as to warrant *his* following in the same course, and accordingly he established at Orbiston, Lanarkshire, a community founded on Mr. Owen's principles, and in which he ventured the whole of his own capital, and induced friends to venture theirs. The sum expended in the erection of houses, schools, and workshops amounted to £36,000, and accommodation was provided for three hundred persons. Clearly, Mr. Combe was strongly impressed with Mr. Owen's success.†

---

* "Threading my Way," R. Dale Owen, chap. iii.
† This scheme ultimately failed, from various causes, after Mr. Abram Combe's death. See "Social Systems," by Mary Hennell, p. 136.

Meanwhile, Mr. Owen had been hampered in his work at New Lanark by the unsympathetic nature of business partners, who aimed principally at financial success, and disliked the diminution of profits arising from the wider aims of the noble man, who sought to combine the moral elevation of the labourers with the material interests of the employers of labour. When he freed himself from this element, another of an equally hampering kind presented itself. A benevolent Quaker became his partner, whose conscientious earnestness proved obstructive. Between Mr. Allen and Mr. Owen there was much unity of feeling, but also much disparity of thought; for Mr. Allen held strictly to the rules set forth in the text-book of early Quakerism—a book which says, "games and sports, plays, dancing, etc., consist not with the gravity and godly fear which the gospel calls for;" whereas Mr. Owen believed in the beneficial action of innocent recreation of *every* kind, and especially desired to promote music and dancing. Mr. Owen's enthusiastic zeal for the benefit of his race made him impatient of restraint, and the irritation caused by the constant opposition to his plans offered by men who only partially understood and sympathized with him was a large factor in deciding him to withdraw from New Lanark and embark on fresh schemes, which were prematurely and somewhat hastily organized, and ultimately resulted in discrediting his principles and bringing them into public disfavour.

The offer of a village ready built in a country where the expression of thought was free was an overpowering temptation to a man whose *ruling desire* was a vast theatre on which to try *all* his plans of social reform, unhampered by adverse opinions and by unsympathetic coadjutors; and the young life beside him threw weight into the scale. "Well, Robert," said Mr. Owen to his eldest son, "what say *you*—New Lanark or Harmony?" to which the lad answered without hesitation, "*Harmony;*" for novelty had its attractions, and the unsophisticated youth was propelled towards the new country by an inward force of which he then said nothing, but in later years wrote, "I thought if our family settled in Western America it would facilitate my marriage with Jessie." The purchase of the Harmony property in Indiana was completed, and a society formed there, which, as the above-mentioned Robert tells, included "waifs and strays—men and women of crude, ill-conditioned, extravagant notions; nay, worse, vagrants looking out for pecuniary gain or a convenient cloak for immoral demeanour." * No wonder, indeed, that what Mr. Owen himself called "the great experiment to ascertain whether a large heterogeneous mass of persons collected by chance can be amalgamated into one community" proved a downright failure; and Mr. Robert candidly tells that his father made a grave mistake in changing the scene of his experiments from New Lanark, Scotland, to the United States.

Later, an experiment was made on the estate of Tytherly, in Hampshire, England, but without success; and Mr. Owen died in 1858, without having permanently effected for any great group of individuals the change in the social and industrial structure of their life which *he* conceived to be necessary. Nevertheless, his work at New Lanark proved that association, gentle treatment, and the introduction of self-discipline will work wonders in the elevation of the masses; that children may be well trained and educated without recourse to punishments of any kind, and without the artificial stimulus of rewards; that, given a market for their produce, the labour of a group of steady, industrious human beings is sufficiently profitable (without pressure of long hours or overexertion) to supply, first, for the adults, all the necessaries and comforts of life and recreational pleasures,—second, for the children, abundant attention and complete education,—and third, an equitable return for the use of capital to carry on the primary industry.

Since Mr. Owen's day, the above propositions have been tested and proved in the case of the Ralahine Co-operative Association and Farm, established in 1831, and successfully conducted by Mr. Craig until the gambling propensities of the landlord brought ruin upon himself, and caused the estate to be sold off and the happy community to be broken up. For information as to this experiment, I must refer my readers to Mr. E. T. Craig's published work.† And at the present time, when unhappy Ireland is again causing alarm and rousing within our breasts mingled indignation and sympathy, the picture given there of the nature of the Celtic race, its docility under gentle treatment, and the marvellous change effected by co-operation in labour and close association in intercourse, is in itself a valuable lesson, which thoughtful English minds would do well to ponder.

Turning, however, now from these great experiments, in

* "Threading my Way," p. 228.
† "A History of Ralahine," published by Trübner and Co.

which hundreds of social units are implicated and enormous sums of money are necessarily risked, we inquire what has the British race done in a much smaller way to give fresh vigour to its daily existence and improve the structure of its social life? Spontaneity is of the very essence of a high civilization, and originality is *certain* to be manifest in any nation brimful of healthy vital energy. What we see, however, within our own nation speaks but little for either its high civilization or its vital energy. I should like, for instance, to know how many individuals have feared to enter into married life for the simple reason that the *customary* habits in that relation appeared unsatisfactory, and yet in how few cases has any trifling alteration in method been adopted!

The noble Mary Wollstonecraft and her philosophic husband, William Godwin, give us one little example of originality in the *minor* arrangements of domestic life, on which so much of individual comfort and happiness depend. Mary Wollstonecraft's extreme sensitiveness and quickness of temper were likely to be tried by Godwin's bachelor habits and far more callous nature. To guard the precious love that held them together from too great a strain in the early days of wedded life, before habit had brought about mutual adaptation, was to their bright intellects the wisest policy. When they married, therefore, "Godwin took rooms in a house about twenty doors from that in the Polygon, Somers Town, which was their joint home." To these rooms he repaired as soon as he rose in the morning, rarely even breakfasting at the Polygon, and here also he often slept. Each was engaged in his or her own literary occupations, and they seldom met, unless they walked together, till dinner-time each day.* "We agreed also," says Godwin, "in condemning the notion prevalent in many situations in life that a man and his wife cannot visit in mixed society but in company with each other, and we rather sought occasions of deviating from than complying with this rule." Mrs. Godwin's sentiments are expressed in a letter to her husband during his first absence from home with a delicious frankness and childlike simplicity. "I am not fatigued with solitude," she says, "yet I have not relished my solitary dinner. A husband is a convenient part of the furniture of a house, *unless he be a clumsy fixture.* I wish you from my soul to be *rivetted* in my heart, but I do not desire to

* "William Godwin, his Friends and Contemporaries," vol. i. p. 233.

have you always at my elbow, although at this moment I should not care if you were. Yours truly and tenderly, Mary."

The universal custom in this country of setting up a separate and often expensive establishment at marriage has this twofold effect. Thousands of young men, who desire at two or three and twenty to marry, delay for years, and in the interim suffer so greatly as to hurt the health, temper, and disposition, and render them at last unfit to become open-hearted, generous-minded, warmly sympathetic husbands. On the other hand, a girl is taken from a domestic life without responsibility (perhaps out of a merry group of sisters and brothers), and planted in her separate establishment, where, if her husband is a commercial or professional man and out all day, she feels very desolate. The management of servants, to which she is unaccustomed, worries her, the hours of solitude depress her, and when the married pair come together in the evening, the mental atmosphere surrounding them is one of fretfulness, vague discomfort, or nerve fatigue. This may seem an exaggerated picture. But I allege that thoughtful middle-aged married ladies have frequently remarked to me, that the *first years* of *married life* are always *more or less miserable* to a *girl*, and have spoken as though *this* were a *necessity* of *nature* to be calmly faced and *acquiesced in*.

During my whole life, I have only in one single instance seen a new method adopted. A young man was encouraged to marry early and bring his bride into his father's large and commodious dwelling-house, where the young couple remained until their family had become numerous, and the pecuniary position admitted of a separate establishment. Whether this arrangement was satisfactory I cannot tell. That would depend, of course, upon the dispositions of the various members of the household. In associated homes, happiness is sure to be *proportionate* to the *absence* of jealousy, dominancy, and all anti-social emotions, and the *presence* of the kindly, gentle, and sympathetic feelings. But to my mind, the experiment was a noble one, indicating elevated character on the part of father, mother, son, and daughter-in-law; and such experiments are in the direct line of social progress. If the *drawbacks can* be overcome, conceive how *great* are the advantages of such associated homes. Marriage would be made possible at an early period of life, without improvidence and without overwhelming care and anxiety. A young man's whole emotional nature might be satisfied, his tender domestic qualities exercised before the shadow of remorse or disappointment had touched him, or any hard struggle for existence had embittered his nature. A girl would be protected from worry and dreary monotony in the first years of her married life, and would have leisure to fill up her education, and adapt her intellectual studies to her husband's tastes, prompted by affection and the desire to think and feel *as he* does; she would have leisure also to devote herself to all the sacred joys and duties of motherhood. These advantages, weighed in the balance against the mere honour and glory of separate establishments—the highly extolled British manliness that must be master in its own house—are *worth* contending for, worth putting forth our highest efforts to obtain.

Experiments of the kind may fail. One single semi-barbarous individual may mar the harmony of an entire household; nevertheless, it is in the above direction that reason and hope will continue to point; for observe, as humanity becomes more and more civilized, the anti-social emotions *all* tend to disappear, and happiness results from the exercise of all *social* feelings, including the emotional, the intellectual, the æsthetic. Now, the exercise of these is absolutely dependent upon intimate intercourse and association; *ergo*, associated homes are in the line of social progress. Again, as humanity becomes more and more civilized, the social feelings grow stronger, the craving for sympathetic relations of every kind more intense, and solitude more painful and depressing; *ergo*, to persist in a line of action which produces *isolation* and *desolation* is contrary to the laws of our own being, it is adverse to progress, it tends to individual misery and to social decay.

But to return to experiments. The custom of a tour immediately after marriage, a honeymoon spent in wandering from place to place, is all but universally practised by the British, and all but equally universally condemned! by some individuals on the ground of its discomfort, by others on the ground of its expense. The Americans frequently discard this practice and spend their honeymoon quietly at home, delaying travel till adaptation to the new life has been established, and a thorough knowledge of each other's tastes and habits fit the married pair for encountering the little irritations and discomforts of an extended tour. With what result is prettily, and I think truly, told in Mr. Howell's charming little sketch of "A Wedding Journey."

But John Bull, in his manly independence, keeps steadily to the prevailing fashion. Custom with him is always *best*, and to alter an old habit implies admission that John Bull is capable of error. It hurts his manly pride—a pride which, alas! makes him callous to individual suffering under the galling thraldom of the tyranny of custom. The British race has bowed the neck and bent the knee to Mrs. Grundy. She is the Moloch upon whose altar John Bull is ready to sacrifice the happiness of his children and his children's children. And when the breath of unsatisfied desires disturbs his equanimity, the impression is momentary; he will not let it seriously impair his self-complacency or destroy his peace, still less will he alter one iota of his practice. He hugs to himself the fond belief that British ways are admirable ; that the habits and customs of British social life are perfect, or nearly so ; that national wealth proves national prosperity and health ; and he, John Bull, may laugh to scorn the warning given by his high-souled, pure-minded son, whose presence has passed from us—" that so few now dare to be eccentric marks the chief danger of the time." *

* Essay on " Liberty," chap. iii.

CHAPTER XIV.

LEGISLATIVE *VERSUS* VOLUNTARY METHODS OF REFORM.

"Though the truth may not be felt or generally acknowledged for generations to come, the only school of genuine moral sentiment is society between equals."—J. S. MILL.

AT the close of the seventeenth or the beginning of the eighteenth century, a fencing-master named Machrie introduced into Scotland the old Greek game of cock-fighting. In England, it was a fashionable amusement at the period, and Henry VIII. had built a cock-pit at Whitehall. James I. was accustomed to amuse himself with cock-fighting twice a week. It was not regarded as more inhuman than hunting, coursing, or shooting, and Machrie was looked upon as a benefactor to Scotland for having started a new, cheap, and innocent amusement. He had written, "in 1705, 'an Essay on the Innocent and Royal Recreation and Art of Cocking,' in which he expressed his hope that 'in cock-war village may be engaged against village, city against city, kingdom against kingdom, nay, the father against the son, until all the wars of Europe, wherein so much innocent Christian blood is spilt, be turned into the innocent pastime of cocking.' The Welsh main, which was the most sanguinary form of the amusement, was exclusively English and of modern origin. . . . As many as sixteen cocks were sometimes matched against each other, at each side, and they fought till all on one side were killed. The victors were then divided and fought, and the process repeated till but a single cock remained. County engaged county in cocking matches, and the church bells are said to have been sometimes rung in honour of the victor in the Welsh main." *

* "History of England in the Eighteenth Century," Lecky, vol. i. p. 554.

The Royal Academy Exhibition for the year 1881 contained a little picture which depicts this "innocent" amusement of the past. Three men, whose laced and ruffled garments betoken high birth, are eagerly watching two miserable cocks, engaged in furiously pecking each other to death. The contrast between the elaborate costume of these fine gentlemen and the purely savage instincts expressed in attitude and face is very striking, and as I stood before it I realized the great change that has taken place in humanity itself during the lapse of a single century. Machrie was probably right. The thirst for blood was better appeased in witnessing the suffering of brutes than the suffering of mankind; and in those days, the appetite was too strong and deep to dwindle into nothingness by simple *disuse*. Rather like a pent-up torrent, it would have broken barrier and forced its way on some destructive course. *Cocking*, therefore, was as a safety-valve for the protection of mankind.

Our position *now* is wholly altered. Except in the case of isolated individuals, the race is comparatively free from bloodthirsty cruelty. Its destructiveness and all predatory instincts are slowly dying out, and *inaction* is concerning them, the proper and masterly policy of the nineteenth century. It is true that some great teachers of the present day think otherwise, and, as I have already pointed out in Chapter II., would move upon Machrie's lines still, and, in the case of little children, do "evil that good may come." The error is a grave one. To excite a child's anti-social emotions over the spectacle of a cat destroying a mouse, and thereby interest him in a reading lesson, is training of a *vicious* kind in reference to his future social life, and unworthy of wise men later than the *eighteenth* century, when outward and inward conditions differed widely from present conditions.

But in reference to anti-social instincts, it is not only in childhood that the policy of *inaction* should be carefully observed. In adult life, warlike tendencies exercised injure character; therefore, society ought to relieve individuals from all necessity to fight for protection of their rights; and with growing rationality it will, in time, be generally recognized that this policy condemns the whole competitive system on which our industrial life is almost universally based. Keen competition means the rousing of a man's rapacity and greed, and frequently a belligerent attitude of mind towards fellow-competitors whose lines of pursuit converge with his. It means vicious joy upon the overthrow of competitors, and sharp desire to drive them one and all from the field. It means the growth of selfish monopoly; and what it displays is a pitiable illustration of the old saying, "Every one for himself, and the devil take the hindmost." As a result of the system, we have a million paupers fed at public cost, and about half a million of criminals who feed themselves or are fed at public cost. We have also a numerous, powerful class of human beings whose success in industrial life has degraded their humanity, who, in short, by the constant exercise of barbarous instincts, have been kept at the level of eighteenth-century social units.

Is there a practical, decisive remedy for this evil state? Does any *one* force exist competent to sweep away competition, with every condition adverse to man's integrity, and establish institutions which would at once secure the material prosperity and moral well-being of an imperfect, an unequally developed race? The force of legislation is very generally regarded as a powerful factor that will remove mountains and accomplish what man, in his strong faith, requires of it. The present age is pre-eminently a political one, and it is to government measures that all classes alike look for amelioration of their miseries and for aid in social reform.

The lower classes, but recently roused from the lethargy of ignorance, look upon *bad laws* as the fundamental cause of suffering, and in every emergency they clamour for fresh legislation to remedy an evil. Those of the people who call themselves "radical reformers," mean by the term reformers of law and government. Their special aims are the suppression of perpetual pensions, of laws of entail, and other abuses; in short, the clearing out of survivals of a past epoch from our legislative code, and adapting our code to present conditions. All great secular leaders of the people have been politicians, and although the extension of the Franchise, the Free-trade movements,* and other important legislative reforms have in a great measure disappointed public expectation, and clearly shown themselves to be only very slightly beneficial, the tendency persists to anticipate redress of *every* grievance and reception of every benefit through the direct action of government.

* For twenty years previous to the repeal of the corn laws, the average price of corn was 56s. 4d. per bushel. The average price for twenty years after (1850-1869) was 52s. 7d. Showing that, despite the very large increase in the quantity of corn imported, the pressure of population has kept up the price of food to almost the same amount.

A large class of individuals, calling themselves " socialists," have grasped the fact that co-operation is destined to supersede competition, and form ultimately the great spinal cord of the industrial social organism of the future. Now, the majority of these socialists* rely upon government action for the execution of their purpose. The old-world method of *external authority* is what they trust to in the inauguration of their new-world *régime*. In them, no less than in the ignorant populace, this mental tendency exists, to look to Government as the good fairy which can reduce disorder to order, and, touching with her finger the secret springs of happiness, can compel them to burst forth in fresh life-giving streams. If *this* is a delusion, how comes it that the delusion is almost universal, and has its hold on active, energetic natures as well as on inert slow-moving ones?

Looked at in relation to a perfect Humanity, external government and legal restraints are, of course, in their nature, simply *evil*. They are at best clumsy mechanism for the strengthening and regulating of a society which has latent capacity for self-regulation and self-restraint, and which will ultimately, by internal, not external, forces, hold its units in a firm embrace and keep itself upright. Scaffolding, everywhere conspicuous in the process of building, disappears when the edifice is complete; and evolution, in raising man out of barbarous degradation and lifting him step by step to a high, noble civilization, pursues a course destined to give him stability, independent of external support, and finally to abolish repressive government and the whole machinery of arbitrary authority off the face of the earth.

It is absolutely essential that we should perceive the path of evolution, if we propose, with open-eyed intelligence, to aid evolution. I have in previous chapters used a figurative expression, viz. that we may take the forces of evolution into our hands, and apply them to our purpose in hastening the rapidity of their action. But we can never alter the path or direction of advance, and unless our action is in harmony with, or (let us say) sets in the same direction as, the main force, we shall only foolishly oppose nature and stultify our own pigmy efforts. The *main force* of progression is, and must long continue to be, purely automatic—a mighty irresistible move-

* The English Co-operation Union differs from continental socialism. "Our co-operators," says Mr. Thomas Hughes, "do not ask the State to do anything for them, beyond giving them a fair field and standing aside while they do their own work in it, in their own way."—" A Manual for Co-operators," preface, page xiii.

ment of unconscious evolution, tending in a definite direction. It is the part of conscious intelligence, as an *evolution agency*, to ascertain this direction and study carefully the numerous, varied subsidiary forces that play important, although unconscious, parts in the general phenomena.

By study of our national history, we can clearly see the order of evolution in the government and legal institutions of a slowly civilizing race. In England, the government has passed through the purely monarchical and aristocratic, or oligarchical, forms, and is now steadily approaching the democratic form.

The government of the aristocracy ceased at the Reform Bill of 1832; and now, in *theory*, the people govern themselves, whilst in reality the people are mere puppets, forming part, it is true, of the outward mechanism of government, but answering unconsciously to the pull of the strings held by their political leaders, or by the wealthy and often the unscrupulous, who have voice enough to render them simply popular. "The conclusion in regard to the republican form of government which I draw," says Mr. George Combe, " is that no people is fit for it in whom the moral and intellectual organs are not largely developed, and in whom also they are not generally and extensively cultivated. The reason is clear. The propensities being all selfish, any talented leader who will address himself strongly to the interests and prejudices of an ignorant people will carry their suffrages to any scheme which he may propose, and he will speedily render himself a dictator and them slaves." * Again, the bribery and corruption that occur occasionally at elections, and the self-interested votes of subservient electors, disclose the secret but potent rule of *money* in the government of our country.

Notwithstanding all this, "Democracy must advance. We cannot put the clock of time back; we must go on, and *this* is becoming generally recognized. 'Is it credible,' asks De Tocqueville, 'that the democracy which has annihilated the feudal system and vanquished kings will respect the bourgeois and the capitalist?'" † No, it is *not* credible. The order of unconscious evolution is definite and clear. From despotism and irresponsible authority, through responsible authority to rational freedom; from tyranny and coercion to liberty of action and self-control, moves on the great stream of things;— and those reformers who would extend and intensify State govern-

---

* "Moral Philosophy," p. 375.
† "Science of Man," Charles Bray, p. 260.

ment, who would, in its present unscientific condition, invite its further interference with private life and give it more of arbitrary control, are setting themselves against, not with, the stream. They may retard, but they will certainly not hasten progress. State socialism may be tried. It will never succeed. And the Comtist system is also but an abortive effort, noble in conception, beautiful in imagination, but opposed to the irresistible flow of unconscious evolution going on around us.

Comtism rejects the force of co-operation and rests its scheme of life, social and economical, upon the force of domination. In this system, the cultured are to rule with gentle but despotic authority. Wealth must not even tend to equality of distribution. It must be concentrated in a small number of holders, who will regulate production, so that the existing amount may somewhat more than reproduce itself, and so that a share sufficient to permit of healthy life falls to the lot of every producer. Capital will rule, but the capitalist, cultured, refined, and morally inspired by the "religion of humanity," will wield his sceptre with the wisdom of Socrates, the justice of Aristides, the tenderness of Christ. Now, domination of the good is no doubt a lesser evil than domination of the unscrupulous and bad, in the same way as cock-fighting is a lesser evil than war. But, as I have shown in regard to cock-fighting, we are now advanced beyond the necessity to replace a major by a minor evil, and if we consciously strive to replace the despotism of kings, or of a corrupt oligarchy, by the despotism of the very salt of the earth—the *élite* of mankind—we are acting similarly to Machrie in the eighteenth century, and out of all fitness to the more advanced and rational epoch in which we live.

Apart from Comtists and Socialists, there are reformers who build *their* hopes for humanity upon an equal distribution of the land, and who long for State interference in this matter, and would not hesitate to sanction such an exercise of dominant power by Government as the abolishing of landlordism at one fell swoop.

No American book on social problems has been more widely read in this country than "Progress and Poverty." In addition to its literary merit, the striking truthfulness of Mr. George's powerful and feelingly rendered picture of the tremendous evils of our present state, and the social miseries that everywhere surround us, and that seem the *natural*, because the as yet invariable accompaniment of a nation's growth in wealth, explains its popularity. But Mr. George's argument throughout is *vitiated* by his denial of the law of population and his palpable ignoring of the *effects* of that law of nature; and the fact that, in spite of this, he makes disciples amongst the thoughtful and intelligent, can only, I think, be explained by the *bias* of the general mind—that is, the *political tendencies* of the age. His remedy of State interference kindles enthusiasm with a magnetic touch, and the slow march of evolution is forgotten in the keen excitement of an honest but misdirected public spirit.

It seems a gross contravention of the ethical principle of *equivalence* of *rights* that £66,000,000,* produced by the labour of one million six hundred and fifty-seven thousand of our fellow-countrymen, should annually pass into the pockets of a small class of landlords, whilst one and a half million of agricultural labourers remain poverty-stricken, or are crowded out of Great Britain to seek a livelihood in some less densely populated land. To remedy this injustice and promote social equality (in which lies, he thinks, general happiness), the State would be justified in seizing rents and applying them to the relief of taxation. The moral position involved in this scheme let us here neglect, and grant for the moment that the injustice to landlords is swamped in the justice it awards to the far wider class of lowly toilers, who dig and plant from day to day in the sweat of their brow. The economic result, however, we must not overlook, and I give it in the words of an evolutionist, who looks more deeply into social questions than the brilliant, earnest, but superficial American. "In the present state of education and individual prudence, were the cultivators of the soil to begin upon a new system and in a high state of comfort, they would increase their marriages, and increase their children, so that in a few years poverty would come back and so many would be living on each farm that all their means would be used, as at present, in satisfying the most urgent cravings of nature." † I need say nothing in support of this position, since, in Chapters VII. and VIII. of this work, the question of population has been fully discussed. But this is not all—a relapse into poverty is not the *only* sad outlook unavoidably attached to Mr. George's scheme of reform.

Looked at in the *abstract, landlordism*, i.e. absolute posses-

---

* "In the opinion of a well-known statistician, the annual rent of the agricultural land in this country is about £66,000,000."—H. FAWCETT, M.P.

† G. A. Gaskell, in a lecture to working men.

sion of the soil of a country by a very select number of the social units born into that country, is essentially and peculiarly unjust, and the evolutionist steadily and confidently anticipates for the human race a *better* time in the future, when landlordism and all other arrangements of social *injustice* will cease to be. But turning from the absolute to the relative, landlordism has been in the past, and *is still* to a very great extent, an important factor of civilization; and its *sudden* removal would cause national ruin, because the people are not as yet sufficiently advanced to *value* civilization and to prudently preserve it *for themselves.*

The peasant proprietors in parts of France, where the state of education is low, have neither books nor prints in their cottages; civilization amongst them hardly exists. On the other hand, those (and there are undoubtedly some) who live up to a certain level of comfort and refinement are stimulated by an education which proceeds from another state of society than their own.

In this country, the sweeping measure of so-called reform advocated by Mr. George would tend to the *decrease*, perhaps the actual *loss*, of science, art, literature, and the whole movement would be a retrograde, not a progressive one. In a social system such as ours, where education has no hold upon the substratum, *i.e.* the masses at the base of our society, and culture has not even permeated the great middle section, for the present, and probably for many generations to come, the wave of evolution, deep, strong, widespread, *must* carry in its bosom the class distinctions and social gradations that at this moment prevail. The policy that is coercive and would cause a break or rent at any point in our present system is a mistaken policy, and these political tendencies of the age are a source of national weakness, not strength. The reason of their universal prevalence is not difficult to understand. In the order of evolution, that which is on the surface is always first perceived; and it is only when superficial measures are found to bring results that are trivial and disappointing that the hidden springs of life are diligently searched for, in the *deeper*, more *vital*, stratum, where they abundantly lie.

"It is clear, that in taking into consideration the institutions under which man can be made most happy, we must include his political as well as economical and social institutions." *
But, in the order of importance, the *first* stands *last* and

* " Science of Man," C. Bray.

the *last, first;* and apart from *all* institutions, the habits and customs of the people, the domestic and individual life, and the ordinary usages of society form the sphere in which *reform* is *most urgently required*, and where every effort put forth in a right direction will *unreservedly tell*. Therefore, instead of attacking classes and governments, if we, who see the truth, do our utmost in spreading knowledge of the causes of evil, and showing how and where our individual habits, our social life, our educational principles, may be improved, we shall be doing that which must hasten the advent of a happier state.

An enlightened, vigorous, warm-hearted teaching of the people is required. But a teaching of what? Not of theories that give a transcendental view of existence—not of an asceticism which by self-repression dulls or even kills faculties of human nature, that are endowments calculated to render existence innocently pleasurable ; but a teaching of the science of pure and happy life, mundane, not ethereal or saintly, but life healthfully egoistic, controlled by reason and by sympathetic instincts, and tending towards regeneration of race and the moral elevation of humanity.

A few pages back I asked this question—Does any one force exist competent to sweep away competition and other adverse conditions of life, and introduce a thorough radical reform? And since asking it, I have been engaged in showing that the Government action, State interference, legislative measures, etc., to which we all more or less instinctively turn an expectant gaze, are only broken reeds to lean upon, empty cisterns from which we shall draw no living water.

But now there is a force which, rightly directed, may accomplish much, and that is the force of *public spirit*. As an element in our present social life, that force is minute in quantity and ineffective in action, through diffusion and frictional waste. It wants nursing into vigorous strength and turning into channels of healthy exercise. But so long as politicians are all at cross purposes, and science has not even entered the domain of popular politics, there is no safe channel for public spirit in *that* direction, and we need hardly regret the small amount of the force. Our social units who are advanced in science are singularly devoid of *public spirit*, and yet this public spirit is essentially a form (if I may so express it) of the Religion of Humanity which must eventually *dominate* society. Science, in the first instance, attracts or dominates the human nature that is phlegmatic and cold in temperament. At a later stage only does the human nature that is warm and sanguine imbibe its spirit and adopt its methods. In the interim progress lingers. Scientists are contemptuous of the public spirit which prompts to action of an irrational, although benevolent and unselfish kind; and they omit the educating of that force in the childhood of their offspring. Whilst, on the other hand, individuals of enthusiastic temperament and unscientific intellect, in whom at present this precious force of public spirit chiefly centres, look timidly askance at science, seated at a high altitude and breathing an atmosphere of cool indifference. When these *two*, viz. science and public spirit, are closely, firmly united, progress will no longer linger—the halcyon days of man's future will commence.

The union I speak of has been visible for years in the life and work of Monsieur Godin, of Guise, that eminent Frenchman who, in building up his own fortune, has steadily ministered to the well-being of his fellow-countrymen ; and in climbing the hill of difficulty that leads to wealth, has carried with him, step by step, the labourers without whose aid he could have accomplished nothing; and has finally placed at least a thousand of them in surroundings of comfort, security, and independence. At the age of seventeen, Monsieur Godin laboured as an apprentice in the iron trade, from five in the morning till eight at night. "I saw," he says, "then, the misery and wants of the workman laid bare, and I resolved, that if ever I rose above the position of workman, I would endeavour to make the working man's life more tolerable and satisfactory to him, and to raise labour from its degradation." Here was the prompting of a *public spirit* which, guided and controlled by a rational and scientific intelligence, ultimately realized the generous, glorious aspirations of youth. My reader will find a detailed account of the Familistère, and a history of its founder, in a lecture, published by Macmillan and Co., London, called "Associated Homes," by Edward Vansittart Neale. But the principles set forth by Monsieur Godin, I must refer to here.

Moral order, Monsieur Godin considered, is ultimately bound up with material organization ; and so long as institutions fitted to conduct men to the practice of goodness do not exist, *neither will moral order*. If a poor man's wife is to be a model of home virtues, cleanliness, tidiness, etc., she must not be compelled to wash and dress linen in the one little chamber where she cooks and eats her daily food. If she is to be a model of good temper, never slapping or abusing her children, she must not be subjected to irritating interruptions at her work. The rich man's nursery performs a function which is equally essential in the poor man's case. Children have to be safely and happily disposed of, whilst parents of toil, as well as parents of luxury and leisure, are otherwise occupied. Again, if the poor man is to behave like a gentleman, he must, like the latter, have surroundings of comfort and happiness within his home, and some security from the wear and tear of constant anxiety regarding the daily bread of himself and his family. In short, the working class, although not rich, ought to have the equivalents of riches, or, in other words, *advantages analogous* to those which fortune grants.

Now, *is this possible?* Monsieur Godin has not simply asserted, he has *proved* that this is possible, and to be carried out by two means—first, the use of capital (which is not *lost* to the capitalist); second, the formation of a unitary or associated home. In this home, every appliance which the wealthy man possesses for his own individual service is also at the service of the poor man, but is not possessed by him individually, only, in common with a group of fellow-workmen and by mutual accomodation, applied to his individual requirements.

The Familistère was built gradually, as Monsieur Godin could afford the outlay. Not being able, as he said, to change the hut or garret of each workman's family into a palace, he has placed the dwelling of the worker *in* a palace. The Familistère is nothing else. It is the Palace of Labour—the Social Palace of the future. The industrious poor man has there a convenient dwelling with the same resources and advantages with which the abode of the rich is provided, and with common institutions which replace those services that the rich derive from domestic servants. The necessary space for each family, the necessary heat along with thorough ventilation, the sufficient water supply, the appliances for rapid disposal of refuse, and so on, along with general baths, washhouses, and laundries, general day nurseries, infant schools, etc., general hospital, and staff for tending the sick, general appliances for amusement and relaxation—all these under the same roof, or close at hand, may be seen at the Familistère, and are obtainable elsewhere without detriment to the capitalist, and without the labourer differing in any essential particular from the great mass of our own industrial population.

The east wing of the Familistère was fully occupied in 1861, and the central building was completed and occupied in the year 1865. In 1874 Monsieur Godin wrote: "For the edification of those who believe that the working classes are undisciplined or undisciplinable, I must say that there has not been in the Familistère since its foundation a single police case, and yet the palace contains nine hundred persons: meetings in it are frequent and numerous, and the most active intercourse and relations exist among all the inhabitants.*

Monsieur Godin has taken a part in the political life of his country, but he has lived in his Palace of Labour. His scientific intellect applied to the improvement of the "*grande industrie*" on which his economic success depended, made various important discoveries and inventions, and in the *benefit* his workmen have always *shared*. He was head and brain of the *grande industrie*, they the hands or limbs; and as in a material organism these members are vitally united, and suffer or prosper *together*, so in this social organism vitalized by a pure and noble *psyche*, viz. *public spirit*, all the members have alike reaped health, comfort, and happiness through the associated action of their distinctive and varied functions. The animating psyche has extended to every member of the social body, and good order is maintained by no despotism or tyranny—no external, arbitrary authority. In Mr. Neale's words, "The life of the Familistère is one of carefully guarded individual liberty, which is prevented from degenerating into license simply by the influence of public opinion among its inhabitants, who, administering their own internal affairs as a united body, exercise a disciplinary action upon each other." † Behold, then, a realization, in a small degree, of the forecast given in my previous chapters, of orderly social life resting upon principles of abstract equity and individual liberty—a life which, made happy in its freedom, blossoms into personal dignity, warms into sympathetic jealousy for the rights of all, and cheerfully pays the price of liberty in an eternal vigilance.

The administration is conducted by committees elected by general suffrage. Thus, there are committees for provident purposes, for medicaments, for medical attendance, for the firemen (who, by the completeness of their organization, and their residence in the building, make a large fire almost impossible), for music, for festivals and pleasures, for the club, the library, the claims of workers. In addition to these committees, there are two general councils of twelve men, elected by all the

* "Associated Homes," by E. V. Neale, p. xi. † Ibid.

males, and twelve women, elected by all the females above sixteen; and the functions of these councils are described as initiatory and observant, with no special limit assigned to either of them. The council of men busies itself principally with questions of improvement of work, the provident institutions, the organization of festivals, and other matters; whilst the council of women quite naturally occupies itself with the domestic functions, the quality of articles sold at the Familistère, the care bestowed on infancy, the general cleanliness, and all improvements which may assist the household management. Both councils note, either zeal, or exactness, or forgetfulness and negligence in those employed in any office. They hold meetings whenever they think fit. They give advice on the conduct of whatever is going on. They serve as a lever of moral influence, and greatly promote the regular course of the general interests both in the factory and in the unitary home.

Attention was called in this country to Monsieur Godin's noble experiment as early as the year 1865. The *Times*, the *Social Science Journal*, *Review*, *All the Year Round*, and other public prints, opened their columns to descriptions of the Familistère. The twenty years that have since elapsed, however, have seen no similar attempt in England, and this seems strange when we call to mind our national benevolence, our numerous wealthy capitalists, our energetic social reformers. True, we have not, I think, amongst these, men of genius such as Monsieur Godin; for, as I have pointed out, the weight of conventional custom in Great Britain—the slavish submission to the tyranny of fashion, "the colossal uniformity, the mechanical motion of that inexorable London," and of the innumerable minor provincial social rings, that swing round in exact conformity with, and imitation of, that inexorable London! —all these, I say, tend to crush originality, to repress spontaneity, to render the birth and development of genius *rare* and *difficult*. To the genuine John Bull, of haughty deportment, and of submissively conventional mind, Monsieur Godin's action of living in close proximity to the labouring class, and assuming that the labour of guidance and direction is, although different in kind, in no way worthy of *greater* reward than the manual labour, without which *it* would be worthless, savours of eccentricity; and what the British rich man *hates* is eccentricity.

Our capitalists amass fortunes, in some perhaps undesirable manufacturing district, or perhaps in distant India, and ultimately drift to London, to live in luxury, and to aid in keeping up the weary mill-round of frivolity and selfishness, which in Great Britain passes for a life of social independence, ease, and happiness. There are in London thousands of houses, at rents of £300, £500, £1000 a year, or more; and when one traverses miles covered with the dwellings of the rich, and not a single poor man's abode apparently within reach, one perceives how far apart the social classes are, and what an enormous amount of the yearly toil of the working classes goes for the support of the idle rich, who live in magnificent mansions, whilst they remain miserably lodged and fed. Palaces for the rich is our British method, but no palaces of labour for the poor. Instead of Familistères there are—I use Miss Ellice Hopkins' words—"extensive districts in London, Liverpool, and all our large towns, where our people are living in little more than half the area of ground required for a corpse, and which could be claimed for them if they were dead, in tenements which are the graves of all decency and chastity." *

But now let us estimate in a rough way how much labour of other people each rich man absorbs year by year according to the income that he spends. Professor Leone Levi estimates that the average wage of a man in this country is 24s. a week, while that of a woman is much less. Let us, however, put it at 25s., which amounts to nearly £65 a year, and behold we find that fifteen persons are labouring for every £1000 spent per annum by the rich man on his domestic living. But an income of £1000 is a very small one in the West End of London. £5000 might be an average, and £5000 means no less than the entire labour for a year of *seventy-five* persons absorbed by one man and his family.

This calculation may seem strange to a reader who does not understand the intricacies of our economic system, but it is no exaggeration, and let us not think that the idle rich man gives any adequate return for the poor man's labour. We have undoubtedly some legislators and rich men who are working for the general good, and *these* repay in *kind* the poor man's labour; for they give to society personal exertion, the work of the brain in the sweat of the brow. But the British estimate of all such labour as compared with the manual labour of the poorer classes differs essentially from Monsieur Godinas The British view is this. All labour that is intellectual or highly responsible and trustworthy, deserves at all times.

* "Social Wreckage," *Contemporary Review*, July, 1883.

reward of greater happiness than the simplest and most non-intellectual labour. Observe, however, it is the intellectual and trustworthy themselves who are the zealous upholders of this opinion. It sways, in fact, a society in which the *others*, I mean the unintellectual poor, are obscure and dumb, and judged of in their absence. Is this opinion, then, to be trusted? Is it the unbiassed outcome of the civilized intellect in union with the moral sentiments appropriate to our epoch? An evolutionist must unhesitatingly answer *no*. The delicate susceptibility to quantitative relations, the moral sense of justice which pulsates in an embryonic state within humanity, but which will by-and-by expand and strengthen, till it breathes in vigorous maturity, and dominates the feelings of mankind— that infant virtue will shrink, turn from, and utterly repudiate the opinion.

Justice *feels* that the clod-hopper working heartily for an hour at digging potatoes, is as much deserving of happiness, as the philosopher, who for an hour strains every nerve to solve a problem in social phenomena. Justice *sees* that "all progress, all diminution of misery and increase of happiness, is in direct proportion to the utilization of the various sorts of capital—physical, intellectual, and moral—land, money, muscles, brains, hearts, which we possess."* Our philosopher's work is intellectual—*he* supplies us with brain power. Our clod-hopper's work is muscular—*he* supplies us with material force. Both are alike necessary to progress, therefore society ought not to mete out to one a share of happiness represented by pounds, to the other a share of happiness represented by pence, in return for one hour of conscientious hearty work according to the ability of each.

Nature, to be sure, has endowed the one with a high order of capacity, and the other with a low order of capacity, and their enjoyment also is of a different order. The philosopher cares for "intellectual ambrosia," whilst the clod-hopper thinks his beer and porridge quite as nice! But there is no reason why society should increase these inequalities. Its nobler mission is to reward with impartial justice and scrupulous equality the labour of all men which is conducive to the general well-being.

Again, the wealthy merchant's time is spoken of as "very valuable," the operative's is treated as of comparatively insig-

* "The Responsibilities of Unbelief," Vernon Lee, *Contemporary Review*, May, 1883.

nificant value. The latter for his ten or twelve hours of labour earns the average wage of twenty-four shillings a week. The former for *his* three or four hours of daily work often reaps, and always considers himself entitled to reap, a far higher reward, and this sentiment is in a great measure the cause of the frequent failure of co-operative production schemes in this country. The leaders *want more pay* than the men. Monsieur Godin required a fair return for the use of his capital. But the value of the time he spent in the directing and guiding of operatives and in the exercise of his scientific faculty (through execution of the high order of work which suited his personal capacity) was computed by him in accordance with a dictum of unselfish *justice*. His action has been to apply the profits of associated work for the equal benefit of *all* the associated workers.

The sentiment which controls the action of ordinary men, which tends to unequal distribution of wealth, and which I have been engaged in examining, has had its distinct use in evolution. But just as, in Chapter IX., I showed that the sentiment of jealousy had in the past aided in evolving *constancy*, and yet, in its essential nature, is thoroughly anti-social, therefore destined to disappear, so we have here a transitory form of human thought, a mood of human feeling, certain to give way and be submerged, as the gradual upheaval of humanity takes place from the lower level to the nobler elevation, the purer atmosphere of a region where reason dominates and social morality sways the daily conduct of mankind.

And now the subject-matter of this chapter is easily summed up. Enthusiastic philanthropy is doomed to disappointment if its hopes of remedy for all the great evils of society are placed in acts of Parliament, State interference, and State superintendence. "The people," says Mr. Herbert Spencer, "will be morally injured by the State doing things for them instead of leaving them to do things for themselves. In proportion as the members (of a nation) are little helped by extraneous power they will become self-helping, and in proportion as they are much helped they will become helpless. And this helplessness shows itself in a retardation of all social growths requiring self-confidence in the people—in a timidity that fears all difficulties not before encountered—in a thoughtless contentment with things as they are." If my reader feels stirring within him the great soul of our future social life—a pure, unselfish public spirit—let him beware of retarding, nay, let him aid in the creation of fresh vigorous forms of social institutions—institutions based on equality and freedom, capable of spontaneous growth, of untrammelled development from within. If he do this, then, by certain laws of evolution, as sure in their action as the law of gravitation, he will aid the elevation of humanity through growth of moral sentiment; for, as Mr. Mill truly says, "The only school of genuine moral sentiment is society between equals."

CHAPTER XV.

HOME, SWEET HOME!

" Two faces o'er a cradle bent:
Two hands above the head were locked;
These pressed each other while they rocked;
Those watched a life which love had sent.
      O solemn hour!
      O hidden power!

" Two parents by the evening fire:
The re l light fell about their knees,
On heads that rose by slow degrees
Like buds upon the lily spire.
      O patient life!
      O tender strife!

" The two still sat together there,
The red light shone about their knees;
But all the heads by slow degrees
Had gone and left that lonely pair.
      O voyage fast!
      O vanished past!"

                              GEORGE ELIOT.

To the ordinary British mind there is no more sacred institution than that of *home*. We are proud of our domestic life, and ideally we worship the British home for its purity, privacy, and peace; its comfort, respectability, and independence; its venerable name and nature. John Bull is not a sentimental fellow, yet his eyes will fill with tender tears as he listens to the melody of " Home, sweet home! there is no place like home." Valued friends have said to me, "'Take care how you approach with cruel criticism our noble conception of the home;" and my own life has its far-off memories of a happy childhood's home, that would alone suffice to guard me from all hasty and flippant treatment of this hallowed subject. I am brought to it, nevertheless, in the course of my examination of our social condition, and although the ground I tread is holy, I can but advance, since I hold intellectually the firm conviction that rational argument will never injure truth, but, on the contrary, it will weaken error and strengthen truth.

The necessary elements of an ideal home are a father and mother of sound constitution and domestic habits, a large family of brothers and sisters affording companionship to one another, and wealth enough for comfortable or luxurious maintenance, abundant recreation, occasional travel, and the exercise of hospitality. Where these conditions exist, and, as age permits, the young inmates are launched into active life with suitable careers, these homes are sending forth the right material for repairing the waste of British life, and building up the nation. But alas for the home itself as the years roll on! If daughters marry its brightness becomes overshadowed. The rooms are emptied one by one, until at last out of that merry group of brothers and sisters perhaps but one remains to minister in the desolation of loneliness to parents whose infirmities of age give them a peculiar claim upon the tenderness and love of *all* their adult children. Observe, the typical British home makes no systematic provision for the solace of old age. It leaves that to haphazard; and although the genuine John Bull would scorn ideally to neglect his aged parents, the actual fact is, that in the hurry and bustle of life, he gives them but a few fragments of his time, and a faint fraction of his tenderness; and they too often pine and languish for the daily presence of the beings on whom their life's best affections have been centred. Even this ideal British home, then, is unstable and imperfect. It fails in respect of a paramount duty of life, viz. the duty incumbent upon the generation that is in its prime, to console, support, and cherish to the utmost the generation that is tottering on the brink of the grave.

But, turning from the ideal to the real, let us consider for a moment what proportion of the actual British homes make any approach to this ideal which we have described. Mr. Bright, in his rectorial address at Glasgow,* stated that in that wealthy city no less than "forty-one of every hundred families live in a single room, and that beyond these forty-one, thirty-seven families out of every hundred live in two rooms." More than two-thirds, then, of the British families in our flourishing

* Delivered in March, 1883.

Glasgow are without such necessary elements for comfortable living as space and air. We may safely surmise that to them home is not the abode of purity and peace.

Dr. Anderson, in his report to the Police Commissioners of Dundee on the ravages of fever, states that there are in Dundee 8620 houses of only one room each, in which there is a population of 23,670 persons, and 16,187 houses of two rooms, into which are crowded 74,374 men, women, and children. Among the population living in one or two rooms, 183 persons per 10,000 were attacked by fever, and rather more than 16 per 10,000 died during the year 1882. The "Bitter Cry of Outcast London" was sent forth to proclaim the abject condition of the homes of the poor, and the cry was re-echoed from Liverpool, Edinburgh, and other large cities. Family life in the dwellings there described is impossible upon lines that answer to any worthy ideal of a home; and the facts compel us to recognize that in outward conditions and inward sentiment a change is taking place, which indicates the approach of, and the necessity for, a *new social order*.

In the year 1878 the Committee of the American Social Science Association meeting at Cincinnati prepared a paper descriptive of the rapid and startling changes brought about in the United States, since the commencement of this century, by the use of labour-saving machinery. What is said there in reference to home life I shall transcribe. "Not long ago the farm found constant employment for all the sons of the farm, and many of the children of the city. Now, the farm furnishes employment for but a very small number of its sons, and that for a very few weeks or months at most in the year, and for the rest work must be had in the cities and towns, or not at all. In the time of our mothers, they, with all their daughters, had an abundance of employment in their homes. Throughout our country every farmhouse possessed its looms and spinning-wheels. From the sheep reared on the farm was the wool taken and carded by our mothers ready for spinning.... The women-folk, day after day, week after week, month in, month out, for fully or more than one half of the year, were all constantly employed in carding, in spinning, and in weaving the woollen and linen cloths that clothed the family, or were traded at the store for tea or coffee and sugar. The household music of that time was the hum of the spinning-wheel, the steady flow of the tones of the flax-wheel, the rattle of the shuttle. These operations were in constant progress in all the farmhouses, and a very large portion of the town houses; and the never-ending labour of our grandmothers must not be forgotten, who with nimble needle knit our stockings or mittens. The knitting-needle was in as constant play as their tongues, whose music ceased only under the power of sleep. All, from the youngest to the oldest, were abundantly employed, and all decently clothed. Now all this is changed. Throughout the length and breadth of our land the hand-card, the spinning-wheel, and the hand-loom are to be found only as articles without use, kept as curiosities of a past age. The occupation of our grandmothers also is gone; no more does the knitting-needle keep time to the music of their tongues, for the knitting-machine in the hands of one little girl will do more work than fifty grandmothers. The consequence is, there is no more work at home for our farmers' daughters; they also must seek the towns and cities, where they find their sisters equally idle, and in thousands are found upon the streets, spinning yarn or weaving webs, the warp of which is not of wool, neither is the woof of linen. In conclusion, labour-saving machinery has broken up and destroyed our *whole system* of household and family manufactures as done by our mothers, when all took part in the labour and shared in the product to the comfort of all, and has compelled the daughters of our country and towns to factory operations for ten or twelve hours a day in the manufacture of cloth they may not wear, or to the city to ply their needles for eighteen or twenty hours a day in hunger and cold, or to the street in thousands spinning yarns and weaving webs that become their shrouds."

Now, a somewhat similar change to that here described must have occurred in the past history of our race; and the conception of *home*, that lies vaguely in the intellect, and embues the emotions of the British people, reflects a *former* social condition, but neither points to a *future* that is possible, nor to the *present* domestic state of the great mass of our over-crowded population. In these old times, doubtless, the life was a poor one, even at its best. Redundant numbers were continually swept off by famine, pestilence, war. But social conditions were more favourable to the union of survivors than now. No facilities for migrating existed, and sisters and brothers, for the most part, lived and died side by side, whilst the family sentiment was, on the whole, strong and enduring.

The parent-stock from which sprang Thomas Carlyle, was planted on a little farm where father, mother, brothers, sisters, worked cheerfully together. Thomas drifted to the town, and soon attracted thither his brother John. Home life on the farm was ultimately dissolved—the units scattered. But in the private history now brought so fully before the public, strength of kinship, warmth of filial and fraternal love, stand out as notable features characterizing this remarkable family. These features have excited wide observation, admiration, and comment, which, I think, shows that the mind of the public has almost ceased to look for strong and abiding love in kinship! Ardent family love, and personal devotion to family interests are *not prominent* in the unstable domestic groups around us; and since the actual presses upon and tends to dim our ideal, we have come to approve and even to reverence the family life that appears *tolerably united*, although it is all but destitute of the appropriate sentiment—deep and ardent love of kin.

As regards the poor, then, rural life in the past was favourable to the growth of the family sentiment; but town life, as at present ordered, is wholly unfavourable to that sentiment. Town life separates the domestic circle prematurely, whilst within the home it miserably circumscribes the pleasures and freedom of youth. There is there no airy hay-field in which to hold merry-makings, and rustic festivals at harvest-home; there is the one room, or the two rooms only (spoken of by Mr. Bright), where children's gambols are an inconvenience, and the cramping limitations fret and irritate, till what results is anti-social bitterness instead of pure domestic love. When factory life begins, brothers and sisters no longer consort. Companions of the factory or workshop are better known, and naturally are preferred. A girl has her lover. Can she bring him home? Alas! no. "There is not room enough, and mother is too busy to care to see him." So the theatre or tavern becomes the nightly resort, and by-and-by a room is taken in a distant narrow street, and another so-called *home* established, where the girl-wife, ignorant of all domestic management and work, scrubs and scours to the best of her ability, but to the young husband seems ever in a muddle. Life goes on somehow, and in a few years *that* room too is filled with children, whilst the parents, prematurely old, are bowed down with care and with a weight of almost joyless responsibilities. What wonder is there that every now and then the public mind is startled, and the heart of the people torn by some horrible tragedy, such as the mental overthrow of that unhappy labourer, who (as I write this) lies in prison for the murder of his five children ? *

But now, one question remains. We still have rural life for many of the poor. Are social conditions in the agricultural districts altogether favourable to the stability and purity of the British home ? In the *Nineteenth Century* magazine for October, 1883, the Rev. Dr. Jessop writes thus: " The immediate future of our agricultural population seems to me to be gloomy. The rustics are not happy; they are sullen, discontented, averse to labour; they are on the alert for any grievance, they are ready for any form of rowdyism . . . they have lost all belief in kindliness or disinterested motives; they disdain to submit to such restraints as religion has a tendency to impose. Hovels which the raggedest tramp would shun . . . hovels which the local board of any borough in the kingdom would condemn in a week as unfit for human habitations, are, in a thousand instances, the only places that our country people can lay their heads in. There has been, or there is, a constant drain of the best men from the villages to the towns. Physically and morally a steady deterioration in the quality of our Arcadian swains has been and is going on. . . . From the parish in which I write, thirty-one sons of the soil have been enrolled as London policemen in thirty years. What does that mean? It means that these young men, who were the very pick of the parish, have been taken from us never to return. Why should they return? They will be fathers of families elsewhere. . . . We retain the sediment—the vicious, the immoral, the sickly, the dissipated and profligate, the roughs who would have been poachers in the days when poaching paid." †

Passing from the lower to a higher social stratum, we must now glance at the average home life there. Poverty prevails, but in a new form, for there is no squalor. There are rooms enough within the dwelling, which is situated in a dark, narrow street. It is closely surrounded by other dwellings whose inmates strive to keep themselves concealed from neighbours, whilst conscious of an irrepressible desire to spy upon, and take a kindly interest in the doings of neighbours. Gossip

---

* I refer to the case of William Gouldstone, at Walthamstow. " No motive can be assigned for the terrible crime than that which is indicated by complaints that his wife had too many children."—September, 1883, Correspondent of the *Scotsman*.

† " Clouds over Arcady," the *Nineteenth Century*, October, 1883.

is disapproved, but gossip more or less is indulged in. In this home there is great refinement which daily struggles to express itself in an environment threatening its destruction. Purity and nicety of detail in all domestic arrangements are necessaries of life, and when the cost of these is covered, there is no surplus income to expend on the thing of beauty which would be a joy for ever in that home. The very children have an innate sense of beauty, causing them to long for pretty playthings, and in gazing at shop windows they proffer prayers for this and that, which the poor mother sighs to refuse, yet dares not grant.

Within the last few hundred years, the shops in all large cities have become a wide sphere of brilliant beauty for æsthetic taste to revel in. Rare gems are no longer hidden away by Jews from public gaze, but in the open light of day attract to the jeweller's counter. Lovely pictures flood the market, at prices that appear an insult to real art, rather than its due reward. And the china-ware and glass shops are filled with objects that are beautiful in form, and of true artistic value. To the rich all this seems gain. It proves rapid development in industrial arts and national wealth. Great Britain can afford to apply a large proportion of her capital to the production of the beautiful, and in doing so she educates public taste and ministers to happiness. Newspapers and public prints written for the wealthy, and of course reflecting the opinions of the wealthy, are embued with this sentiment. It is possible, however, to regard the question differently. We have tens of thousands of individuals amongst us, whose taste is already considerably educated whilst their pockets are empty. On every side the beautiful attracts them, but admiration is not the only emotion called forth. Their human nature is complex. It contains a sentiment of acquisitiveness or love of property which is apt, nay certain, to glow at the sight of the beautiful, that lies within the range of human possession. This sentiment we must remember is highly approved by public opinion. The lack of it is what we are constantly deploring in the thriftless pauper class ; and the general mind is tolerably conscious of the fact, that to it we owe much of the energy and industrial progress of our race. We do not then wish to depress it unduly, nevertheless we blindly rejoice in social conditions that tend to the turning of this sentiment—love of property—into a source of depression, irritation, even acute suffering, in the case of myriads of our fellow-countrymen.

To an impecunious young man of taste, the sight of lovely things he may not prudently acquire, instigates to debt, or frets the nature into morbid discontent. To the woman of refinement whose delicate organization thrills a response to every line of outward beauty, the joy of seeing lovely things is frightfully impaired by contrast with surroundings that are poor, bald, mean. Unless she holds the master-key of money, and can embellish her home in accordance with her own pure taste, the æsthetic drawing-room of a more fortunate friend, the shops filled with *bric-à-brac* and *objets de vertu*, are like the Dead Sea apple which turns to bitter ashes in the mouth. Dickens has given us a fancy picture of poverty with contentment in his Tommy Traddles and his young wife Sophy. "When it's fine," said Traddles, "and we go out for a walk in the evening, the streets abound with enjoyment for us. We look into the glittering windows of the jewellers' shops, and I show Sophy which of the diamond-eyed serpents coiled up on white satin rising grounds I would give her if I could afford it; and Sophy shows me which of the gold watches she would buy for me if *she* could afford it; and we pick out the spoons and forks, fish-slices, butter-knives, and sugar-tongs we should both prefer if we could both afford it; and *really we go away as if we had got them !*"

Traddles and Sophy, however, are in the heyday of young love, and Dickens's creations are not altogether the flesh and blood human beings of our everyday experience. No! humanity is not truly represented there, and if in judging of this matter we extend our view until it embraces every social class, the poor of every grade as well as the rich, we must, I think, perceive that the amazingly rapid growth in national production of *objets de luxe* does *not* minister to national happiness in any broad sense, but only to the happiness of the few. We are bound to put into the scale, the painful emotions engendered, the longing desires, the envy, the discontent; and without reference to harsher results in malice or crime, we find, since the poor are *far more numerous* than the rich, that the balance descends upon the wrong side—not happiness, but pain, is so far the national result.

To return to the home in which poverty and refinement are in close alliance. The children are educated at day schools, where they mingle with all sorts and conditions of boys and girls, and their young ideas take shape in a mental environment of which the parents know literally nothing. A vulgar phrase will startle them and be corrected superficially, but the coarse thought it covered sinks into the boyish mind, lies undetected there, and germinates to undermine in manhood the delicate refinement of the inherited nature. The parents would willingly know all their children's friends, but to keep open table is expensive, and would entail domestic service beyond their means, so only a favourite companion is brought in now and then, and the range of her children's feelings, as well as thoughts, passes from the mother's ken. A girl of ten is asked in holiday-time to visit a companion whose home is in the country. Fresh air for the child who droops in summer in the town, country freedom, keen enjoyment, these advantages are great. Consent is given, although not one member of the group she enters is known to her parents, and the social forces dominant there may counteract the moral teaching they have striven to impart. The boys of the family are literally let loose upon the public. I mean—as there is no billiard-table at home, no lawn-tennis ground, no smoking-room, they cater for amusement elsewhere, and in my chapter upon youth I have sufficiently described the life that is led. Its dangers and drawbacks could scarcely be exaggerated, and the profound anxiety that tender mothers feel is a deplorable source of suffering. Yet *no* mother, whose home is in a town, can possibly escape this anxiety, so long as social life is what it is, and the constitution of the British family retains its present form.

Nor is suffering mitigated when sons depart one after another to begin the individual industrial career. One goes to London, where the temptations of city life are greater still, and a solitary lodging is the poor substitute for home. Others to New York, China, India, Australia. Backwoods or diggings, or a rough peasant's life in some wood shanty planted in the solitary pampas, are the alternatives to youths pushed far afield by social pressure and industrial competition in the overcrowded land that gave them birth. And daughters too, move on. The home even for them is no abiding-place. Do they marry? Alas! no. The young men they could love are far away, and marriage everywhere brings such expense, responsibilities, and cares, that contrary to her theory, the mother's heart rejoices over lives predestined to be mulcted or mutilated in the tender ties that give completeness, roundness, happiness to womanhood.

Two of the girls are exceptionally intellectual, but since

strict economy has to be practised, no suitable sphere save tuition is open to them. One passes through a normal school curriculum; takes many prizes; and in a few years is settled for life in a village in the north of Scotland. A tiny cottage attached to the school is her home. She has no society, for the villagers are uncultured. She pines for mother and friends, but can only write to them affectionately, visit them once a year, and for the rest plod on as best she may. The other scholar of the family has great musical as well as intellectual talent. The stage would be her natural sphere, but the stage is so surrounded by dangers to youth, that nothing would induce her tender parents to make dramatic art her calling. She goes as private governess to Russia, sees society, and mixes in a wholly different world from that of her sisters, with whom her correspondence becomes lax, indifferent, almost cold. How can it be otherwise? She has become an accomplished woman of the world, but the literature she delights in, the people she meets, are alike unknown to her family at home.

One sister there has entered a shop; millinery is her daily occupation. Her fingers are agile, but her brain is not active. Her mental conceptions deal only with the ordinary doings of her neighbours, and kindly gossip is the main interest of her life. She is affectionate and good to the poor parents, now left almost desolate in the old dwelling. One daughter only with them still, but not one son. The daughter, feeble in health, has little spring of youth within her. A narrow life, the want of breezy freshness in her home, with the despondency of old age overshadowing it, tell upon the nerves and spirits, and at the early age of thirty, she is practically a wreck—sweet, gentle, devout, but sickly in body and mind.

The trio are annually transplanted for a month to seaside quarters; the daughters at a distance send money for this purpose. But the variety in their monotonous life has come too late for pure enjoyment. In summer the restless spirits of this restless age are all abroad. The whole world travels. A little rise of wages occurs in the iron trade, immediately the operatives rush off to the coast. With money in their pockets, monotony and smoke become intolerable; they must have change at any cost. A servant in the midlands gets a ten days' holiday. Regardless of thrift, she cheerfully spends two of these precious days in travelling, and a third of her year's earnings, for the pleasure of visiting Scotland, of which she has heard her master or mistress frequently speak. The wave of general movement is at its height in August, and it is not surprising that the thoughtless poor are caught upon its crest and whirl about in the *maelstrom*. This expansion in traffic, however, is not satisfactorily provided for by railway companies. Cheap trips, excursion trains, etc., are organized hastily, and bustle, confusion, noise, and wear and tear of nerve at stations, the overcrowding everywhere, makes travelling to the aged too great an effort and toil. Our trio emerging once a year from still life in a familiar groove, are shaken out of all composure by a railway journey, and when they reach their destination, chosen because some sixty years ago it was a rural retreat of charming simplicity and quiet, behold, it too is transformed. It is a bustling, thriving, vulgarized watering-place, where, roaming on the sea-beach, may be seen company as gaily dressed as on a city promenade.

In Great Britain it is difficult indeed to find a *haven of refuge* for the old, or to place around them in winter and summer those varied conditions of life which are essential to health in the years of declining strength, when life's active work is approaching the close. The great Mrs. Somerville maintained health of body and mind to the last; but she forsook her native land and made Italy her home. At the advanced age of eighty-one she found solace in grief (the grief of Mr. Somerville's death) in writing a new book on Microscopic Science. The undertaking was a great one, but, "instead" (she says) "of being discouraged by its magnitude, I seemed to have resumed the perseverance and energy of my youth, and began to write with courage, though I did not think I should live to finish even the sketch I had made." She completed this work, however, and lived on to the age of ninety-two, when "her pure spirit passed away so gently that those around her scarcely perceived when she left them. It was the beautiful and painless close of a noble and happy life." The same witness, her daughter, writes: "My mother's old age was a thoroughly happy one. She often said that not even in the joyous spring of life had she been more truly happy. Serene and cheerful, full of life and activity as far as physical strength permitted—to the last her intellect remained perfectly unclouded; her affection for those she loved, and her sympathy for all living beings, as fervent as ever. . . . She always maintained her habit of study. . . . Mathematics delighted and amused her to the end. Her last occupations, continued to the actual day of her death, were the revision and completion of a treatise on the "Theory of Differences" and the study of a book on "Quaternions." *

Now, my reader may regard this sketch of Mrs. Somerville's history as a wide divergence from my subject—the British home—and fail to see that I can base any argument concerning the latter upon personal facts relating to a woman so illustrious and exceptional. I admit that such minds as hers are rare, but all minds, even the most commonplace, are immensely influenced by surroundings up to the last moment of existence; and whilst Mrs. Somerville had in old age an environment favourable to happy contentment, and healthful mental activity, and was usefully productive to the end of life, I maintain that amongst us the great majority of aged ones have on the contrary an environment that promotes sadness, tends to the relaxing not the bracing of bodily and mental powers, and is in every way unfavourable to healthful life. The happy activity of old age is, in each generation, eminently desirable for *national welfare*. We suffer, I am convinced, an enormous national loss by premature suppression of a vitality, which (if I may so express it) bears within its bosom, a wide knowledge, and an intimate experience of life.

Mrs. Somerville's life was sustained by all the natural ties which normal humanity requires, and consciously or unconsciously craves. She had a devoted husband, with nature akin to her own, whose society was freely bestowed upon her up to the hour of his death. She had the companionship of daughters who lived at her side till she passed peacefully away. She had a large circle of intellectual friends, with whom she held converse upon a platform of mental equality. She had in sunny Italy the warmth that old age requires. She had variety, the frequent change of air and scene that keeps the mind vigorous; and this without the hurry, bustle, and nerve exhaustion which travelling in Great Britain commonly entails. From Venice, she writes to her son: "This place suits us, to the life, constant air and no fatigue; I never once have had a headache." And in another letter, "This place is well suited for old people, who require air without fatigue." † Alas! how many of our town-dwellers at the age of sixty can obtain air without fatigue? "Nay," says my reader, "gondolas of

---

* "Personal Recollections of Mary Somerville," by her daughter.
† Ibid, p. 259.

course are out of reach, but there are private carriages." And what are they amongst so many? and who drive in them? Often the young women of fashionable society whose health would be vastly improved were they obliged to walk! Many young men, to be sure, succeed in mercantile or professional life, and straight proclaim it to the world by setting up (as the phrase goes) their carriage. But it is the wives and children, not the aged parents, as a rule, who benefit thereby. These sit contented (the world supposes) all day long in easy chairs at a desolate hearth, and are compelled (unlike Mrs. Somerville) to narrow their mental as well as physical range.

To return to my typical British family. Several of the sons lie in distant graves, and the thought of death-beds unsoothed by the love of mother or sisters makes the hearts of the aged parents ache. But two of the number attain what the world reckons enviable success. They make considerable fortunes and return *home* to marry and settle in London; for observe, *home* to them now, means simply their native country, without reference to aged parents and the dwelling of their childhood. Filial and fraternal love have no warmer colour than what prompts to the gift of a little money now and then if required. The tender chords of affection that bound them in childhood are broken and silent, and if an emotional life is to be re-awakened in beings who have had no domestic relations for many years, perhaps the forming of new ties is a wiser course than seeking to reunite ties that have been severed

The daughters, however, with the pure idealism of womanhood make the experiment. When the parents are dead and they alone in the world, with the meagre capital that steady toil has enabled them to accumulate. they resolve to live together. Alas, the power to do so *happily* has passed away. They jar and irritate one another, and the home is full of discords, not of harmony. Shall we call these sisters unnatural, and blame them? By no means, for the whole phenomena are pre-eminently natural. Social units buffeted by adverse forces become misshapen or deformed, and when that process has been completed, to adopt a life that requires the welding of these units into solidarity is to demand of humanity what is directly *opposed* to nature. The sisters separate again, for their dream of family life and a restful home after the struggle for existence was over, has proved a mockery and delusion. Sadly each one pursues a solitary path, feeling utterly forlorn,

T

although closely surrounded by a race that daily boasts of its venerable institution, the *British Home*.

And now in reference to our highest social stratum, I draw my picture of domestic life there, from Mr. Escott's work on England, and quote his words. "The domestic life of England has undergone a complete metamorphosis. For the nation is only an aggregate of households. Modern society is possessed by a nomadic spirit, which is the sure destroyer of all home ties. The English aristocracy pass their existence in a perpetual round of visits. They flit from mansion to mansion during the country-house season; they know no peace during the London season. They seldom endure the tranquillity of their own homes in the provinces for more than a month at a time, and then they temper their rural solitude, by a succession of visitors from the great city. Existence for the fashionable and the wealthy, is thus one unending whirl of excitement, admitting small opportunity for the cultivation of the domestic affections, no time for reflection, or the formation of those virtues which depend upon occasional intervals of thought and seclusion." * And again, "The influences of the time are not favourable to domesticity, and in our progress towards cosmopolitanism, the taste for the family life which was once supposed to be the special characteristic of England, has to a great extent been lost. The claims of society have continually acquired precedence of the duties of home."

Assuredly it cannot be denied, that the whole tendency of modern industrial life, of modern society, and of modern philosophy, is towards a weakening of family relations. "It is not, that there will not be in the future," says a thoughtful evolutionist, "as in the past, a prevalent disposition to form strong attachments within the family group. But, the mere fact of being children of the same parents will not be held as sufficient ground for life-long intimacy and favouritism. 'Behold thy mother and thy brethren stand without desiring to speak with thee,' said some one to Jesus. But he answered and said unto them that told him, 'Who is my mother? and who are my brethren?' And he stretched forth his hand towards his disciples and said, 'Behold my mother, and my brethren! For whosoever shall do the will of my Father who is in heaven, the same is my brother and sister and mother.' This is," says the evolutionist, "an exact statement of the

* "England, its People, Polity, and Pursuits," T. H. S. Escott, vol. ii. pp. 9 and 13.

frame of mind of the man who is religious in the best sense, whose every thought is subordinated to the welfare of the sentient beings by whom he is surrounded. 'What to me,' he virtually asks, 'are the accidents of birth? If any men and women are by preference my brothers and sisters, they are those who manifest most brotherly and sisterly feeling for other men and women.' . . . Love of kin as such has had its day. As a step towards a wider and more spontaneous sympathy it has been useful, even necessary. But the family has now ceased to be the social unit. The sceptre is fast falling from the hands of *paterfamilias*, and it needs no Daniel to read the writing on the wall which pronounces the doom of the primal despotism." *

Disintegration is the process that unconscious evolution is year by year forwarding in the British family and home. And unconscious evolution is, as I have already pointed out, a vast force, impossible for man under present conditions to control. He must conform to the general drift of the onward stream. But he may nevertheless voluntarily contribute new forces to that stream, forces that will annul or counteract many of the evils that prevail. If the contracted sentiment of love of kin has had its day, there are new emotions ready to take its place, — so soon as man destroys the barriers that restrain these new emotions, and prepares for the life that is to come.

"The wine of life is on the lees," sang our great old poet with a melancholy ring. But, "New wine in new bottles," is the true theme of inspiration for our nineteenth-century singers. Let us turn to face the future rather than the past. Let us renounce old ideals that are incomplete or false, and openly condemn the system which leaves cultured minds to stagnate in solitude, and tender hearts to be bruised and broken in a world teeming with life. Family home life never was, never can be, a perfect or ideal life. Frailty characterizes it. Frailty belongs to it as a necessary defect. It is venerable, and, like all institutions of the past, has played its part in evolution; but it is no longer adapted to the present environment, and moreover in its actual form it is incapable of adaptation.

By emigration alone may the British race still approach its ideal of the good old family life. In a new country sparsely populated a large family group may co-operate and keep together, although employed in various industries, provided

* D., in *National Reformer*, No. 1113.

there is sufficient capital for maintenance until the group is firmly established. In Great Britain itself no such life is possible. Have we then reason to look back regretfully, and mourn continuously the good old days that are gone? Says Mr. Holyoake : * " Let unthinking and unknowing people who talk of the good old days which are gone, understand that there never were very good days for the poor. There has always been a golden age for opulence, but no age save a desolate one for indigence." Yes, the burden of redundant numbers has in every age brought sorrow to the poor. When they migrated less, they had to suffer the shocking alternative of a higher death-rate, and a shorter life. The golden age that will embrace all, lies before us, not behind ; and a devout reverence for human happiness must constrain us, sooner or later, to abjure our reverence for the venerable but unstable and sadly defective domestic institution—*the isolated Family Home.*

* In his " History of Co-operation," vol. i. p. 12.

CHAPTER XVI.

THE EXPANSION OF OUR DOMESTIC SYSTEM.

"Home must be a sanctuary of exhilarating enjoyments, as well as an abode of peace. The labours of every day must be relieved by the constant return of tranquil pleasures, and heartfelt delights."—ISAAC TAYLOR.

THE importance of possessing true ideals is little understood by the majority of mankind. The British are intensely practical. As a rule they have no patience with a theorist or speculator (unless his speculation be one of money-making), and to have new views upon any subject that is important is regarded as sometimes dangerous, and always eminently *unpractical*. Since the commencement of the industrial epoch our marvellous success in the creation of wealth has resulted in the steady, rapid growth of population, producing in the lower classes an unceasing struggle for existence; in the higher a mental restlessness, a constant emulation in the spending, if not the making of money. These conditions have created outwardly a social atmosphere of hurry, bustle, confusion, and inwardly a love of action for its own sake, which precludes the calm, deliberate forming of noble ideals, or pushes them aside for some practical outcome inspired by *expediency* alone. When in isolated cases doubts and vague suspicions arise that things are not as they might and should be, two factors—the fear of Mrs. Grundy, and the outward and inward clamour for action—stifle the doubt.

"They dare not, one man out of ten,
To think their thought thoroughly out;
The practical plucks at their sleeve,
And they're frightened to shock and to grieve." *

* "Hilda among the Broken Gods," Walter C. Smith, D.D.

Nevertheless, speculation is necessary. Ideals are of the utmost signification, and our first step towards the thorough reform of our domestic system is, that in the minds of a considerable number of individuals there should shape itself a conception of home life, different from and superior to the home life that is everywhere around them. The conception must be at first purely ideal, but as it strengthens, tentative efforts will be put forth to realize it again and again, until at length the noble ideal is practically attained. If action, however, precedes the full development of the mental conception, and the thirst to be practical hurries society into premature efforts, the tide of reform will be caused to ebb rather than flow.

Domestic relations and individual home life form, as touching happiness, the most vital spot of Humanity, and in initiating reform here, evolution must pass from the purely reflex, automatic stage into a conscious state, in which reason and elevated moral sentiment are bound to play a conspicuous part. The formation of a superior domestic and social system is like the weaving of a fabric of highly complex structure and exquisitely delicate quality. It can only be accomplished under certain conditions, and the primary condition is fitness or adaptability of the raw material, out of which the superior structure must be elaborated. By raw material I mean, of course, in this case, the human nature to be welded into the social organism, the mental and moral calibre of the individuals destined to combine in close domestic union. The raw material chosen in Robert Owen's experiments was unfit, and how unfit we may gather from this one significant fact—at Queenwood it was found that some of the members had not even read the community rules, and everywhere there had been grave infractions of these rules! In every well-ordered isolated home household rules are acted upon, although they attract no public attention, and are not necessarily read and adopted by the family. Parents arrange domestic matters to suit their own comfort and well-being. Servants and children are trained to be cleanly, orderly, punctual at meals, etc., and the rule of the house is followed without any written legislative code. A few oral hints and the force of example are generally sufficient in the simple primitive social organism of the family.

But with a complex social organism of, say, twenty families combined, the position is entirely different. In order to secure happy individual relations and free interaction of the social units without jar or discord, simple but clear and definite rules have to be thought out by *some*, and by all members have to be strictly observed. For instance, that no one shall enter the private room of another, except by special request; that no one shall smoke where it interferes with the comfort of others; that no one shall talk in a reading-room; that dirty boots in carpeted rooms are not permissible; that punctuality at the dinner hour and other set times is essential; and so on. Such matters in our present disorganized, loose, and shambling system of life appear to many individuals the merest trifles, not worth considering; but, in reality, they are extremely important, and their significance will be understood—so soon as humanity perceives the true position of the *order of domestic life*, and exalts it to its right level as a noble, invaluable *art*.

"But," says my reader, "if such trifling matters as these have to be punctiliously attended to, where is the superior freedom that a superior life should accord?" I reply, the freedom lies in the banishment of useless etiquette, in boundless liberty to approach our fellow-creatures with confiding warmth, to enjoy with them intimate intercourse, and enter spontaneously and without reproach into every kindly social relation that is mutually pleasurable. It lies in the absence of all necessity for self-protection, the serene enjoyment of a sphere where belligerent attitude is never called for, where no one trespasses on another's rights, and personal dignity needs no defence. It lies in power to spend one's hours of recreation in such pursuits as minister to pure taste and æsthetic joys. "How so?" my reader asks; and I reply, whereas in family life, individuals who are musical can only have music occasionally at home, and now and then attend a public concert; in associated homes, the means for regular and frequent indulgence of all innocent and pure tastes would be universally provided.

By the present system, twenty families of the genteel order have each a small parlour or dining-room and a small drawing-room, in all, forty sitting-rooms. Yet the varied family group is confined to one of these two sitting-rooms, where the long winter evenings, spent in regular succession, are extremely monotonous and dull. If chess is played, all must sit silent; if piano and violin are played, whist must be relinquished by the elders, and the young man who loves reading must lay down his book. Unless all the members of the group have similar tastes, the liberty of some is necessarily restricted; and if the environment of the home is a town, the young men seek pleasure outside (where and how parents know not, and scarcely dare inquire), whilst the sisters are left in possession of the domestic hearth, whose monotonous dulness drives them to seek excitement in novels and day-dreams that sadly enervate the mental powers.

When the twenty families combine, the forty small rooms are exchanged for, perhaps, two large well-heated and ventilated but uncarpeted refectories, a music-room, an art-room, a smoking-room, a dancing-room, a chess and whist room, a library, a mechanic's room, some school-rooms, and two nurseries—in all, perhaps twelve or fifteen fine rooms, furnished comfortably if not luxuriously. Every member has a wide choice of where, how, and with whom he will spend his precious hours of relaxation. By strict rule, each chamber is devoted to a special purpose. Intrusion or mixture of occupation is not permitted. Within each a small group of happy individuals is, under a law of natural selection—affinity of nature—harmoniously and pleasurably occupied. No one is compelled to talk who prefers to read or write; no one needs check his own desires to accommodate his neighbour. The home itself gives facilities for following the bent of every innocent inclination, and this freedom to spend our evenings just as we like is a great part of *true liberty in life*.

"But," says my reader, "you infer that all humanity is social and gay, and bright evenings spent in the company of others is what every individual longs for?" By no means. There are, I am well aware, gentle, lonely souls to whom society is often oppressive, whose natures, congenial and responsive on the whole, need an ever-recurring solitude to breathe freely. Unless privacy and all the necessary conditions for solitary thought and calm reflection can be secured to every member, the unitary home would not be what I assert it will be—a social organism superior to the family in its power of promoting human happiness. Observe, there are two methods by which solitude can be secured and maintained, for each individual at the moments of desire. The sleeping-chambers may be furnished according to the continental system. In each is placed a writing-table, a couch, an easy chair, with facilities for lighting a fire; and when the inmate chooses to be alone, he is there safe from all intrusion, and in the comfortable surroundings necessary for study or happy thought. If, however, in the Unitary Home the system of large dormitories has been adopted, some ten or twelve small boudoirs are provided for purposes of solitary study, and by means of a small piece of mechanism placed at the door, whoever enters, desiring to be alone, can indicate to any one approaching the door that the room is occupied, and his privacy will remain undisturbed.

"But will there be no bold spirits eager to intrude and utterly indifferent to wishes so expressed?" In the open world there are undoubtedly plenty of social units ready to trespass on their neighbour's freedom in matters small and great, and without one particle of the sweetest modern sentiment of civilization—a jealous regard for the rights of others; but these will probably adhere by choice to the social system of the past, or if they enter into the new system, it will either adapt them gradually, or reject them as material unfit for a social organism whose healthy life depends upon delicate sensibilities, of which they, alas! are wholly destitute. An advanced social organism can only exist permanently by the exercise of advanced principles in a medium of advanced sentiment. If units wholly unfit intrude, they will in time become again extruded; or if they preponderate in the mass, the organism will disintegrate and ultimately break up.

We must not conclude, however, that brilliant intellect is required for the superior life, and that individuals of ordinary mind have no capacity for an associated home. The general and particular rules devised in regard to the well-being of the community may be perfectly simple and clear. They will be carefully inculcated, all the children of the home will be trained to observe them, and should they be followed mechanically (that is, unintelligently), that *alone* would prove sufficient. It should never be forgotten that habits once formed are perfectly easy, and that good habits are quite as easy to follow as are bad habits, rather, I should say, easier; for good habits in good surroundings minister so directly to happiness that they suffer no check, whilst bad habits cause discomfort to others, and bring to the perpetrator more or less of resistance and punishment. If my reader will carefully study the natural laws that lie at the basis of social life, he will discover, I think, that in the true order of nature all such unitary homes as are destined to embrace family life, to supersede isolated homes and to survive into the distant future are certain to be schools of *good habits*, where new graces will be evolved for the adornment of humanity, but where the frivolous rules of a senseless etiquette will as certainly have no existence.

In our present social state, the mingling of etiquette and bad habits is bewildering, thoroughly uncomfortable, and often intolerable. To give an extreme instance taken from life. A young man will enter society dressed like a Beau Brummel. He has expended anxious thought over the fit of his garments, and is sensitive to a degree upon the delicate subjects of his necktie and button-hole. In the drawing-room, if a lady drops her pocket-handkerchief, he hastens to pick it up; if she approaches a door, he flies to open it. He hates music, yet he compliments and thanks the girl who has blundered through a sonata, and his politeness excites general admiration. At home, his habits are irregular and slatternly. When he leaves his bedroom, his well-fitting garments litter the floor. An individual belonging to the sex he so politely waits upon in society (possibly a mother or sister) performs the menial office of picking up, folding, and replacing them in his wardrobe. As he leaves the house he calls brusquely to his mother that he will be home at seven o'clock to dinner. A game of billiards, however, detains him, and he makes no apology when he enters his home at eight, but grumbles openly if his food is not served to perfection. His sisters are dependent upon him for escort to places of public amusement. He readily promises his escort, but quite as readily breaks his word, and disappoints their expectations. He is never exact to any appointment made with the inmates of his home. He keeps them waiting for him without compunction. His presence is often a source of irritation to feminine nerves. It is his habit to light his cigar in the hall; to leave the sitting-room door open, or close it with a loud bang; to make no response when called in the morning, or ignore the breakfast bell. On Sundays he lies in bed till eleven or twelve o'clock, thereby disarranging household work, to the annoyance of parents, the disgust of servants. Yet when remonstrance is made, he fails to see that these habits are other than insignificant trifles; and he conscientiously believes that his general behaviour is thoroughly consistent with the character he assumes—a civilized man of the world and perfect gentleman.

He is a curious anomaly, a combination of laboured gentility and striking ill-breeding; a model of fastidious elegance, yet an insusceptible human machine; a creature craving society, but miserably deficient in all the delicate qualities that render social life easy and pleasant;—in short, a narrow and artificial, not a simple, genuine, product of an advanced civilization. Nevertheless, we must judge him as we judged the pauper and the girl of the period—without censure to the individual. Two broad factors have produced the type, viz. the false system of hollow conventional society that encircles our domestic system ; and, within the latter, the isolated home, in its unstable equilibrium, ever tending towards the " doom of the primal despotism"—the loss of parental authority. The important question is, Would this type be favourably affected by a new domestic system ?

Had Mr. John Stuart Mill lived, the world would probably long ere this have received from his pen a complete work upon Socialism, containing valuable opinions concerning its defects and its worth as a system of life. In one of his articles on this subject, published in the *Fortnightly Review* in 1879, these words occur: "What is incumbent on us is a calm comparison between two different systems of society, with a view of determining which of them affords the greatest resources for overcoming the inevitable difficulties of life." He expressed the hope that "the best parts of the old and new systems might be combined in a renovated social fabric ;" and he made this striking observation : " in communist associations private life would be brought in an unexampled degree within the dominion of public authority." *

Now, I have shown how the tendency towards individual freedom throughout the community has told fatally upon our masculine youth. The forces for its regeneration undoubtedly require to be disciplinary—capable of inducing social purity and order ; and humanizing—capable of inducing sympathy and ready responsiveness to all the amenities of a gentle life. The conventional type of young man, with his courtly manners and private bad habits, is no subject for a commune. Yet the authority necessary for the ennobling of his character would there (according to Mr. Mill) exist in an unexampled degree. And there also, I venture to assert, would be found homely influences wide enough and warm enough to develop in his nature the genial domestic qualities of which he stands in need. I do not, however, deem it likely that an only son, for instance, brought up in an isolated home, petted and spoiled in childhood, and in manhood permitted to dominate the household, would submit to communism, and be moulded by it to a higher form. Individuals may cease to be plastic. But let us ever

* "Chapters on Socialism," by John Stuart Mill, *Fortnightly Review*, March, 1879.

remember that the race is plastic, and adaptation step by step is the *modus operandi* in the path of progress. In pursuit of personal happiness, the race at present is struggling to adapt itself to a weak domestic system and a vicious social system, and in the struggle it is degenerating. If the domestic and social systems are remodelled, and attention paid to the laws of heredity, the inevitable happy result will be a regenerated race.

In the "Manual for Co-operators" it is well pointed out, that organized combination for the purpose of realizing a life consistent with the ideal held, *i.e.* the conception of the true end and purpose of life, is no new thing. Our Catholic ancestors systematically and successfully established the monastic or ascetic form according to *their* ideal of the highest and noblest life possible. Since their day the ideal has changed. Let us slightly alter the well-known lines of a poet of the Church of England, who stands midway between the past and the present, and adapt them to the purely modern ideal.

> "We need not bid, for cloistered cell,
> Our neighbour and the world farewell;
> The trivial round, the common task,
> Will furnish all we ought to ask;"—
> With happy heart and gentle mind
> To cling more closely to our kind.

Are we able, like and yet unlike our Catholic ancestors, systematically and successfully to create a social order in harmony with this ideal? The task is no easy one.

Voices from America tell of social dangers similar to ours, the escape from which can only be by the birth of a new social order. A recent writer says: "I am afraid that the influences at work among us, while competent to produce a good deal of individual refinement, single specimens of humanity characterized by a sort of hot-house rapidity of growth and delicacy of structure, are adverse, and growing more adverse, each year to gathering and organizing these fine units into anything deserving the name of a social order." *

A *new social order* is, indeed, the pressing want of civilized man. It speaks to us in all the restless music of this age. There is all around us a voiceless wail of humanity, waiting for the birth of a new order, which as yet gives no sign. Blindly waiting, for in the mind of our civilized humanity there is no

* Miss Preston, in the *National Review*.

distinct conception of what the new order shall be, although deep within its tender heart, the emotion of desire shapes itself to the masterly problem of Morelly, viz. " to find that state of things in which it should be *impossible* for any one to be depraved or poor." A social order that does not embrace the future of the poor, as well as of the rich, that does not promise to elevate a_ll_, to make happiness and goodness possible to *all*, will never satisfy the noble cravings of a socially developed race.

Does it logically follow, then, that State socialism is the order required? I think not. A system established and upheld by force could have no stability in a civilized nation, and the use of force inevitably injures the steady growth of moral sentiment. Voluntary association is the only instrument available in the organizing of a new order; and neither advanced communism * nor socialism will at present be voluntarily assumed. A community where intellect has been sharpened by competition and made preternaturally keen in the pursuit of wealth is not prepared for any levelling system, that in securing the well-being of the whole, would banish distinctions often bravely fought for and boldly won. In the yet far distant future, when another stage of life's history has been reached, when men know the happiness-value of material wealth and turn from its too eager pursuit to seek higher and purer joys, socialism may become possible. Meanwhile for us the step to take is a less ambitious one, and in aspiring less we shall accomplish more.

Must we, then, turn from the various systems proposed by Fourier, by Robert Owen, and others, to the far less sweeping measures of reform adopted by the members of the Co-operative Union, and advocated by them as practical efforts rather than utopian theories ? Voluntary co-operation in the production and distribution of wealth, I have spoken of elsewhere as eminently desirable; and wherever success is achieved, the results will be valuable in two directions—economically, in the saving of labour and avoiding of waste ; educationally, in the diminishing of anti-social feelings that are fostered by competition, and the increase of those gentle social qualities that are indispensable to progress. Nevertheless, so long as co-operation is confined to industrial life, and leaves untouched our

* Communism implies the absence of private property. Socialism usually implies that the land and implements of production should be the property of communities, or associations, or governments.

unstable homes and defective domestic system, it will accomplish comparatively little. The direct path to rapid advance lies, in my opinion, the other way. Robert Owen went too fast. He tried to make people live and work together before they were fit to simply live in harmony. We call for union in *domestic life*, and when people can live together happily, then, but not till then, they will take the other great step of progress, widely and effectively. Let motion of a healthy expansive kind begin in the very centre of our life—the *home* itself, and the wave will roll outwards in ever-widening circles, till our social life, our industrial life, our national life, are thrown into harmonious movement.

But again I repeat that true *ideals* must precede practical effort. Some individuals in the community must know clearly and definitely what is wanted and why, before healthy movement can arise. It may be that initial efforts will take the form of societies for promoting improvement in the social relations of life—societies in which meetings will be held for discussion of questions regarding the varied evils of our present domestic system, and for consultation upon the methods by which these evils may be mitigated or overcome. The scientific principles of harmonious social living would be brought forward and adopted, and finally the rational evolving of an advanced and stable unitary home would be the outcome of intelligent constructive *art*.

Rational discussion must soon make prominent what elements are *essential* in a Unitary Home. The primary elements undoubtedly are, that it shall afford greater comfort, ease, and happiness to adults than their present condition affords, and such facilities in training the young as will give promise of a still better and happier future. In reference to the first of these two essentials, it is a very remarkable fact that, premature as was the Tytherly experiment, fatal the mistakes made in the commencement, and miscellaneous the members collected together, superior happiness was certainly afforded to many. Mr. Holyoake tells us, that as the end approached working members said they would rather live on an Irish diet of potatoes than go again into the old world, if that would enable the society to hold on. " Residents—and there were many who were boarders in the community—all regretted the end of their tenancy." Later they spoke with regret of the loss of the happiness Queenwood had afforded them. " Ladies, who are always," says Mr. Holyoake, " difficulties in a new state of miscellaneous association, came to prefer Queenwood life." Some, "who were tartars in their social relations in the old world—just women at heart, but impatient of the crude wayward ways of domestics—there became the most agreeable and honoured of residents. It was not because they had to control their tempers, but because the occasions of natural irritation no longer existed under the happier circumstances of equality of duties and enjoyments."*

Perhaps Mr. Holyoake, like others of his sex, scarcely appreciates the magnitude of the burden so many of my sex bear, in the management of domestic servants. "Why don't you dismiss her at once?" a husband will say, when his wife complains of a servant, oblivious apparently of the fact that to obtain another entails effort, and that the chances of improving matters by this masterly policy are extremely small. The weak, defective domestic system in the homes of the labouring class, causes the problem of domestic service in the middle class to become increasingly difficult. Girls are no longer trained at home by mothers, who were good servants in their day, retaining their first situations up to the hour of marriage. Our girls go to board schools at an early age, and are so occupied with book-learning that learning, how to use the hands and exercise the heart at home, has little or no place in their lives. The outward conditions of their youth mould them to sturdy independence, and to the restless spirit of the age. They hate monotony, dulness, routine, and whereas the nurse whom I loved in my childhood was thirty years a member of my father's household, the neater smarter maids I meet with now, flit rapidly from place to place and never feel their master's house is *home*. But in this matter, as in most others, money gives its possessor an undue advantage. The rich hold butterfly servants by a purely commercial tie, instead of one mingled, as of old, with the sentiment of household affection. They raise wages freely and generously, according to their personal ability, without reference to the inability of others in their social class to follow suit, and every step in this direction increases the difficulties of life to families whose incomes are limited or small.

"English servants," says Mr. Escott, "are not in good repute. They are often idle, exacting, thankless, incompetent, wasteful, and dishonest. There are a few English households in which not a single English servant is kept. . . . There are

* Holyoake's "History of Co-operation in England," vol. i. p. 311.

German and French nursemaids, the cook is Belgian, the parlourmaid Swiss, the footman Italian." *

In the colonies, and all new countries, to find British servants at all is extremely difficult, and in old countries the facility varies with the state of trade. When trade is brisk marriages within the class are more numerous; and other means of making a livelihood presenting themselves, there immediately follows a scarcity of good domestic servants. It is social pressure alone that causes ready consent to domestic bondage, and the inference we are compelled to draw is this. In the future, as social pressure decreases, through the exercise of conjugal prudence, few persons will be willing to become servants, and the only natural and adequate escape from this difficulty lies in the wide adoption of the associated home system. The central truth in reference to this question is, that unconscious evolution has already carried us beyond that stage of our history, when domestic masterhood and servitude of adults, was an appropriate relation. This relation is at the present moment ill suited to the human nature around us, and in the future Unitary Home, domestic servants would prove incongruous. Service without servitude, the banishment of every species of slavery, with the introduction of the voluntary fulfilment of every domestic duty, is an essential element in the new social order, an element, however, not likely to be attained rapidly or suddenly.

Meanwhile the *initial* movements that will lead up to the disappearance of domestic slavery are to be seen in various directions. I know of two families in which the young members have chosen to undertake domestic work, in preference to the alternative of elders suffering from domestic worry; and I have been told of three ladies living in a good house in a London square, who, disliking to have a servant, do all for themselves except rough work, which is left to a charwoman.

But more significant still is the fact that girls of birth and breeding are everywhere sick to death of " waiting to be married." They are shaking off enforced passivity, pressing into fields of labour, and bravely buckling on their armour to enter the arena of active life, and find their true function amidst the " eager, striving, bustling crowd " around them. Mistakes at present are often made. Ambition is taken for

* " England, its People, Polity, and Pursuits," T. H. S. Escott, vol. ii. p. 6.

capacity, and girls thoughtlessly choose occupations for which they are mentally and physically unfit. Change of sentiment, however, will tend to counteract this evil. False ideas of gentility will cease to sway the mind, and when every kind of useful work is perceived to be noble, and if well done, highly creditable, our numerous Fanny Dovers will instinctively seek and find their true places in the order of industrial life.

Experimental Unitary Homes are sure to be of mixed character, in conformity with the mixed outward and inward conditions of life, in a transitional epoch. Some method will be adopted by which servitude may be modified, so as to lose the repulsive aspect of serfdom. If a class of persons, inferior in point of gentleness and training to the actual members of the home, are required for duties that are at present regarded as menial, they will work by contract,* and not on the principle of personal subjection to superiors. And I make this assertion, because it is clear to me, that the only possible stable homes for the future will be those based upon principles of distinct equality, and individual freedom. The difficulties of domestic service, however, may prove much less serious than they look at present. In talking with a refined gentlewoman of those difficulties, she remarked : "I go to my kitchen every morning, and order dinner for eight persons. I could as easily arrange a dinner for forty or fifty, and if in doing so I discharged the duties of perhaps six lady-housekeepers, these ladies would have their nerve force set free for other domestic functions in our Unitary Home." Economy of labour would result, not only from union of function, but from the superior amount of thought bestowed upon the problem of how to work efficiently. Labour-saving machinery may also come into service. In this direction invention has yet done very little in domestic life, because the rich are able to pay for labour, whilst the poor in isolated homes can ill afford the outlay that new methods always entail. By combination the economic difficulty may be overcome and intelligence be stimulated to the discovery of mechanical appliances for the relief of domestic service.

But the great question of economics, and of agreement in the arrangement of economic details, is a fundamental one in reference to the Unitary Home. It was the rock upon which

* This suggestion is made in an article entitled "Home Life of English Dwellings," in the *Westminster Review* for January, 1875.

the ship of the Tytherly experiment struck and foundered. Too much money had been spent upon the hall. The experimental ship was launched with "insufficient capital to last while the new order of life consolidated itself, and the conditions of industrial profit were found."* In the more cautious experiments (as I conceive them) that will ultimately succeed, the stability of the Home will not depend upon the industrial profit arising from association. I mean that individuals not associated in industrial life, will unite in domestic life, and share the household expenses. The apportionment of these expenses will be a matter of somewhat delicate adjustment, to be dealt with only in a spirit of broad, generous humanity. It is clear, that the human nature that is mean and selfish in money matters, composed of misshapen social units in whom acquisitiveness or love of personal property amounts to a deformity, is not the material from which to elaborate a Unitary Home.

At the same time the entire relinquishment of individual property would be unnecessary, and in tentative experiments unwise. Individuals must reserve power to withdraw, should the experiment not increase their happiness, and means to re-establish themselves according to the old domestic system, without detriment to the Home itself. There will, doubtless, occur some failures of adaptation in individual cases; but these, I think, will not prove numerous, since the mixed conditions necessary to progress may now be understood and secured. They are—first, a home life full and satisfactory, in the midst of a large group of individuals, the group consisting of adults and small families of children co-operating in close community of interests; second, an industrial life, outside the home, in which the individual is temporarily subjected to the prevailing system of competition. This environment will tend to the improvement of human nature, whilst not rendering it unfit to grapple with the special difficulties of a transitional epoch. In Isaac Taylor's work on Home Education he thus speaks of the family home: "No disparagement, no privation is to be endured by some of the little community, for the aggrandizement or ease of others. Along with great inequalities of dignity, power, and merit, there is yet a perfect and unconscious equality in regard to comforts, enjoyments, and personal consideration. There is no room either for grudges or for individual solicitude. Whatever may be the measure of good for the whole, the sum is distributed without a thought of

* "History of Co-operation," G. J. Holyoake, vol. i. p. 311.

distinction between one and another. This single circumstance would enable parents to cherish with more success those bland sentiments, the development of which it favours. Refined and generous emotions may thus have room to expand, and may become the fixed habits of the mind, before adverse principles have come into play. Home is a garden, high walled towards the blighting north-east of selfish care." Again, "Within the circle of home each individual is known to all, and all respect the same principles of justice and kindness. There is therefore no need for that caution, reserve, or suspicion, or for those measures of defence and restraint, which in the open world are safeguards against the guile, the lawlessness, and the ferocity of a few, and which are never altogether out of sight or out of mind. But these operate to depress very much the level of the generous sympathies, and to chill or deaden the happiest emotions of our nature. It is otherwise within the republic of home, where the most absolute confidence, and an unchecked good-will may, and ought always to prevail. On this ground we possess . . . a main means for raising the happiest feelings to a high pitch, and for keeping them there."* Isaac Taylor understood the nature of man, and the mental and moral atmosphere essential to elevation of character. But what would he say in our day, of the myriads of homes within which there is no atmosphere of moral purity or genial love? In very sooth, the home he dreamed of, as "a garden high walled towards the blighting north-east of selfish care," has no existence. The bitter wind of anxious care penetrates into every home, and blights the tender blossoms of our noblest life. Man's generous sympathies, the happiest emotions of his nature, do, in the very bosom of our often miserable homes, continually flag and languish even unto death. Isaac Taylor was deemed a practical man, yet his ideal waits its fulfilment now. We of the nineteenth century have to create, to call into existence the true republic of home, which after Isaac Taylor's ideal shall be free from selfish care, free from all suspicion, guile, lawlessness, and capable of raising happy feeling to the highest pitch, and keeping it steadfastly there. The means by which we shall accomplish this are, I believe, as Mr. Mill foresaw, the combining of the best parts of Socialism and of the old system, and therefrom evolving a renovated social fabric.

"Plato's Republic, Sir Thomas More's Utopia," said a

* "Home Education," pp. 33, 34.

practical man of the world to me, "are quite as realizable as your scheme, and we are as little likely to see *your* scheme acted upon as *theirs*. Do you not see that in the hydropathic establishments that dot the country, buildings answering precisely to your picture of what is required, associated life is tried by an immense number of the community and found wanting. You have there meals in common, society in the evenings, recreation rooms, and so on ; yet no one finds the life satisfactory. Ten days of it makes a man sigh for his home with all its drawbacks ; and (this given with emphasis as a crushing argument) hydropathics are for the most part *commercial failures.*" Now observe, to practical minds of this order I have no hope of commending my scheme. Unless there are, throughout our community, minds practical, and yet like Isaac Taylor's mind, *logically idealistic* also, we are not prepared for any great and definite step in social progress. The tendency of the age towards association is seen in the birth of hydropathic establishments, although the purely commercial spirit has dominated the movement, and through competition almost worked its ruin. The growth of social feeling is also seen within these hydropathics. Individuals from various social classes, of heterogeneous nature and interests, are there promiscuously thrown together, yet gentleness and kindness prevail, and it is extremely rare to see the rules of the house disobeyed, or to hear any anti-social feeling expressed. So far from resembling the Unitary Home, however, these establishments are destitute of all the essential elements of a home. Without knowledge of one another's characteristics and antecedents, how can rational beings closely unite, and live together in open-hearted confidence ? As is natural, the inmates of a hydropathic are reticent, self-defensive, and inwardly suspicious of the companions, whom they may possibly never meet again, or possibly may encounter in the world of social and political life as competitors and rivals.

That the hydropathic is a shade less expensive and more homelike than an ordinary hotel, is all that can be said in its favour. The domestic comfort it presents is by no means great, whilst privacy and comfort *combined* are not attainable within its walls. The household management and service are based upon individual interest alone, and without self-assertion the socially weak may have their rights infringed by the socially strong. The principles of socialism have no part in the system. The institution is simply one of commercial enterprise, and pressure in that field of enterprise has brought about the failures which my illogical friend irrelevantly uses as argument against the introduction of a wholly new, or at least an entirely different system!

In the order of unconscious evolution these establishments simply indicate the growing tendency of humanity towards union, and some fitness in humanity for the great step of conscious evolution, viz. the adopting of an Associated Domestic System. Other indications of approaching fitness are from time to time brought before the public, such as when Mr. S. P. Unwin, writing to the *Pall Mall Gazette* from Shipley, of a successful series of social entertainments given to the working classes, says : " I should like to notice the immense improvement in manners, bearing, and social disposition that has taken place among this great manufacturing population during the last few years. Among so much that makes good men and women sad and fearful of the future, it is a good and hopeful sign. Twenty or fifteen years ago it would have been impossible to have conducted such assemblies as those I have been describing; they would have broken up in rude horse-play; but in this undertaking there has been no difficulty, the behaviour of the people to each other has been charming to witness."

This " social disposition " is not, however, of itself sufficient to warrant decisive change of system. Fitness for a superior life implies advanced moral character, rather than mere sociability; and whilst in a well-formed Unitary Home a comparatively *unsocial* individual should find it pleasant to live, there are certain moral qualifications, without which the stability of the Unitary Home could never be ensured. Let me make this point unmistakably clear. The individual fit for a Unitary Home, is one who is prepared to give up liberty in some directions for the sake of special advantages and liberty in others. He must renounce liberty to domineer over others, and liberty to be wayward in living. He must maintain equitable relations, and be wary of trespassing on the liberty of his neighbours. He must be prompt and thorough in fulfilling his set duties and engagements. He must be reasonable and patient in conferring with fellow-members, ready to promote freedom in the expression of opinion, and to view without reproach all mere personal eccentricity in fellow-members. He must be able to take a generous, not a rigidly self-exacting view, of the proportional expenditure. He must be free from class prejudices, and personal pride, and above all, he must abhor anger, jealousy, ill-feeling, and quarrels.

Now, in what social class are we likely to find these qualifications? Not amongst the rich, with whom domination has become the daily habit of the mind, who value the position in conventional circles that money gives, and deem themselves superior, because they can indulge in luxuries and pride. And alas! not amongst the labouring poor, the great mass of whom as yet are hasty in thought, uncontrolled in action, and incapable of any sensitive appreciation of true social relations. The rough training of the children of the poor utterly spoils them for a superior state of life, and even makes that state incomprehensible to them.

The people most likely to be ready for the initiation of a new order of life are to be found, I expect, only in the middle class,—people of gentle training and small incomes who are at once free from vulgarity and class pride. To these, the economic motive will be a strong motor force, prompting to the successful working out of the problem, how by association and mutual help to secure the equivalents of wealth, and such society within the home as will make it an abode of peace, where the labours of every day are relieved by the constant return of tranquil pleasures and heartfelt delights. To these people there is another strong (although negative) motor force to the putting forth of effort in this direction. I mean the impossibility of securing any satisfactory training for their children under the present system. Home education is not within their means, the high-class colleges and boarding-schools are too expensive; the alternative (they have literally no choice) is day-schools of an essentially vulgar order. In these schools whole armies of children are drilled into uniformity in the matter of intellectual acquisition, whilst the formation of character is in no way attempted. The excitement of school life with its competitive examinations, its bursaries, its clubs for every form of child-amusement, is intense, and throws home life victoriously into the shade. To the children home is dull, monotonous, sometimes even hateful, and parents are year by year losing more of their hold upon the young, whilst the sentiment of family life is tending to disappear. There is to my mind, in the exigency of these disastrous conditions, no escape save one—the Unitary Home. There alone will parents of moderate income be able to give their children, without undue expense, the surroundings of happy childhood, the systematic supervision and training that will fit them for responsible existence, the abundant companionship which is absolutely necessary to healthful life in the offspring of a socially developed race. There, children and the one parent at least will be under the same roof—not, as now, separated for many hours of each day. A tender mother, of capacity, will be able to take cognizance of the children's occupations from hour to hour, to watch the evolution of intellect and play of emotion, to direct the desires, control the passions, form the habits, and maintain her rightful position in the hearts of her children, and her sway over the delicate intricacies of their inner lives. The systematic teaching of different branches of education may be supplied by members of the home, married or single, who devote themselves to the fulfilment of that social function by choice, and have carefully fitted themselves for it by a preliminary course of adequate study at Girton College, or elsewhere.

The formation of Unitary Homes is likely, I think, to begin in the neighbourhood of large towns, where the independent industrial life can be secured. The organizing and arranging of domestic matters will be complex, and of course somewhat difficult, entailing the creation of various committees for the regulation of affairs, including amusements. Mr. Nordhoff says in reference to the numerous committees he found existing in the Communes of America : "At first view these many committees and departments may appear cumbrous, but in practice they work well." When we call to mind, how throughout our native land in all large towns there are hundreds of young men who live in lodgings solitary and homeless, whilst in many country districts there are hundreds of girls daily pining for a practical sphere of social activity, and a wider, fuller life of healthy emotion, it is impossible to doubt the existence in our midst of a strong, vigorous social force, capable of overcoming all obstacles, however great, if only it is rightly directed into a happiness-producing path.

The policy attached to the Unitary Home is of a high order. It is, that each member helps toward the happiness of all, in the sure hope that in giving others happiness, the happiness of each will be enhanced, as it never could be by any other policy. The valuable sentiment of home-pride will diffuse itself throughout the dwelling. Each member will take pleasure in enriching and improving, by his inventions and exertions, the home that is dear to him, and a keen zest will be added to this pleasure, through the knowledge of how many tender friends and gentle fellow-creatures will directly benefit in his labour. Loneliness, that intolerable burden to Humanity—that secret ache that drives so many solitary beings to despair, will be banished from home life; so also will be banished the gnawing pain of anxiety and care regarding the ways and means of living; and everywhere the sympathy that civilized man keenly longs for, will flow freely in the channels of a highly sensitive and organic social life. The Unitary Home will have its quiet corners full of comfort for the aged, where, till the last breath of life is drawn, the love and cherishing of many friends will be their portion; and the young will live in closest union, boys and girls allied in study and in play, till mutual experience of their diverse human nature becomes the ground of knowledge, and kindles sympathy to friendship and to love.

Finally, upon the creation of associated homes, three important social changes will ensue. First—the experience of old age will no longer run to waste. It will permeate and leaven domestic life. Second—the ardent warmth, the healthy vigour of youth will be utilized. It will become the stable factor of general happiness. Third—the strength of manhood will prevail to discipline society and everywhere to reduce it to order, purity, and peace.

CHAPTER XVII.

MARRIAGE.

"We may define marriage as that union of the sexes which is most in accordance with the moral and physical necessities of human beings, and which harmonizes best with their other relations of life."—RICHARD HARTE.

MR. ESCOTT is of opinion that of late years there has occurred amongst us a change in the theory of marriage. "The central idea, the very type of marriage with the English girl," he says, "used to be home, but in the higher strata of (English) society, girls marry in a large proportion of instances not that they may become wives, mothers, mistresses of households, but mistresses of themselves, and are often goaded into it by a sense that a fashionable mother finds them inconveniently in the way.... The spirit of feminine independence after marriage ... often asserts itself in a manner comparatively new to English society." "The wife creates for herself a little world of her own, in which the husband only figures as an occasional visitor." "It is impossible to deny that the relations of husband and wife show often an increasing laxity. ... The flirtations of girlhood are perpetuated, or reproduced in what was once, the staid and decorous epoch of matronhood. Nor is it merely that such things are; they are conventionally recognized as existing, and when recognition has been once won for a fact or a custom, it has practically obtained a social sanction."* Mr. Escott goes on to describe the daily life of a modern English matron, but he does not give its companion picture, viz. the daily doings of her fashionable husband. He has his club life into which his wife is not invited to enter. Within their home he has his smoking-room,

* "England, its People, Polity, and Pursuits," T. H. S. Escott, vol. ii. pp. 15, 16, 17.

where he prefers to sit with masculine companions. He is a sporting man. His afternoon resort is Hurlingham. He flirts as freely as his wife—and *equal licence* is the principle of matrimonial harmony.

The explanation of this greater laxity in the social and domestic life of our upper circles, Mr. Escott finds in the close intercourse that now obtains between France and England, and the tendency of the English race to imitate French manners and customs. The French have no doubt in a great measure discarded the pompous, artificial stiffness of the past, and made society easy and pleasant. They exalt social life into an art, whilst they leave domestic life in a corrupt and barbarous condition.

But evolution in each distinct race, or people, moves upon independent lines, and the changes that we perceive to be advancing in our social condition, and that affect every class from the highest to the lowest, are not due to any surface or superficial movement bearing directly on the upper circles alone, but arise from social forces that are national, and deep-seated in their source, and certain to be far-spreading in their ultimate development. When Mr. Escott says, "the very type of marriage with the English girl used to be home," I fail to understand him; and when he expresses himself thus—"the central idea" of marriage used to be home, I fail to participate in his regret that the theory of marriage is altering. If a man is tired of bachelor life, and marries because he wants a home, or if a girl accepts a husband because she would rather be mistress of a household than remain a comparative cipher in her father's house, the union effected has a purely mercantile aspect, and is by no means a relation of life which the instructed can view with unmixed reverence and approval.

But putting aside for the moment the question of theory, and turning from the upper circle, and the plutocracy or wealthy class, I affirm that in the social strata of our great middle class, the conditions of marriage have vastly altered within the last sixty years, and three powerful factors have produced these changed conditions.

First—The growing intelligence and growing prudence of the mass of the people, under pressure of population, checks marriage or causes it to take place later in individual life.

Second—The increasingly great disproportion of the sexes makes marriages fewer in relation to population. Polyandry and polygamy are forms of marriage suitable to a society where the sexes are numerically unequal; but when we have monogamy as the established form of marriage in a society containing a vast preponderance of the female sex, the conditions are anomalous and, as regards the past, unprecedented. In no country of the world hitherto has there been reared, year after year, thousands of female children whose environment in adult life is certain to preclude the possibility of their ever entering into a relation which humanity teaches us to regard as pre-eminently normal, and true to nature. In savage communities surplus female infants are commonly destroyed, and the absence of a civilized sentiment—reverence for human life—is favourable to the happiness of a people whose development has not advanced beyond the stage when the purely physical and sensational enjoyments predominate over all other. *We* are far removed from *that* stage; but with all due regard to the moral, intellectual, and emotional requirements of civilized man, marriage must ever keep its place as the supremely natural, the happiest condition of human life.

Mr. Escott observes that "marriage is and will continue to be, the grand object in life, to every young Englishwoman." I differ from him here. There are plenty of young Englishwomen who perceive that if they make marriage the grand object of their life, they must vigorously compete with other girls for a husband, and with the instinct of a true refinement they turn from the task. In past ages there has been fighting for wives, capturing wives, and various modes of purchasing wives; but competing for husbands is a feature of modern society only; a new position, in which men become coy, and women aggressive.

The third social factor to which I allude is the advancing movement which will lead to the entire emancipation of woman. Independence strengthens character and is productive of nothing but happiness to a free self-governing creature. But, no married woman in this country is free, and the growing spirit of independence in the female sex carries changed conditions, productive of misery and bitterness, into the private relations of married life. Notwithstanding the late amendments to our marriage law, the inferior position of woman is still its fundamental basis, whilst the only possible condition out of which a pure and happy monogamy can arise is perfect equality of sex. The legal status of woman does not of course result from modern tyranny, and in the present generation of husbands a majority, I conceive, are adapted in sentiment to sex equality, which is the true principle of a high civilization. But there remains the large minority of husbands, whose nature is tyrannic in grain; and whereas formerly these marital and paternal despots ruled supreme without danger of rebellion in the domestic camp, now, the rising spirit of independence among women, and the tendency to insubordination amongst the young, are opposing forces that disturb equilibruim and create conjugal discord and social confusion. In the upper circles, as we have seen, an independent life of pleasure is led by husbands and wives, and the relations between them show increasing laxity. In the middle classes, separation or divorce is an increasing outcome of matrimony, and where a life of independence is possible through female self-support, matrimony is often avoided.

Now, under the predominant force of evolution, the three great factors of change, which I have indicated, will cease to act *similarly* on social life. The first and second factors will tend to decrease, whilst the third will increase, although ceasing to create disturbance. Population will not double itself in short periods when humanity is sufficiently enlightened and moral to exercise self-control in a matter which, above all others, affects its happiness; and the disproportion of the sexes will be far less extreme when social pressure no longer pushes our masculine youths to distant lands unmarried and alone. But the force that carries freedom and emancipation in its train, will advance, until the opposing forces of tyranny and injustice give way before it, and peace is won, through universal acceptance of the principle of sex equality.

In the social strata of our lower classes the conditions of marriage are not the same as in the middle classes. Pressure of population does not cause marriages there to be few. On the contrary, they are numerous, and often consummated at a very early age, whilst the birth-rate, embracing many illegitimate children, is out of all proportion high as compared with that of the middle and upper social strata. Along with the sturdy independence which makes the young throw off all ordinary parental guidance, there is, throughout this class, a growing spirit of dependence upon State aid, or private charity. The Poor Laws, and the widespread machinery of benevolence with which we surround the poor, have undermined the sentiment of self-dependence and pauperized vast multitudes. Religion turns, from the study and teaching of dogmas, to alleviate human misery; and the spirit of universal brotherhood which it inculcates has a twofold aspect. In the benefactor it breathes a tender sympathy; in the beneficiary it often begets a hard defiant claim of personal rights beyond the just limits of national welfare.

An utter absence of prudence, and an easy-going life of thoughtlessness, are the natural results of some of the social forces that press upon and mould our lowest classes. Girls and boys marry, or live together without marrying, and have children, at an age when the true welfare of society requires that they should be under strict guardianship, and the rule of domestic or social authority. What matters it, in their view, that no provision for the future of themselves or their children has been secured by personal endeavour? The children will be clothed and educated by the public;* in sickness charitable aid is always to be had, and in old age, there is the House or Poor-rates. No authority or regulative force follows the social units, who become parents when they are mere children, into their isolated homes. Our reverence for the freedom of the British subject forbids intrusion there; and the overcrowded, miserable condition of the poor in every town is now forcing itself upon public attention. In the words of a leading journal of December 12th, 1883: "Things are about as bad as they can be without creating an actual panic. Overcrowding is the rule ... what is the result? It is not so much an alarming death-rate, though the death-rate is often far too high. It is rather to bring down the general standard of health in a way of which no public record is kept. It multiplies the number of weak and ailing children, of workmen liable to sickness, of mothers unable to nourish their offspring; in short, it converts the nation into a feeble folk. It is bad health that leads directly to the intemperance of the poorer working men. If we mean to hold our own in the international race, we must try to elevate the physical and moral stamina of our working men."

On the question of *how* we try to do this, everything depends. If new efforts are put forth only upon the old lines, the moral and physical deterioration will simply advance with greater impetus. One hopeful sign I gladly note. In an account of the "Peabody Town, Bunhill Fields," given in the *Pall Mall Gazette* † I find this paragraph: "It goes almost without saying, that the superintendent watches carefully the increase

---

* I once heard a respectable workman say, "My children get their garments made at our Dorcas. A Dorcas is useful to the ladies : it teaches them to sew."

† Of November 21st, 1883.

of population, and when a family has been so enlarged that the apartment, or apartments, originally taken have become overcrowded, it must either move to a larger tenement, or quit the dwellings." Here, then, we have the attention of these tenants forcibly drawn to the advisability of limiting their families, and this policy is virtuous, although it may appear harsh when viewed in contrast with the vicious * policy of those benevolent ladies who supply all manner of necessaries to thoughtless, improvident mothers, on the birth of an infant.

How ought society to deal with thoughtless early marriages amongst the poor? This question is a fundamental one in relation to national health and well-being. I have shown that in our lowest class the Unitary Home system, with its valuable discipline and guardianship of the young, will not quickly take the place of the old domestic system that is rapidly decaying. The rough training of the poor, their low moral calibre, and tendency to spontaneous, irrational action, makes elevated domestic life impossible to them at present, and failure would follow any attempt to gather them into Unitary Homes. It is true that in the Familistère a large group of the French working class live in harmony and purity, but they have Monsieur Godin, as their inspiring genius and familiar friend. Should society, then, in our case condemn early marriages and by public authority restrain from marriage, boys and girls in their teens? Even if this were possible, it would be unwise. Early marriage is at least respectable, and an increase of sexual immorality would result from its restraint. The lowest Irish, though steeped in poverty and destitution, are yet far above the British in sexual morality, and this is universally imputed to their habit of early marriage.

It is not with marriage itself, but only with its evil consequences, in the too rapid birth-rate, that society ought to interfere. If immature persons marry, deferred parentage is what society has a right to require, and it should exercise moral pressure, through an enlightened public opinion, to that effect, and bring home to all parents the social responsibility they incur by the birth of a child; whilst private action of social reformers should teach the rational methods by which, in this class, the populating tendency may be kept within due limits, and prudence, foresight, and self-dependence be promoted. To other points of social reform within this class I will later refer; meantime I pass on to make this broad state-

* *Vicious*—because tending to evil.

ment—that throughout our whole society, from the upper circles downwards, early marriage is desirable, and that the prudence which restrains marriage in the middle classes is not a factor that tends to social purity and well-being.

The awful fact that in London alone there are upwards of eighty thousand prostitutes, reveals how far distant we are as yet from the national adoption of such sex relations as would be worthy the name of a *pure monogamy;* and this other startling fact, that as civilization advances, this great evil does not diminish but increases, and in this nineteenth century is fatally corrupting the merest children, calls for a thorough sifting of this hideous subject, and an exposing of the forces that underlie the painful social condition.* Miss Ellice Hopkins' work amongst ladies is strikingly significant. It shows that we are rapidly approaching the conscious epoch of evolution, and that even the strong sentiment of delicacy, admirable though it be, will no longer permit humanity to close its eyes to truth, however ghastly. Until the facts of human life are made pure and sweet, a delicate refinement *will* not and *cannot* dominate the whole mental atmosphere of social intercourse. Science has taught us in this age a reverence for truth unknown before. It is training us to the patient investigation of facts. It has led us to reject Puritanism as the best form of human life, and to give sex appetite an honourable position in our theory of life.† But it has not reorganized society, nor shown us how the normal healthful exercise of a general function is to become harmonious with moral and social well-being. It has simply launched the new conception into a society already tottering upon its Puritan foundation, and something like an immoral chaos has resulted.

I have drawn attention in Chapter XI. to the late Mr. W. R. Greg's opinions concerning the old Malthusianism, with its doctrine of late marriage. The strength of human passion in the majority of men and women makes the entire suppression of it in youth unnatural—a sacrifice to the well-being

* Whilst this work is passing through the press, the *Pall Mall Gazette* disclosures of July, 1885, have drawn wide public attention to this subject of the systematic seduction and violation of young girls. Of all the painful lessons that those dreadful revelations teach us, the most important is—that parents should instruct their children in all that relates to sex, and the dangers they may encounter in life ; for, it is well said that " unsuspecting innocence is the prey of systematic vice."

† I refer my reader to H. Spencer's " Data of_Ethics," p. 75. and to Buckle's " History of Civilization," vol. iii. pp. 271 and 297.

of society which comparatively few individuals only are capable of rendering, and further, that this animal instinct is redeemed from animalism by being blended with unselfish affections and ennobled by tender associations.

Neo-Malthusianism, with its power of subjugating the law of population and deferring parentage, is a new key to the social position, and by its use early marriage, judiciously surrounded by new conditions, will lead to national regeneration and the gradual extinction of sexual vice. "A thing to be carefully remembered," says Miss Martineau, "is that asceticism and licentiousness universally co-exist. All experience proves this; and every principle of human nature might prophesy the proof. Passions and emotions cannot be extinguished by general rules."* Let us then, as a community, grant this social science proposition, and set ourselves to purge out our ghastly licentiousness by early marriage, keeping steadily in view that the marriage aimed at must be capable of guarding purity of morals, constancy of affection, and domestic peace.

In my last chapter I pointed out how our poor attainment in the matter of home life had corrupted our ideal of home and lowered the family sentiment, and I showed that if we can create a worthy conception, a noble ideal of what home should be, in the general mind, it would prove a powerful factor in improving and elevating our domestic system. Now, similarly here, a noble *ideal of marriage* is an imperious want in this nineteenth-century civilization of ours. The general mind is utterly confused as to what are the true motives that should lead to marriage—what is the right attitude of mind in the relation of marriage—what consequences of marriage society is bound to control. And this need excite no surprise, for hitherto the whole subject has been wrapped in mystery and carefully hidden away, or coarsely and flippantly played with, or decked out in delusive fairy-like fancies. It is true that romances have altered their character of late years, and many of our best novel-writers are no longer fanciful. They are purely realistic. The scientific spirit of the day controls their genius, and what they represent is, things as they *are*, not things as they should be, and as we should strive to make them.

But in poetry at least, we may surely look for aspiration, for ideals that will stimulate to purer life and mould humanity

* "How to Observe: Morals and Manners," by H. Martineau, p. 169.

to higher forms. Alas! even poetry fails us here. Turn, for instance, to the later works of our Poet Laureate. In "The Sisters" he represents marriage as following naturally and worthily from passion inspired by the merest glimpse of a pretty face. Love at first sight is exalted into an honourable position, and the hero, a young man of supposed civilized mind and manners, basely forsakes the girl who loved him and whom he was prepared to marry, so soon as he has a second glimpse of his first attraction. After the two glimpses only, the hero deliberately soliloquizes, "But could I wed her loving the other?" Observe also, the girl who is forsaken defers to the sanctity and superior claims of love at first sight, and never perceives the flimsy superficiality of the lover. She dies in consequence of her disappointment, and later the worthless hero, posing as the affectionate father of two marriageable daughters, does not desire them to marry in order to increase their happiness, but "one should marry" (he says) "or all the broad lands . . . will pass collaterally." Love at first sight, then, and parental pride of property are motives for marriage, which the poet whom our nation has delighted to honour thinks worthy of his verse. Yet surely my reader will agree with me, that the first of these motives is respectable only in a savage epoch, and the second, during the feudal stage of man's development. Doubtlessly we have plenty of primitive men and women amongst us even now, but in poetry we look for ideals that bear fitness to the highest rather than the lowest type of the present age, and point us to the humanity of the future.

The civilized man has a large development of intellect, of emotion, of æsthetic and moral sentiment. In him the love, that is the only true foundation for marriage,* is no simple passion kindled by the glimpse of a pretty face, or any mere superficial contact whatever. It is, on the contrary, a highly complex emotion, and if it fails in complexity, if it does not embrace sympathy in thought, feeling, and moral aspiration, it will fail to sustain happiness in married life.

If this be so, however, dare we rationally expect that early marriages will result in happiness? and that the immature love of youth and maiden will expand into the fuller development of sympathetic thought and feeling, that complete union in mature life signifies? Under certain conditions, I think we

* In Mr. Mill's "Subjection of Women," p. 177, there is the only true ideal of civilized marriage which I have been able to find.

X

may. In the annals of the past I find recorded that Lady Sarah Cadogan, daughter of William, first Earl Cadogan, was married, at the age of thirteen, to Charles, second Duke of Richmond, aged eighteen. It is said that the marriage was a bargain to cancel a gambling debt between the parents, Lady Sarah being a co-heiress. The young Lord March was brought from college and the little lady from her nursery, for the ceremony, which took place at the Hague. The bride was amazed and silent, but the husband exclaimed, "Surely you are not going to marry me to that dowdy?" After the ceremony, his tutor took the bridegroom away whilst the bride went back to her mother. Three years after, Lord March returned from his travels, but in no hurry to join his wife. He went to the theatre, and there saw a lady so beautiful that he asked who she was. "The reigning toast, Lady March," was the answer. He then hastened to claim her, and their lifelong affection is much commented upon by contemporaneous writers; indeed it was said that the duchess, who only survived him a year, died of grief. Now, here we have a bright little picture of primitive simplicity. Marriage by purchase and arranged by the parents, cemented, however, by passion springing from superficial attractions only, and leading to simple domestic felicity, the human nature involved being comparatively uncomplex or homogeneous.

Would similar antecedents, however, be followed by similar consequents, amid the complexity of life and variety of human sentiments that surround us now? Undoubtedly no. Modern sentiment would revolt from early marriages arranged by parents for purely selfish purposes. The young would revolt from all marriage in which personal choice was not the prominent element; and although a pretty face excites passion, and in early youth especially prompts to marriage, the college-bred husband of the present day demands more than superficial attractions in a wife, if he is to enjoy domestic felicity. His demands are not as yet understood or responded to, and there is much truth in Mrs. Lynn Linton's picture of Bored Husbands. "We can scarcely wonder," she says, "that so many husbands think matrimony a mistake as we have it in our insular arrangements. . . . Few women take a living interest in the lives of men, and fewer still understand them. . . . They are utterly unable to comprehend his (a husband's) pleasures, his thoughts, his duties, the responsibilities of his profession, or the bearings of any public question in which he takes a part. . . . Most men are horribly bored at home, and the mass of them really suffer from domestic stagnation."

Isolated homes and uneducated wives drive men to clubs, and club-life has grown enormously in the last forty years, and further separated the sexes.* Nevertheless, some elements of a nearer approach are plainly to be seen. Billiard-playing within doors, and out of doors lawn-tennis, skating, etc., unite the youth of both sexes; and girls are fired with intellectual ambition, and emulate their brothers' student-life at Oxford or at Cambridge. Here, however, there is no union. Male and female students live widely apart; and the higher education of girls has had no appreciable result in enkindling any widespread intellectual sympathy between the sexes.

But marriage, in my opinion, might frequently precede, rather than follow after, the student-life; and the tutor need not intervene to carry off the husband, as in the case of Lord March. The youthful pair might together attend college, and co-operate in all the serious duties of youth as well as in its recreations. Mental development side by side would promote conjugal union, and mutual knowledge, mutual interests, and mutual tastes, would be immeasurably enhanced. These student years would of course only be a preparation for the full married life with its parental and social responsibilities. Public morality forbids parentage during years of immaturity, and to parents belongs the duty of inculcating moral social restraints. Together the young husband and wife would study physiology in all its branches—the laws of health in marriage and parenthood, and when family responsibilities came to be incurred no ignorant mistakes would mar the health and happiness of offspring.

My next chapter will treat of heredity, therefore I need not pause to touch on that subject now, but rather ask this question: Are parents alarmed at the mere suggestion of mixed colleges? Would they tremble for the safety of daughters studying—not in the conventual seclusion of Girton, Newnam, Saint Margaret's, Somerville Hall, etc., but side by side with a young husband or brother at Oxford or Cambridge; and if so, why? In America immorality has not resulted from mixture of sex at universities. I find recorded in the *Century Magazine* of March, 1883, that " the testimony of the colleges

* A social club for young men *and* women was some time ago started at the Workman's Hall, Bell Street, Edgware Road. The president is the Rev. Llewelyn Davies. This is a favourable sign of the times.

already open to both sexes, of academies and high schools, in the hands of men touching life at more points, with equal uniformity encouraged it. In theory it was averred that the girls would become mannish, or the boys effeminate; that the standard of scholarship would be lowered in concession to feminine limitations; and that sentimentalism would be developed, with a consequent deterioration of morals. In practice it was proved that while the boys acquired finer manners, the girls advanced in truthfulness, sincerity, and courage; that the standard of scholarship was raised, and that the predicted period of sentimentalism, though everywhere overdue, had persistently failed to appear."*

The following evidence concerning Antioch College is also reassuring. "At Antioch College in America the young men and women are educated together; and we are told that the influence of sex upon sex is admirable; the young gentlemen become refined in manner and habits, and their higher ambitions are stimulated; the effect is naturally stronger upon the older students . . . No restraint beyond that of absolute propriety is put upon their intercourse; the recitations and lectures are of course in common, and so are the meals, the refectory being arranged with small tables, at which the students form their own parties of four or six, eating and talking together like acquaintances in a restaurant. Out of study they associate as they please, and often get up picnics or excursions to visit the show-places in the neighbourhood. Preferences and attachments are inevitable; but these, honourably pursued, are not discouraged or interfered with. One instance has occurred of husband and wife taking their degree as Bachelors of Arts at the same time. One pair of best scholars, who had become attached while Seniors, and married after graduating, returned to the college to become professors."

It is precisely the separation of the sexes that makes college life dangerous to youth; and it fills one with profound melancholy to think how many fine young men might have been saved from debasing habits or a licentious life if the sweet and pure influence of mother and sister had been with them at the most critical period of their youth. "By the mere play of the affections," says Mr. Buckle, "the finished man is refined and completed." This thought, which he gives us concerning mother and son, I apply to the marriage relation: "The most sacred form of human love, the purest, the highest, and the

* Lucia Gilbert Runkle, in the *Century Magazine*, p. 685, March, 1883.

holiest compact of which our human nature is capable, becomes an engine for the advancement of knowledge and the discovery of truth; the most truly eminent men have had not only their affections, but also their intellect, greatly influenced by women."

A brother and sister are sent up to Oxford from a happy home. The influences of that home will encompass the lad far more closely than if he were alone; and many a wine-party will be eschewed for pleasures in which his sister and her lady-friends can safely join. There is no greater risk of immorality in mixed than in unmixed colleges. The risk is of early marriage, and sometimes of hasty and foolish marriage: and the question is, have parents cause to regard this risk as a danger too great to incur? In our transition state there are no conditions absolutely safe. The girl who pines for occupation and a wider life at home is in danger of losing physical, moral, and intellectual health, and so marring the happiness of her future husband and children, if marriage ever comes to her. The choice is not between good and evil. It is a choice of evils, in which the bias of established custom may cause parents to mistake the lesser for the greater evil.

Let us glance at the question from an economic point. A young man, who is maintained by his parents at college, loves a girl who is there also, and maintained by her parents. Marriage is approved by the parents, who will continue to meet expenses; and these are in no way greater than before. Neither a wedding tour nor a period of idleness is considered necessary. The young couple are in parental leading-strings, and gently kept to the performance of their duty in diligent study. But passion and affection are satisfied. Love is not mistaken for the whole of life, by either of them. It falls into its true place in sweetening the whole. It lightens and stimulates intellectual effort. It widens experience and broadens sympathy. It comforts in sorrow, and deepens the moral sentiment. It elevates and strengthens the whole nature.*

Now, I dare not assert that in every case the love so begun will continue warm and enduring to the end of life. But I am convinced that the chances of its doing so are great (for in

* I might here dwell also on sexual emulation in self-improvement. When ideals of what are the best qualities in each sex are better understood and taught, each sex will then require of the opposite sex these good qualities, with the result that development in virtue amongst the young will be vastly stimulated.

the period of youth they will grow to each other intellectually and emotionally), and infinitely greater than if the lad at twenty-five, when established in his profession, had married the girl from her home (where the busy idleness of a life of pleasure had been her sole occupation) after a few interviews at bazaars, balls, concerts, and lawn-tennis tournaments.

To parents, then, would it be a hardship to support their children in the first few years of marriage, to guide and control them at the most critical period of life, and launch them into independent careers married rather than single? A mother would have to sacrifice pride. She would not be congratulated upon her daughter's brilliant marriage. But I address myself specially to the parents whose strongest desire is, that the happiness of their children may be secured, and who will pursue that end unbiassed by considerations that are purely selfish. These parents, however, may say to me, "Why suggest our sending daughters to Oxford or Cambridge so long as the men's colleges there are not open to women?" I reply, that my aim is to give what aid I can in educating public opinion, and creating healthy sentiment on this subject of the mixture of sex, and with this aim anticipation of the future is unavoidable. Even now there are in Birmingham, Manchester, London, and elsewhere, educational institutions open to both sexes, where young men and women may pursue scientific and art studies together.

Some change of sentiment in respect to marriage is an all-important step of advance at the present moment. Parents should cease to regard marriage as the ultimate crown of life, which their sons may only hope to attain after years of patient waiting and earnest endeavour. They should regard it as a means to an end rather than an end in itself—a vital and vitalizing means, namely, by which the young will build up strong and elevated character, and achieve for themselves a widely useful and honourable life.

The British race presents to the world the "fair surface of a decorous society." It publicly professes monogamy, and preaches to the young a puritan asceticism; but behind the scenes the drama enacted would disgrace a civilization of the Middle Ages. Amongst the lower classes, wife-beating and wife-murder—amongst the upper classes, the hideous revelations of the Divorce Court—witness to the impurity and misery of our boasted monogamy. Privately we tolerate licence; we condone and studiously conceal the vicious propensities of youth—we harbour in our midst a social evil of gigantic magnitude; we permit hypocrisy to prevail; we instigate the young to form mercantile self-interested marriages; we have, in short, no lofty ideal of sex relations. We are corrupt in our social life—debased in our mental and moral conception of what life *should be*.

Truth, simplicity, justice, must form the threefold basis of a better system of things in which asceticism will be put aside as false to the facts of adult human nature, and *early marriage* will be everywhere encouraged and facilitated. Simplicity will take the place of luxury in domestic living, and happiness be aimed at instead of magnificence and show. Justice will give freedom within the marriage bond itself, creating outward or legal conditions that are neither immoral nor lax, but sufficiently elastic to permit of retrieving the errors of youth, and honestly or openly retracing steps that have converged to mutual dissatisfaction and misery rather than to mutual comfort and peace.

Marriage, as it exists, is, in hundreds of thousands of cases, an artificial sham. If we would make it pure and true we must concede that " the dissolution of marriage shall be as free and honourable a transaction as its formation." From divorce made easy, however, the British mind instinctively revolts, and I venture to assert that the mass of intelligent thinkers in the present day would prefer that the nation should hold its units strictly in bonds that are a sham and delusion, rather than that it should legalize the dissolution of marriages felt to be false. But why do we hold divorce in disgust and disrespect? Is it not because we have ourselves defiled it, and made its conditions odious? When its conditions are made reasonable and pure we shall respect it as necessary to general happiness. Meantime, Mr. Emerson's theory that in the conduct of life the great essentials are, to escape from all false ties, and to have courage to be what we are, is fully admitted, whilst the practical outcome of the theory, if openly advocated, would be fiercely opposed.

Not so in the case of some grave and noble thinkers of the past. John Milton held that one of the four *necessary properties* of marriage was " love to the height of dearness," and where that property did not exist there could not be true matrimony, and the pair ought not to be counted man and wife. " Christ Himself tells," he says, " who should not be put asunder, viz. those whom God hath joined." But "when is it that God may be said to join? when the parties and their friends consent? No, surely. . . . Or is it when church rites are finished? Neither; for the efficacy of those depends upon the presupposed fitness of either party. . . . It is left, that only then when the minds are fitly disposed and enabled to maintain a cheerful conversation, to the solace and love of each other . . . *that* only can be thought to be God's joining. . . . The rest, whom either disproportion, or deadness of spirit, or something distasteful and averse in the immutable bent of nature renders conjugal, error may have joined, but God never joined against the meaning of His own ordinance. And if He joined them not, then there is no power above their own consent to hinder them from unjoining, when they cannot reap the soberest ends of being together in any tolerable sort. . . . Undoubtedly a peaceful divorce is a less evil, and less in scandal than hateful, hard-hearted, and destructive continuance of marriage in the judgment of Moses and of Christ." *

Again, he says : " What thing more instituted to the delight and solace of man than marriage ? And yet the misinterpreting of some Scripture . . , hath changed the blessing of matrimony into a drooping and disconsolate household captivity, without refuge or redemption. What a calamity is this ! What a sore evil is this under the sun ! . . . Not that license and levity should herein be countenanced, but that some conscionable and tender pity might be had for those who have unwarily, in a thing they never practised before, made themselves the bondmen of a luckless and helpless matrimony." † To the objection that individual dispositions might and ought to be known before marriage takes place, Milton replies : " For all the wariness that can be used it may yet befall a discreet man to be mistaken in his choice; and we have plenty of examples. Nor is there that freedom of access granted or presumed, as may suffice to a perfect discerning till too late." ‡ " Who sees not," he exclaims, " how much more Christianity it would be to break by divorce that which is more broken by undue and forcible keeping, rather than to cover the altar of the Lord with continual tears . . . rather than that the whole worship of a Christian man's life should languish and fade away beneath the weight of an immeasurable grief and discouragement !" §

* "The Doctrine and Discipline of Divorce," John Milton. Edited by J. A. St. John, ch. xvi.
† Ibid., bk. i., Preface.   ‡ Ibid., ch. iii.   § Ibid., ch. vi.

And again : "He therefore, who lacking of his due in the most native and humane end of marriage thinks it better to part than to live sadly and injuriously to that cheerful covenant (for not to be beloved, and yet retained, is the greatest injury to a gentle spirit)—he, I say, who therefore seeks to part, is one who highly honours the married life, and would not stain it; and the reasons which now move him to divorce are equal to the best of those that could first warrant him to marry; for, as was plainly shown, both the hate which now diverts him, and the loneliness which leads him still powerfully to seek a fit help, hath not the least grain of a sin in it if he be worthy to understand himself." * "It is a less breach of wedlock to part with wise and quiet consent betimes, than still to foil and profane that mystery of joy and union with a polluting sadness and perpetual distemper; for it is not the outward continuing of marriage that keeps whole that covenant, but whatsoever does most according to peace and love, whether in marriage or in divorce, he it is that breaks marriage least; it being so often written that 'Love only is the fulfilling of every commandment.'" †

Milton wrote the essays from which I have quoted, about the year 1659, and nearly two hundred years later Mr. St. John, in re-editing them, calls attention to the fact that although every question connected with marriage and divorce is there discussed with surprising eloquence, learning, and freedom, these essays had produced no sensible effect on the laws or manners of the country. The Roman Catholic theory of marriage has prevailed ever since, and yet now it is, says Mr. St. John, "repudiated by perhaps a majority of those who are able to think for themselves. Well, however, might Milton inveigh against custom. That which has been long established is usually invested by us with a sacred character; on which account we continue to submit to it though conscious of the innumerable evils of which it may be the cause to us and others. The object of marriage must be admitted to be the happiness of those who enter into it, not their mere worldly prosperity, or the well ordering of their household and families, but in a moral and intellectual sense, their own individual delight and tranquillity of mind. Where this is not aimed at, marriage degenerates into a mere social connection for economical purposes, in which both husband and wife become

* "The Doctrine and Discipline of Divorce," ch. iv.
† Ibid., ch. vi.

subservient to the property they bring together, or may happen to amass." *

Since Mr. St. John wrote these words there has been a reaction in the minds of social units able to think for themselves. They no longer repudiate indissoluble marriage. Individual freedom has brought such social disorganization that further freedom is instinctively dreaded, and intellectually, we would rather submit to the innumerable evils of our present state, than face the unknown dangers that easy divorce might bring. Meantime, industrial life, accompanied by social pressure, has brought us into the exact position that Mr. St. John so well describes. In our upper and middle classes innumerable marriages are mere social connections for economical purposes. Money is the main instigator to marriage, the important element in matrimonial contracts. Husband and wife are subservient to property, and individual delight and tranquillity of mind play little part either as the motives to marriage or as its result. Meanwhile, in the lowest class, promiscuous unions and unlimited child-bearing depress the standard of national health, and weaken the social order of the State.

It is an interesting, although a somewhat humiliating study, to compare our civilization with that of some of the ancient Grecian States. In the Spartan State, for instance, where the higher culture and the higher arts were discouraged, and science was unknown, the common sense displayed by legislators and people is marvellous to look back upon. The prominent idea, in the minds of both, was the strength and permanence of the State, and if individual freedom was adverse to the welfare of the State, or individual interests clashed with the general interests of the community, the former were unhesitatingly sacrificed. The position of Sparta made war frequent and necessary, and the rearing of warlike men a public duty. All efforts, therefore, were directed to this end. Great attention and care were bestowed on the physical training of women. From their earliest days they engaged in gymnastic exercises, and when old enough entered into contests with each other in wrestling, racing, throwing the quoit.

In this way the whole body of citizens knew a girl's powers; there could be no concealment of disease; no sickly girl could pass herself off as healthy. The girls mingled freely with the

* Editor's preliminary remarks to "The Doctrine and Discipline of Divorce."

young men. They came to know each other well. Long before the time of marriage they had formed attachments, and knew each other's characters. All Spartans girls married except the sickly. No sickly woman was allowed to marry, for *offspring must be healthy*. And indeed, if left to consult her own feelings, the sickly girl would refrain from marriage. The age of marriage was fixed, special care being taken that girls should not marry too soon. In all these regulations women were not treated more strictly than men. Men were practically compelled to marry. The man who ventured on remaining a bachelor was punished in various ways. Men were also punished for marrying too late, or for marrying women disproportionately young or old. By the Spartan system the main object of the legislators was achieved.

For four or five hundred years there was a succession of the strongest men that ever existed on the face of the earth, and they held supremacy in Greece for a considerable time through sheer energy, bravery, and obedience to law. The women helped to maintain this high position as much as the men. They were vigorous and beautiful, patriotic, and remarkable for moral courage. The wives were true to their husbands, the husbands fond and proud of their wives, and adultery was almost entirely unknown. Lastly, a legitimate pride in physical health and purity of blood was felt and expressed. A poor maiden was asked what dowry she could give to her lover? and her answer was, "Ancestral purity." *

How few of our poor maidens could with equal truth boast of this endowment! and yet we live in a scientific age, when rational means to rational ends are within the compass of an intellectually developed race. Our ends are wider and fuller than those of the Spartans. We want something more than strong and brave men, healthy and beautiful women, to perpetuate fitly the British race; but as a nation we have not yet had the common sense to recognize, as the Spartans did, that marriage and generation are facts to be looked to as of vital national importance, and dealt with without that secrecy which is the invariable accompaniment of vice. Nor have we like the Spartans trained the sentiments of youth in accordance with national welfare, nor formed our regulations with impartial justice, treating men as strictly as women, or (should that seem the preferable course) women as unstrictly as men.

* *Vide* an article on "The Position of Women in Ancient Greece," by James Donaldson, *Contemporary Review*, July, 1878.

It is a very striking feature of the present age that the action of putting immoral women under the stringent control of the police, and young girls under protection of the law, is regarded as the best possible remedy for social evils in which the other sex is deeply implicated.* The Spartans appear to have been before us in the knowledge of this truth: that national happiness and virtue are not likely to spring from unequal treatment of the sexes, or from injustice in any form.

"The Romans admitted three kinds of marriage—the '*confarreatio*,' which was accompanied by the most awful religious ceremonies, was practically indissoluble, and was jealously restricted to patricians; the '*coemptio*,' which was purely civil, and derived its name from a symbolical sale; and the '*usus*,' effected by a simple declaration" of marriage.† This last form became general in the empire and had this important consequence—the woman so married remained in the eyes of the law, in the family of her father, and under his guardianship. Another important consequence resulted from this form of marriage. Being looked upon as a civil contract entered into for the happiness of the contracting parties, its continuance depended upon mutual consent. Either party might dissolve it at will, and the dissolution gave both parties a right to re-marry. Now, observe, this gradual loosening of the marriage bonds in Rome was synchronous with the dissolution of public morals, and the setting in of a period of great corruption. To the casual observer of history, therefore, the experience of the Romans appears unfavourable to the wisdom of a policy that facilitates divorce. Mr. Lecky, however, clearly shows that there were independent causes for the vast wave of corruption that flowed in upon Rome, and these would, under any system of law, have penetrated into domestic life. He lays down the proposition that, "in a purer state of public opinion a very wide latitude of divorce might probably have been allowed to both parties without any serious consequence." To this he adds a significant observation, viz. "Of those who in Rome scandalized good men by the rapid recurrence of their mar-

---

* I denounce the system as irrational as well as unjust. Nothing so rapidly demoralizes people as to show them no respect, and be continually harassing them. So long as we have prostitutes at all, we are bound to preserve in them as many good qualities as possible, whereas by the above system we destroy the good and bring out all the bad qualities of their nature.

† Lecky's "History of Morals," vol. ii. p. 322. Second edition.

riages, probably most, if marriage was indissoluble, would have refrained from entering into it, and would have contented themselves with many informal connections, or if they had married would have gratified their love of change by simple adultery."

To the student of social evolution the present position of the British Empire is not analogous to that of the Roman Empire approaching its fall. Bad as our social state is, false and impure as are the relations of sex amongst us, public opinion and sentiment are so far firmly fixed. Monogamy is felt to be the *one* form of marriage suitable to an advanced civilization, and we need fear no relapse into polygamy, polyandry, or any of the barbarous forms out of which the civilized man has emerged, and from which his moral nature revolts. The purification of our monogamy, not its dissolution, is the goal to which we shall advance; and under the rule of a pure public opinion *latitude of divorce* is certain to prove a direct means of the purification which we are bound to achieve.

If my reader will refer to Mr. Spencer's "Principles of Sociology," vol. i. p. 788, he will there find described the natural course of evolution in marriage. When "permanent monogamy was being evolved, the union by law (originally the act of purchase) was regarded as the *essential* part of marriage, and the union by affection as non-essential; and whereas at present the union by law is thought the more important, and the union by affection the less important; there will come a time when the union by affection will be held of *primary* moment, and the union by law as of secondary moment." Then occurs the condemnation and reprobation of marital relations in which affection has melted away and the union regarded as sacred has naturally dissolved. In some civilized countries this stage has been entered upon.

Let us glance at results in two of these countries which are in advance of ourselves in respect to the sacred institution of marriage. In the United States, although divorces are numerous, it was publicly declared at a meeting held at Piccadilly, London, in June, 1883, that there is no country where the relations of husband and wife are more respected and tender, and, I think, American light literature gives ample proof of the truth of this statement. In France, the law makes no provision for divorce, and even legal separations are few. But if we compare French romances with American novels, it is in the former, not the latter, that we find pictured a domestic life that is corrupt, and sentiments that are immoral, on the subject of marital relations.

In an article on the increase of divorce in America, in which Mr. Gladden examines the statistics for the different States, and compares one with another, he observes : " It is not true . . . that the multiplication of divorces is accompanied by a corresponding increase in crimes against chastity." And again : " We trust, therefore, that the figures we have been studying do not indicate an increase of immorality corresponding to the increase in the number of divorces. The trouble (that is the increase of divorce) is institutional rather than ethical. It is not the vice and corruption of society that are assailing the family." * One of the remedies suggested by Mr. Gladden is, that means should be taken to prevent ill-assorted marriages, for these and rash marriages furnish, he says, the soil from which the many divorces spring.

Switzerland holds a high place amongst civilized countries. The simplicity, industry, contentment, and moral purity of her people have long been recognized, yet in words taken from a letter to the *Times*, of September. 1883—" Switzerland is the most *divorced* of European countries. In only one city— Berlin—are divorces relatively so frequent. They are less frequent among the easy classes than among small traders and working folks, for in good society divorces give rise to scandal, and are not regarded favourably. But amongst the masses no scruples are felt, and young married couples at the end of a twelvemonth frequently join in an application for divorce on the ground of irreconcilable incompatibility of temper. As the Courts accept a *joint* application as sufficient proof of the alleged incompatibility, the young couple can always obtain the loosening of these bonds."

Now, amongst us, how fare the young couples who find after a year's experience that matrimony has been fatal to happiness? Our masses are quite as ignorant and perhaps more impulsive than the Swiss, therefore they are quite as liable to make mistakes. If poor, there is no divorce open to them. Divorce with us is the privilege of the rich. Their history may be traced in the police court—the annals of crime—the fearful statistics of prostitution and drunkenness that mark our progress, and annually give evidence of individual misery and wretchedness. Public opinion unaided by law checks divorce in good society amongst the Swiss, for scandal is disliked and

* The *Century Magazine* for January, 1882, p. 414.

avoided, whereas in our good society (notwithstanding our more stringent marriage laws) scandal is by no means disliked. The society journals are literally filled with gossip and scandal, with details of the private life of members of the royal family and individuals of the upper class, and the more discreditable are these details, the more are they certain to be read. Prurient curiosity and vulgarity of mind are not absent but widely present in our upper circles, and the hideous revelations of the Divorce Court furnish ample pabulum for the growth and nourishing of these noxious social weeds.

It is not only in countries where divorce is made easy that divorce has increased. In England, says Mr. Gladden, the divorce laws have not been essentially altered since 1857. But whereas, in 1860, the petitions for divorce and legal separation were one to every six hundred and twenty-eight marriages, they had risen in 1875 to one for every four hundred marriages. Now, when we call to mind the only cause for which divorce is obtainable, and the nature of the details of private life that are scrutinized and actually set forth in public newspapers, to meet the eye and contaminate the pure mind of youth, surely we must admit that the path chosen by Germany, Switzerland, America, in this matter of divorce, is quite as likely to promote national dignity and purity of thought and action, as the path which we continue to pursue. Also we must admit that there was far-seeing wisdom in Milton's thought that "peaceful divorce was the least evil and the least in scandal; that it were better far to part with quiet consent betimes, than to pollute and profane that mystery of joy and union that marriage ought to be." Divorce will decrease, not by the forcible holding together of units that do not and cannot combine, but by the elevation of humanity—the growth of intellect and moral sentiment—and, as Milton expressed it, "that freedom of access granted, or presumed, as may suffice to a perfect discerning before it is too late."

Society may do much in the prevention of ill-assorted unions; but it can do literally nothing in the way of healing, or curing, when the affection, which made union sacred, has dissolved. Dissolution of marriage before parentage occurs ought to be a very simple affair, requiring only the expressed desire of both parties before a responsible public officer or registrar, and so inexpensive as to be available for the poor as well as the rich, thereby removing all temptation to illegal or illicit connections, and also repressing the social tendency to inquisitorial interference with the strictly private matters of domestic life.

The really important *social* epoch of marriage is at the birth of the first child; and after that event, divorce should be more difficult and carefully guarded by society, in the interests of the child or children. These interests, however, may often be far better secured by parental divorce, than by enforced parental union, and private friends, chosen by the parties and approved by the legal court, should be charged with the duty of private investigation and settlement of the questions at issue. Arbitration of mutual friends acting under the dictates of common sense and from experience of the human nature around us, is a social force likely to adjudicate more justly and wisely, in reference to these delicate relations of life, than public officials applying the antiquated regulations, which are survivals of an old *régime*, neither well adapted to the conditions of modern life, nor to the complexity of modern humanity.

The law concerning parental guardianship, which efforts are now being made to alter, is a manifest survival from the epoch when man ignorantly believed that father and child were more closely and vitally related than mother and child! We are surely sufficiently advanced to perceive that the parental relation does not so differ, and to demand that for the two sexes equality of parental rights and duties should be recorded in the statute book, and justice be made the basis of our marriage laws.

Legal changes are required in *two* directions, viz. towards greater freedom as to marriage and greater strictness as to parentage. The marriage union is essentially a private matter with which society has no call and no right to interfere. Childbirth, on the contrary, is a public event. It touches the interests of the whole nation. The simplest form of marriage is the best. We should do well to follow the ancient Romans, and adopt the *usus* with its natural consequence—that the wife if young remains under the guardianship of her parents. But as regards childbirth we must become Spartans, and with firmness and fidelity assume censorship, and, to some extent, control over action that affects the order of society, and the entire well-being of the coming race.

In reference to both, our customs at the present moment are destitute of common sense. We sanction, by religious services, the mercenary marriages that in our hearts we condemn. We emphasize with social pomp and ceremony the commencement of unions that in our present evil social state may speedily terminate in private misery and public disgrace, whilst we view with patient serenity and without public comment or rebuke, a careless, heartless assumption of parental responsibilities and an utter neglect of a great social duty, viz. that of transmitting only to the incoming generation ancestral purity and physical, mental, and moral health.

In a scientific age man is bound to recognize physiological reasons for early marriage and physiological reasons for delayed parentage. Of the former, I have here to say that an early moderate stimulation of the female sexual organs (after puberty is reached) tends, by the law of exercise promoting development of structure, to make parturition in mature life easy and safe; and that the healthy functional and emotional life of the marriage union is the best preventive of hysteria, chlorosis, love melancholy, and other unhappy ailments to which our young women are cruelly and barbarously exposed, and which, I do not hesitate to say, make them in many cases feel their youth to be an almost *insufferable* martyrdom. There are no less serious sexual evils which overtake masculine youth, if chaste and debarred from early marriage, *i.e.* persistent and mentally irritating cravings, self-abuse, etc. To these I need not, however, further allude.

Conscious evolution demands that the eyes of the public shall be opened to the *necessity* for marriage, and the great difficulties that lie in the way of true marriage, in order that all obstacles to human happiness, save such as are *inevitable*, may be bravely grappled with and overcome. The ineffectual gropings of unconscious humanity propelled by blind necessity are touching to enlightened minds. That civilized men and women, who for mutual happiness require the closest unions of affection, should ever have invented the miserable device of *matrimonial agencies*, and that these should be commercially profitable,* speaks volumes on the unsatisfactory state of our social life. Many newspaper advertisements under the heading "Matrimony" are genuine, and genuine also was a poor man's written appeal to the superintendent of a workhouse: "Sir, I am in lodgings and have no home. Could I have a tidy woman from your institution which could make me a tidy wife?"

Not so much with the uneduca'ed, however, as amongst the educated and those of gentle birth is there lacking, that free-

* The *Matrimonial News* journal is here referred to, and there are various matrimonial agencies on the Continent.

dom of access which might suffice for that discerning which is undoubtedly the only sure antecedent to *true* marriage. There are in the present day thousands of men situated as was Anthony Trollope during seven critical years of life. He lived in solitary lodgings in that great London, where he says, "There was no house in which I could habitually see a lady's face and hear a lady's voice. No allurement to decent respectability came in my way." How was he to spend his evenings? The alternatives before him were, a life *not* respectable, or in solitary lodgings, to sit for hours "reading good books and drinking tea." Frankly he tells us, "It seems to me that in such circumstances the temptations of loose life will almost certainly prevail with a young man. . . . The temptations at any rate prevailed with me."* Later in his "Autobiography" he speaks of his marriage, and adds, "I ought to name that happy day as the commencement of my better life."

A social system that, on the one hand, leaves thousands of educated women without daily intercourse with educated men, and therefore renders them unable to comprehend a man's pleasures, thoughts, duties, the responsibilities of his profession, or the bearings of any public question in which he takes a part. and, on the other hand, leaves thousands of gentlemen utterly unable to "habitually see a lady's face and hear a lady's voice!"—such a system is a failure, and worse than a failure in respect to necessary qualifications for promoting the happiness and moral well-being of man.

Let all social reformers aid in *radical* reform, viz. in the evolving of a *new system* which will bring men and women together in close relations of a frank, pure, and healthy character—friendships, that are intimate. affectionate, faithful, to the benefit and enjoyment of all, and true marriage based on discernment of individual nature for all who desire to marry in so far as disproportion of sex will allow.

Wealth must no longer figure as the chief promoter of human happiness. Wealth must be discrowned, and many products of wealth will have to be renounced. It is not in pomp, glitter, or show, in luxurious and extravagant living that the *new system* will be evolved. It will declare itself in simplicity of domestic and social life. by sympathetic combination in all the sacred duties and varied enjoyments of life. and by the clear enunciation of the principle that in the sphere of the emotions man finds his highest and purest solace, in universal kindliness and love the satisfaction of his being.

* "An Autobiography," by Anthony Trollope, vol. i. p. 69.

CHAPTER XVIII.

HEREDITY.

"The first step towards the reduction of disease is beginning at the beginning to provide for the health of the unborn."—Dr. RICHARDSON.

"The laws of heredity constitute the most important agency whereby the vital forces, the vigour and soundness of the physical system, are changed for better or worse."—NATHAN ALLEN, M.D., LL.D.

WHERE individual life is not menaced by poverty or destitution, *disease* is the bane of existence, the great obstacle to human happiness. Calm resignation and even an exalted condition of inward peace are compatible with disease; but bright spirits, elastic exuberant joyfulness, are unattainable without some approach to sound health. And alas! how few of us have any permanent possession of sound health. Sickness and pain dog our footsteps from infancy up to maturity, and onwards to death; and that in spite of medical science, of sanitary protection, of enormous strides made during the last hundred years in the knowledge of pathological conditions, and of vast resources now at our command for subduing and mitigating every form of physical evil.

If I turn my thoughts to the life-histories of the friends I know well, everywhere I perceive wounds and scars left by our common enemy disease. I see mothers whose light-hearted buoyancy died out when the children whom they had guarded with the fondest care were carried to the grave; husbands deprived by death in child-birth of the chosen companions of their lives—the young wasting in consumption—the middle-aged bowed down by chronic ailments, the aged martyrs to rheumatism, paralysis, gout. In every direction there is disease made visible, or poison germs lying latent, ready to awake and carry sickness and sorrow into wide circles of human life. Nor is this direct physical suffering even, the weightiest evil in the case. Its indirect effects in sympathetic pain and sorrow, in the torments of mental anxiety and nervous fears, in the anticipation of unutterable griefs, these are beyond calculation, and yet so real, as to appal the mind that for one moment strives to conceive and dwell upon them.

The undeveloped nervous system of the savage, his weak feelings and callous nature, enable him to live his life at a tolerably even level. The height, depth, and breadth of emotion, in the civilized man, is to him unknown. Civilization has altered the nerve structure, developed and revealed its latent powers, and transformed a gross thick-skinned humanity into a vital instrument of innumerable strings, ready to thrill, to vibrate, and to throb, responsive to every wave that moves within its wide-reaching sphere. The waves, however, that circle persistently round us are those of sorrow. The delicate instrument of man's emotions vibrates to mortal anguish, and thrills to accents of despair. At no time in the history of our race has the capacity for happiness and the subjective need of happiness been so great; whilst at no former time has human misery been so openly revealed and keenly felt. In view of the complexity of man's nature, and of his civilized life, and in view of the intimate action and interaction of physical pain and mental sorrow, the great problem of how happiness is to be achieved overwhelms one with bewilderment and dismay. But hope persistently whispers, human reason has latent powers, science has resources as yet untried, and the application of *these* to the very mightiest of tasks *may*, nay *will* sweep the world forward to happy issues which we can but dimly foresee.

There is an assembly, says Dr. Richardson, of learned men, of earnest men, bent on understanding to the full human failures from health. These men spare no pains; to gain a spark of light they will labour like miners in a mine.* Observe, the subject-matter of these men's study is, " human failures from health." In other words, behind and below actual disease in all its forms, the *cause* or *causes* of disease are being diligently searched for. The world is at last beginning to "awaken to the fact, that the life of the individual is in some real sense a prolongation of those of his ancestry. His vigour, his character, his diseases are principally derived from theirs; sometimes his

* Unfortunately we have no earnest band of women devoted to this cause, although this field of inquiry is one in which science specially requires female aid.

faculties are blends of ancestral qualities. . . . The life-histories of our relatives are prophetic of our own futures." *Eugenics* * forms the new field of inquiry into which enlightened reason is now carrying its penetrative investigations, and in that field undoubtedly the great battle with disease must ultimately be fought out.

As yet, what medical and statistical researches have principally brought to light are some painful facts and a curious anomaly. The facts are the extreme prevalency of inherited disease, an increase of insanity, idiocy, and suicide, a deterioration taking place in physical stature, a degeneracy of the structure of the teeth,† and clear evidence that, as Mr. Galton expresses it, "our human civilized stock is far more weakly through congenital imperfection than that of any other species of animals, whether wild or domestic." ‡ The anomaly may be stated thus : "Whereas amongst savages the weak in body and mind are soon eliminated, and those who survive commonly exhibit a vigorous state of health," § an opposite condition results from our civilization. Physical weakness and degeneration predominate, and as regards the mental and moral state, I again quote the words of Mr. A. R. Wallace : "At the present day it does not seem possible for natural selection to act in any way so as to secure the permanent advancement of morality and intelligence, for it is indisputably the mediocre, if not the low, both as regards morality and intelligence, who succeed best in life and multiply fastest."

Now, I have in previous chapters explained what are the moral, emotional, and social factors which have counteracted the law of natural selection and deprived us of its benefits in the survival of the fittest. I have shown that we have long been strong enough as a nation to withstand adverse forces without, whilst protecting weak members within our circle.

---

* In the work "Inquiries into Human Faculty," from which, at page 44, the above quotation is taken, Mr. Galton points out that we greatly want a brief word to express the science of improving stock, and he proposes the adoption of *"eugenics"* from the Greek *eugenes*.

† The *Lancet* endorses an opinion recently delivered by Mr. Spence Bate, F.R.S., and long since maintained by the few thoughtful dentists, that human teeth among the cultivated classes are in process of degeneracy. The dentine is becoming deteriorated, interglobular spaces not found in the savage races making their appearance, while the enamel is becoming opaque."--*Spectator*, September 18th, 1883.

‡ "Inquiries into Human Faculty," F. Galton, p. 23.

§ "The Action of Natural Selection on Man," by A. R. Wallace.

We can afford to be generous, and accordingly we have become so. The cruelty that prompted our savage ancestors to destroy the feeble, the maimed, the incapable amongst them, has no existence in our nature. On the contrary, there has been developed in us both a reverence for life that will not permit us callously to let a fellow-creature die, and a benevolence that has impelled us to the discovery and use of innumerable methods for partially subduing disease and warding off its final catastrophe. Finally, there has grown up within us a moral sense of justice that *revolts* from selection by the blind mechanical forces of external nature, and teaches us to oppose to these the vital, tender forces of a universal human sympathy. Natural selection has given way to sympathetic selection, in which man has played an important part, and indiscriminate survival has taken the place of survival of the fittest. All this has occurred in the epoch of unconscious evolution. But now we stand upon the threshold of a new epoch, and conscious evolution is awakening us to the fatal consequences of our own unconscious action. The truth is dawning upon us, that pure benevolence and moral impulse have forged the weapons wherewith we frustrate our own best designs. Whilst earnestly desiring the happiness and goodness of all mankind, we have kept alive and generously supported the miserable amongst us—the physically, intellectually, and morally weak. These have propagated their kind, extending thoughtlessly the action of their disorganized frames and tainted blood. Population has too rapidly increased. The intelligent and thoughtful have refrained from parentage, and the general result comes out in deterioration of national health, an appreciable loss of vigour and vitality in the white races of mankind.

The diseased condition of the stock from which we spring first presented itself to my mind through a remark made by an old, experienced medical man. I was relating to him how, in a family which he knew well, two members had fallen victims to the same organic disease. With regret shadowing his face he said gravely, "Well, in all my large practice that family was the one I believed to have the greatest purity of blood." Can it be possible, I thought, that impurity of blood is the rule and purity only the exception? Health is, I am convinced, the basis of happiness. How can man be happy if all the varied springs of his existence are poisoned at the source?

But now let me ask, Is this melancholy reflection supported by acknowledged facts? In so far as great pestilences and loathsome diseases are concerned it appears almost certain that no civilized people has suffered *less* than we do now; and the average of life during the last half-century and more has been steadily, though slowly, increasing. Progress in sanitation and in medical science, however, accounts for this. More care is taken of life; hence there is an increase in life periods. When all due allowance is made for the more favourable *external* conditions, we are compelled to admit that the internal forces are less vigorous and less efficient than they were a century ago. Our progenitors were more healthy than we are; our great grandmothers had stronger muscles and firmer nerves. They had fewer doctors, but they needed the doctor less.

Let us think for a moment of the multitudinous advertisements of medicines in our day—the innumerable appliances for relief of suffering. "Ah!" says my reader, "these in many instances are not genuine remedies." True; nevertheless the fact that society pays for these advertisements, and all they represent, proves to demonstration the prevalence of disease, the sharpness of a suffering that seeks relief on every side and prompts to wild experiment when hope is flickering or slowly quenching in despair. Mr. Holloway alone expended £50,000 a year in advertisements! and this large sum, in addition to the cost of production of his drugs, and the enormous profits he reaped from his business, has been borne by invalids and their friends. Let us think of the number of doctors supported by invalids and their friends, and the subdivision of labour within their order; of the number of dentists with flourishing practices, proving the diseased condition of human teeth, an enfeebled power of withstanding adverse forces that penetrate and break down the dental structure; of the huge hospitals and infirmaries that dot the country (St. Thomas's Hospital in London alone embraces six large buildings); and the numerous asylums for the insane. One is driven to the conclusion that, although pestilence and other scourges of the past no longer sweep away or thin the ranks, humanity has made *no gain* in that sound health which *must be laid*, the only sure foundation for man's happiness.*

* On the 15th of February, 1885, Dr. Hime, B.A., the medical officer of health for Bradford, stated publicly that no less a sum than £200,000 is spent on disease in a year in Bradford. This did not include loss of wages or cost of funerals. He gave his method of computation, and said that his calculation tallied closely with one made on an entirely different basis by the great Dr. Farr, whose estimate was, however, higher. Dr. Farr calcu-

"In point of health, our children in these times," says Dr. Richardson, "are our reproach. Where is there a healthy child? . . . You may put before me a child in all its innocence. It has done no wrong that it should suffer; it may show to the unskilled mind no trace of disease, and yet I know that if I or any skilled observer were to look into the history of the life in question it cannot be found intrinsically sound. . . . It is sure to have some inherited failure." Now observe, Dr. Richardson speaks of the "history of the life;" but surely a babe has no life-history of its own? It is only a flower in the bud, an aggregate of possibilities. It is precisely these possibilities, however, to which he refers. The potential, though not the actual, adult life lies there. The character of the future man— his physical, mental, moral qualities—are transmitted from his parents, and knowledge of the stock from which he springs gives, not exact, but approximate knowledge of what his chances are; that is, whether he is most likely to suffer failure or achieve success in life. As regards physical health, Dr. Richardson's view is firm and pronounced: "We are as yet," he remarks, "unacquainted with all the phenomena of disease that pass in the hereditary line, but we admit the following as proved:—phenomena of scrofula or struma, of cancer, of consumption, of epilepsy, of rheumatism, of gout, etc. It would be wrong, however, to limit the hereditary proclivities of disease to the above list. . . . The further my own observations extend of the present, from experience and experiments; of the past, from historical reading; the stronger is the impression made upon my mind that the majority of the phenomena of disease have a certain hereditariness of character." * Sir James Paget, on this subject, stated before a committee of the House of Lords, in 1882, "We now know that certain diseases of the lungs, liver, and spleen are all of syphilitic origin, and the mortality from syphilis in its later forms is every year found to be larger and larger, by its being found to be the source of a number of diseases which previously were referred to other origins." Sir William Jenner dwelt on the transmission of syphilis to the offspring of a diseased parent.†

lated that "in England during 1871 there were 1,029,758 persons constantly sick, a number sufficient to fill 10,298 hospitals, each holding 100 patients, and the numbers were sustained by an annual influx of 12,357,090 patients, each ill for a month on an average, of whom 11,842,217 recovered and 514,879 died."—From *Bradford Observer* of February 17th, 1885.
  * "Diseases of Modern Life," p. 38.
  † This account is copied from the *Times* of August 11th, 1882.

In America, as well as in this country, an advancing deterioration of race is noted and commented upon. "A gradual change is taking place," says Mr. Nathan Allen, M.D., " in the organization of our New England people. . . . At the present day we have a larger class of diseases arising from general debility. . . . The framework of the body generally is not so large . . . the countenance is paler, the features are more pointed, and not so expressive of health, though more so of intelligence. The nervous temperament, with all its advantages and disadvantages, is becoming too predominant for other parts of the body. As one of the consequences, we have more diseases of the brain and nervous system, more sudden deaths from apoplexy, paralysis, and also from diseases of the heart. This change of organization has occurred principally within the last two or three generations." It is imputed by Mr. Allen to various social causes, which have, he says, " exerted a pernicious influence upon the *female* constitution, and so upon the laws of inheritance."* Amongst these causes he places, "the increasing use of alcohol in its various forms, and especially of tobacco," and he remarks, "Could the evils of alcohol, tobacco, and opium, as transmitted by *hereditary influences*, be fully realized, what more powerful motives could be presented for a reform in their use, or for their absolute prohibition?" The gravest aspect of these laws of inheritance relates to the transmission of no positive disease, but of a feebleness of vitality, a general incapacity to meet the conditions of life. " In sound healthy stock " (I am again quoting Mr. Allen's words) " we have developed in a far higher degree the recuperative powers of nature, and generally one single disease at a time to combat; whereas, in case the original constitution is feeble . . . diseases of almost every kind become far more complicated, and their treatment more difficult, as well as doubtful in result. . . . We venture the assertion that all permanent improvement or progress in the civilization of any people or nation is more dependent on *these laws* (the laws of inheritance) than upon any other agency whatever." †

Now, the whole theory concerning heredity, and its marvellous influence for good or evil, is a nauseous draught for mankind to swallow. No wonder we revolt instinctively from a doctrine that charges tender parents with transmitting an evil

---

* " Medical Problems of the Day." The annual discourse before the Massachusetts Medical Society, June 3rd, 1874, by Nathan Allen, M.D., LL.D. † Ibid., p. 34.

heritage to the offspring whom they passionately love; and although many important books draw attention to the facts, so far as they are ascertained, these momentous facts have as yet made no impression on the general mind. The extremely varied conditions under which the physiological law of inheritance acts, makes it comparatively easy for an unwilling mind to disregard the logical conclusions that ought to be deduced from it, and applied to the regulation of conduct. Disease often becomes transformed in passing from one generation to another. The failure of healthy organic function, which in a parent shows itself as paralysis, may, in a child, be scrofula. A mother's consumption may become insanity in her son.* A family tainted with disease may show a generation comparatively sound, and in the next there will reappear the morbid action of inherited weakness in, it may be, every member of the group.

Biological phenomena do not display the sameness and precision of mechanical phenomena, although equally determined by unalterable laws. If, however, we could perceive all the conditions, outward and inward, and take them into account, the line of causation would be clear to us, and disease would be found no more an accident than the storm that breaks upon the seaboard, or the volcanic flames that burst from the mountain-top. That we cannot perceive the whole conditions makes no difference to the truth of this thesis. The antecedents in every case are there, and the largest proportion of these antecedents are unquestionably some heritage of weakness transmitted from parents, some disabilities for healthy life, resulting from a *bad descent*.

But let us look more closely at this matter; and in order to understand disease, we must first of all enquire, What is life? "Life in all its forms, physical or mental, morbid or healthy, is a relation; its phenomena result from the reciprocal action of an individual organism and external forces; health, as the consequence and evidence of a successful adaptation to the conditions of existence, implies the preservation, well-being, and development of the organism, while disease marks a failure in organic adaptation to external conditions, and leads,

---

* "Watching the decay of a family, it is often seen that phthisis and insanity are of frequent occurrence amongst its members; and when extinction of it occurs, when the last of the family dies, he not seldom dies insane or phthisical, or both."—" The Physiology and Pathology of the Mind," H. Maudsley, M.D., p. 204.

therefore, to disorder, decay, and death." We shall better understand these relations of organism and environment, by glancing first, for a moment, at the disease of madness. "When it is said that mental anxiety, produced by adverse circumstances, has made any one mad, there is implied commonly some inherent infirmity of nervous element which has co-operated. Were the nervous system in a state of perfect soundness, and in possession of that reserve power which it then has of adapting itself, within certain limits, to the varying external conditions, it is probable that the most unfavourable circumstances would not be sufficient to disturb permanently the relation, and to initiate mental disease. But when unfavourable action from without conspires with an infirmity of nature within, then the conditions of disorder are established, and a discord, or madness, is produced." *

It is the outward circumstances of life that often decide the nature of a disease, whilst inherited infirmity is nevertheless its primary cause; and we perceive that a man, liable to madness under mental anxiety, may, under temperature too severe, or fluctuations of temperature too great, develop, not madness, but consumption, the heritage of weakness appearing in a breakdown of the structure of the lungs, rather than in a breakdown of the still more delicate structure of the brain. "Though the child is not the exact algebraic sum of his parents, yet both are constantly expressed in him." † If both or either of the parents are diseased, the child may or may not inherit the particular disease of the parent, but he will certainly inherit a constitution liable to some "kind of morbid degeneration, or a constitution destitute of that reserve power necessary to meet the trying occasions of life." ‡

The word disease, as here applied, must also be extended to the moral as well as the mental and physical condition of man. I mean that moral defects are *distinctly transmissible*, and as they pass from one generation to another may tend to physical and mental degradation. I have already pointed out the massive proportions of our national acquisitiveness, or love of property. I have shown that our industrial age has swelled this sentiment to undue proportions, and in many individuals it has become a moral deformity. The millionaire, the bold speculator, may be morally diseased. All men, in fact, are so

* "The Physiology and Pathology of the Mind," p. 199.
† "Blood Relationship in Marriage," Arthur Mitchell, A.M., M.D.
‡ H. Maudsley, M.D.

whose sole aim is to become rich. A life spent in the pursuit of wealth alone, saps the altruistic element in human nature. It makes a man egoistic or unsympathetic, and in his person humanity has relapsed or deteriorated. What is the result? "If," says Dr. Henry Maudsley, "one conviction has been fixed in my mind more distinctly than another by observation of instances, it is that it is extremely unlikely such a man will beget healthy children. . . . In several instances in which the father has toiled upwards from poverty to vast wealth, with the aim and hope of founding a family, I have witnessed the results in a degeneracy, mental and physical, of his offspring, which has sometimes gone as far as extinction of the family in the third or fourth generation. . . . I cannot but think, after what I have seen, that the extreme passion for getting rich, absorbing the whole energies of a life, does predispose to mental degeneration in the offspring, either to moral defect, or to moral and intellectual deficiency, or to outbreaks of positive insanity under the conditions of life." *

Now, in our present state of general confusion of thought, it is not the members of our lowest social classes only, who are mechanically and instinctively bringing children into the world without regard to regulative principles. In no social class is there any systematic attention paid to *eugenics*, that is, to improvement of the human stock. Ignorance and prejudice are holding mankind in the old paths of ordinary custom, although these paths lead onward to misery as the portion of coming generations.

Intelligent thought in every age, however, is in advance of popular custom, and Mr. Galton has already brought before the public a scheme for race regeneration, which we have now to consider. Mr. Galton's problem seems to stand thus. How shall man be able to elevate the intellectual, moral, and physical standard, and keep up the numbers of his race, by increase of the superior breed, and decrease and gradual elimination of the inferior breed, until the former shall finally crowd out the latter? His policy is based upon facts collected by Dr. Matthews Duncan in regard to relative fertility in early and late marriage. He shows that a group of a hundred mothers, whose marriages and those of their daughters should take place at the age of twenty, would, in the course of a few generations, breed down a group of a hundred mothers, whose marriages and those of their daughters were delayed until the age of twenty-nine.

* "The Physiology and Pathology of the Mind," p. 206.

Let us then, he reasons, promote, by every means in our power, the early marriage of human beings of superior quality, whilst we discountenance early marriage in those social members who are less favourably endowed. And "few," he says, "would deserve better of their country than those who determine to live celibate lives through a reasonable conviction that their issue would probably be less fitted than the generality to play their part as citizens." *

In examination for official appointments he would have attention paid to a candidate's ancestral qualifications as well as his personal ability. The man of inherited sound constitution and average ability should be preferred to the man of superior ability who belongs to a delicate and short-lived family. The former will in all probability become the more valuable servant of the two. Some scheme should be devised by which to bestow marks for family merit, to put, as it were, a guinea stamp to the sterling guinea's worth of natural nobility; and this, he conceives, might set a great social avalanche in motion. It would open the eyes of every family, and of society at large, to the importance of marriage alliance with a good stock; it would introduce the subject of race as a permanent topic of consideration, and lead to a careful collecting of family histories, and noting of those facts which are absolutely necessary for guidance in right conduct. Late marriage, as advised by Malthus, Mr. Galton utterly condemns. The prudent alone are influenced by that doctrine, and it is, he says, a most pernicious rule of conduct in its bearing upon race. His policy, then, is early and fruitful marriage for the best specimens of our race, and widespread celibacy in the case of those less highly favoured, whilst everywhere the sentiment should prevail, that *eugenics*, or the improvement of the human stock, is the primary consideration in marriage, and the guiding principle in sex relations.

Now, this theory I hold to be contracted, or one-sided, therefore defective; and the policy built upon it is misleading, and to some extent false. Mr. Galton ignores the fundamental principle of social life, viz. that the happiness of all at all times should be the aim and object of rational man, and he mistakes the quality of human nature in highly civilized man. To demand celibacy of men and women, whose defective organisms it is not desirable to perpetuate, would be in hundreds of thousands of instances to sacrifice unnecessarily

* "Inquiries into Human Faculty," p. 336.

present happiness to future gain—to build up the comfort and enjoyment of coming generations at the expense of the comfort and enjoyment of our own generation. The sentiment of justice as strongly repudiates this action, as it condemns the reverse position of a reckless, self-indulgent populating to the deterioration of the human stock; whilst reason distinctly shows that individual liberty in respect of marriage is a social necessity perfectly compatible with the well-being of all. True theories concerning marriage are essential in the founding of a pure and elevated social state, and physical regeneration of race will not be achieved by an overstrained morality, that does violence to the emotional human nature of the normal and average man.

Marriage and parentage are not necessarily conjoined. It is with the latter alone that society properly deals. As guardian of public health, and of the coming race, it is entitled, nay required, to forbid the propagation of disease; but since, by careful use of artificial checks, parentage, and hence the propagation of disease, may be avoided, all adult individuals of ordinary public morality or conscientiousness are free to marry, as spontaneous impulse dictates. *Eugenics* is *not* the primary consideration in marriage. If philoprogenitiveness is in some men the natural prompter to marriage, these men are at least rare. To put forward, therefore, as the correct motive to marriage the bringing into the world of healthy children, is to oppose the laws of human nature, and render more difficult the attainment of social purity and peace. In the normal man and woman, love is the true motive to marriage. The object of marriage is the happiness of the two who are united—their own individual delight and tranquillity of mind—and where not this but something else is primarily aimed at (even if that something else be no mean and despicable aim, such as money or social position, but the nobler aim of bringing healthy children into the world), marriage is degraded into a mere social connection, a tie that is not normal but in a measure false.

Love in a highly civilized race is not the sexuality of the savage. It differs in this respect that it is widely diffused. In Mr. Lester F. Ward's words: " It were foolish to deny that the seat and radiating source of all love between the sexes is the generative system, yet in a highly organized individual this sexuality is so thoroughly diffused that perpetual and enduring pleasures flow from the entire nervous system which are regarded as spiritual emotions."* The passion of love in its elevated form is properly the object of man's highest reverence. It must *not* be thought of as subservient to *eugenics*. It must take its true place in our minds, as the *essential condition*, in most lives, to individual happiness—a condition, moreover, of which society has no proper means of judging, and concerning which society, therefore, has no voice. Already men and women who love most, and most faithfully, are surrounded by difficulties and confronted by social barriers in forming a true marriage. Let us beware of a misleading doctrine which would increase further the obstacles to pure conjugal union, by regarding it as inextricably involved with, and indeed resting upon, the social forces of reproduction. This divergence from my subject of Heredity was unavoidable. Mr. Galton is our most advanced teacher in the field of *eugenics;* and it is important that my reader should clearly apprehend the points wherein his system of social reform differs from that scientific meliorism which it is the purpose of this work to set forth, and advocate.

The systems are alike in teaching that it is in man's power, therefore it is clearly his duty, to improve the physical, intellectual, and moral structure of his race, by intelligent forethought and careful action, in exercising the function of propagating his kind. Public opinion is exhorted to express abhorrence of all thoughtless, immoral inattention to the first principles of *eugenics*, or stirpiculture. Population must not be kept up by consumptives, or by persons whose pedigree is tainted by insanity, or any other disease known to be hereditary.† Such persons, Mr. Galton thinks, should live in celibacy; whilst I consider celibacy as in itself a vital evil, destroying individual happiness, and tending to social disorder. Wherever love in its highest form is a bond of union between two individuals, marriage is eminently desirable; but if either or both be afflicted by hereditary taint, the sacrifice demanded of them is to carefully abstain from giving birth to children. This sacrifice is not great to intelligent minds, because they foresee and shrink from the certain alternative of transmitting some organic weakness which would cause misery to offspring, and become to sensitive parents a perpetual and stinging

---

* "Dynamic Sociology," Lester F. Ward, vol. i. p. 610.
† Temporary illness ought also to be considered. It is when parents are in their best state of health only, that they are morally justified in bringing children into the world.

reproach. Men actuated by rational thought or moral feeling will sit in judgment on their own case and act in accordance with their own best interests, in so far as these do not conflict with the interests of the human race to which they belong. The ignorant, however, must be led into the right path by various social forces, such as prohibitory and attractive legislation, the influence of example, and the well-directed control of an enlightened public opinion.

Mr. Galton is not the only social science thinker who affirms the possibility of universal action which will slowly but surely eliminate unsound constitutions and eradicate the germs of hereditary disease, whilst steadily raising the intellectual and moral standard of the race. An American writer has lately said : "If there is one social phenomenon which human ingenuity ought to bring completely under the control of the will, it is the phenomenon of procreation. Just as every one is his own judge of how much he shall eat and drink, of what commodities he wants to render life enjoyable, so every one should be his own judge of how large a family he desires, and should have power in the same degree to leave off when the requisite number is reached."* It will not be by means of the miserable lonely lives that celibacy entails, but by the exercise of an enlightened self-control, that civilized men and women will refrain from having any family at all, because in their case "the reasonable conviction is, that their issue would probably be less fitted than the generality to play their part as citizens." I admit that human ingenuity has much to do as yet in solving the problem, how to bring this social phenomenon completely and universally under control of the will. Ingenuity, in fact, has not been exercised upon the subject, for the simple reason that there has been no frank and dignified drawing of public attention to the vital importance of the problem, and to the profound changes that its solution will inaugurate, in respect of individual happiness and social well-being. In the savage epoch of our history, the force of natural selection produced survival of the fittest. From that epoch we have long since passed into a humanitarian semi-civilized epoch, in which sympathetic selection produces a miserable state of indiscriminate survival; and now we wait the solution of the above problem, to pass onwards to a rational, wholly civilized epoch, when intelligent selection will systematically secure the birth of the morally, intellectually, and physically fit.

* "Dynamic Sociology," Lester F. Ward, A.M., vol. ii. p. 465.

APPENDIX TO CHAPTER XVIII.

I AM enabled to place before my readers the following interesting correspondence.

November 13*th*, 1878.

Charles Darwin, Esq.,

SIR,
You have so often invited correspondence on the subjects treated of in your most valuable books, that I trust you will pardon this liberty I, a perfect stranger to you, venture to take, of offering for your consideration some thoughts mainly originated by your writings.

For many years I have been accustomed to think sadly of the present condition and probable future of the human race. The works of Malthus and J. S. Mill, your own works, and some others, have so clearly pointed out the evils under which man strives, and how slow and cruel in their action are various forces that tend to better his condition, that it is with a great feeling of relief I have quite recently been brought to believe that there are forces at work of which I had previously little conception, which will in a comparatively short time, and in a wholly admirable manner, bring about that state of things which is so earnestly to be desired.

You say (in "The Descent of Man"), "It is impossible not to regret bitterly, but whether wisely is another question, the rate at which man tends to increase;" and further, that man "has no right to expect an immunity from the evils consequent on the struggle for existence." In regard to this last, with all respect, I am glad to be able to say, I, in great part, differ from you. I think, from the advance of civilization, which is so much a conquest over nature, and the growth of altruism, we have reason to hope for this immunity; and as I now think we can have it without any deterioration of race and decline of virtue, I am free to think it wise to regret the continuance of the pressure of population on comfort and subsistence.

It is my duty to be concise in what I have to say, in order to take up as little of your time as possible, in case my ideas should be worthless; but I hope in the very short statement of the main

z

results at which I have arrived, I shall still be able to make myself understood.

I believe I can point out, as now in action, two important laws of Race, to add to the one already so fully displayed by yourself. They are both naturally destructive of the action of the first law which is Natural Selection ; and the last, which is now in the first stages of evolution, annuls as it grows the action of the two preceding ones.

They each have existence for the same reason, viz. that they tend to greater adaptability of race with conditions, or greater strength against the forces which environ.

I summarize these laws as follows :—

The three great laws of Race Preservation in their natural order of sequence in evolution are—

First, the Organological Law—Natural Selection, or the Survival of the Fittest.

Second, the Sociological Law—Sympathetic Selection, or Indiscriminate Survival.

Third, the Moral Law—Social Selection, or the Birth of the Fittest.

These three laws arise naturally and gradually out of the conditions which precede each.

The first is the Physical Law, which governs all organisms in which no form of sympathy is yet developed ; it tends to greater strength in the unit or more adaptability of the individual to its conditions.

The second is the Psychological Law, which necessarily arises with the growth of sympathy, and is the natural opponent of the first, which it gradually supersedes. It tends to greater strength and adaptability in the aggregate, but to less strength and health in the unit.

The third is the Judicial Law, evolved as a rule of conscience for well-being. It gradually annuls the preceding laws while combining their beneficial results, on the basis of tending to greater strength and health, both in the aggregate and in the unit. It is the final outcome of Human Evolution in the order of forces governing race propagation. It is necessarily evolved in the mind by the interaction of reason and sympathy, and its development proceeds on the fact of artificial birth-control, unopposed to the force of sexual passion which otherwise would, with the weaker individuals, most certainly be too powerful to permit its action.

Of the first of these laws I need say nothing, except that I have been so bold as to name it " Organological."

Of the second I may say, I have formulated it from a consideration of much in your writings, especially of chapters iii., iv., and v., in the " Descent of Man," of portions in the writings of Mr. Herbert Spencer, Mr. A. R. Wallace, Mr. F. Galton, Mr. W. R. Greg, and others. Natural Selection was evidently defeated, and yet species continued to flourish ; so it seemed evident to me that a new law had been evolved, and this I set myself to discover. The word *sympathy* I have used in a wide sense, and as the quality meant has, as you point out, been most probably developed through natural selection, it exists in varying degrees of strength.

Of the reality of the third law there will be most dispute. That its evolution is proceeding, I cannot myself see reason to doubt; and that it is destined to act a most beneficent part in the future of mankind, I firmly believe.

As instance of its evolution I may mention the growing opinion that it is wrong for consumptive people and persons inclined to insanity and epilepsy to marry; the opinion, becoming more and more prevalent, that it is wrong to have more children than can be brought up well; the opinion that celibacy is an evil, and that asceticism is absurd; that the sexual passion is at the spring of much that is noble in life, and is nothing to be ashamed of, but requires only to be regulated; the inference that in no case is it wrong to apply knowledge to guard against natural evils, so long as no injury devolves on others by so doing; the conclusion that procreation is perhaps of all social actions the most important, and ought therefore to be most seriously regarded, and effected only under moral conditions; the opinion that tendency to vice is hereditary, and that it would be best for society if confirmed criminals were "put compendiously under water." And finally I may refer to the present painful conflict between reason and sympathy relative to the preservation of the weak and incompetent while they propagate their stock to the injury of posterity.

I think the extending force of the practice of the arts preventive of conception is in proportion to the capability in these arts of increasing adaptation to conditions within and without the human organism.

If it is a fact that they do increase this adaptability, it appears to me certain that their practice will increase to the extent of society.

The prejudice against them founds itself on the belief that they are in themselves immoral, or of immoral tendency, because social instinct is against them. But social instinct has, as you justly point out, been developed in favour of the general good of the species; it follows, then, that if the general good conflicts eventually with an instinct, instinct will in time have to adjust itself to the new conditions.

A physiological fact having relation to man and society is one among other factors in the determination of morals. The concealment cannot be defended; and if the knowledge of it is of use, it is hopeless to expect any attempt at concealment to be effectual.

If it be true that these arts do not increase adaptability to conditions, I see not how their manifest spread can be accounted for. I think their action is rapidly becoming a sociological fact of the gravest importance, which cannot be left out of consideration in any speculation on social tendencies. I need but refer to France and its extraordinary statistics of births in relation to marriages.

I gather that you fear much reduced social pressure would result in indolence. I submit that indolence is more a physical weakness than an acquired habit, and cannot, I think, be increased under "Birth of the Fittest." To those who love children will be left the task of bringing them up. This love is hereditary, and will increase by survival, and become a presiding force. It may not be utopian to expect that some day a medical certificate may be required to define the rectitude of adding a new member to society. The weak in body or mind may be cared for and protected so long as they conform to the social mandate not to continue their race. They may, to use Professor Mantegazza's words, "love, but must not have offspring."

In conclusion, I submit, the birth of the fittest offers a much milder solution of the population difficulty, than the survival of the fittest and the destruction of the weak.

I feel I take a liberty in speaking of any subject about which you must know so much more than I do. If I have been so fortunate as to make a true generalization, you will see it as such without many words from me.

My present intention is to further develop these ideas as long as I think them true.

I am, sir, with much esteem,
Yours truly,
G. A. GASKELL.

Down, Beckenham, Kent,
November 15*th*, 1878.

DEAR SIR,
Your letter seems to me very interesting and clearly expressed ; and I hope that you are in the right.

Your second law appears to be largely acted on in all civilized countries, and I just alluded to it in my remarks to the effect (as far as I remember) that the evils which would follow by checking benevolence and sympathy in not fostering the weak and diseased would be greater than by allowing them to survive and then to procreate.

With respect to your third law, I do not know whether you have read an article (I forget when published) by F. Galton, in which he proposes certificates of health, etc., for marriage, and that the best should be matched.

I have lately been led to reflect a little (for now that I am growing old, my work has become merely special) on the artificial checks to increase, and I cannot but doubt greatly, whether such would be advantageous to the world at large at present, however it may be in the distant future.

Suppose that such checks had been in action during the last two or three centuries, or even for a shorter time in Britain, what a difference it would have made in the world, when we consider America, Australia, New Zealand, and South Africa! No words can exaggerate the importance, in my opinion, of our colonization for the future history of the world.

If it were universally known that the birth of children could be prevented, and this was not thought immoral by married persons, would there not be great danger of extreme profligacy amongst unmarried women, and might we not become like to "arreois" societies in the Pacific?

In the course of a century, France will tell us the result in many ways. We can already see that the French nation does not spread or increase much.

I am glad that you intend to continue your investigations, and I hope ultimately may publish on the subject.

I beg leave to remain, dear Sir,
Yours faithfully,
CH. DARWIN.

P.S.—This note is badly expressed and written, but I have not time or strength to re-write it.

November 20th, 1878.

DEAR SIR,

I beg to thank you for your most courteous and encouraging letter. I shall devote particular attention to the points you raise, which are most important, though extremely difficult to deal with.

The very strength of the popular fear lest these new checks should lead to immorality, gives me some confidence that the human mind, so long trained in favour of that which tends to social order, will be able to withstand the greater license of new conditions without relapse.

Social change being evolutional is gradual; such disorder as may be prompted must therefore arise in detail, while social order obtains in the mass; disorder is disorganization, destruction of itself. I cannot conceive of the present order not being able to withstand the small corroding tendencies of disorder *met in detail*: surely it will outlive them.

The "arreois" societies are societies for death, not life; they are social suicides. The libertine and selfish natures, in furthering their own ends, will, I trust, further their own destruction, and so be eliminated from society, while order survives.

If I could conceive disorder to arise at one time from numerous centres, and grow in corrosive power until the combination of order should be destroyed by it, then would I fear the extinction of the human race; but disorder is of fitful growth and crumbles as it grows.

Without, I hope, overlooking the importance of colonization, there is much, I think, in what Mr. W. R. Greg says in his essay on "The Obligations of the Soil."

Colonization if slower would have one advantage—that it would be less painful. There is something about colonization at present, which reminds me of a panic in an assembly, where the people get jammed in the doorway. Subsistence is so difficult, that is, food is so dear, that emigrants may often view the fertile land they cannot cultivate for want of capital, or a year's provisions, and so be forced to turn away and starve. High pressure sometimes defeats its own ends.

There is certainly one great danger in lessened fertility of some races, viz. that the pressure of other races upon them might extinguish them. The lessened fertility commences in the races which are stronger socially; I trust they will endure. The nations guided by reason, could not long submit to having their standard of comfort lowered or their means lessened by the influx of an inferior race. I trust little to legislation, but its most useful action may some day be to preserve a civilized nation against the social encroachments of an uncivilized.

It was only that I could not find that what I call the law of sympathetic selection was formulated, that I ventured to draw attention to it. It is, as you point out, alluded to in your writings, and I am glad of the confirmation you give me.

The sympathetic are protective of their kind : the unsocial are left less protected. The law which might be called the survival of the sympathetic (the fittest socially) is a law of *protection* and survival, conducing to the compactness of the social organism, and therefore to existence. Natural selection is a law of *destruction* and survival.

I am hopeful that dispassionate study may help us to the resolution of several important questions. What I submit to you, I submit with much diffidence. I beg you will not let any feeling of courtesy lead you to reply to this letter ; I should be sorry to seem to give you this trouble, and much regret the state of your health.

I beg to remain, dear Sir,
Yours truly,
G. A. GASKELL.

Charles Darwin, Esq.

CHAPTER XIX.

EDUCATION *VERSUS* CULTURE.

" Next in importance to the inborn nature, is the acquired nature which a person owes to his education and training; not alone to the education which is called learning, but to that development of character which has been evoked by the conditions of life."—DR. H. MAUDSLEY.

To exalt culture is a common tendency of this age. The higher classes amongst us decidedly overvalue it. Nevertheless, there are many individuals who perceive, clearly enough, that culture is by no means the one thing needful, and that it may exist in respectable quantity and unimpeachable quality, where education is markedly deficient. A man may be an excellent Greek or Latin scholar, he may have classics at his finger ends, so to speak, and even a smattering of science frequently hovering on the tip of his tongue; he may be extremely æsthetic in his dress, in the choice of pictures to grace his walls, and colouring for his dining-room carpet, and yet fail each day, and many times in the course of the day, in the exercise of various useful faculties with which nature has endowed him.

Education is a term of wide signification. It implies the acquisition of knowledge, and especially of such scientific knowledge of phenomena as will lead to right action and the wise conduct of life; but that is not all. It also signifies the exercise of all non-injurious, useful faculties, physical, emotional, intellectual, and moral, in due direction, order, and succession; until, in the maturity of their strength, the individual applies them freely, in response to every stimulus from without, and lives a full, satisfactory life, harmonious with his own inward capacity and the environment, inorganic,

organic, domestic, social, and universal, of which he is the centre. We are each the centre of an environment more or less extensive, and to meet that environment all round with healthful, easy, pleasurable response constitutes the happiness and usefulness of human life.

In the beginning of life the faculties and environment are alike limited, nevertheless the healthy babe in his nursery breathes easily and pleasurably the air of the well-ventilated room; he crows with delight when the sun's rays stream in at the window; he jumps by impact with the floor when safely held in nurse's arms; he fingers and examines with concentrated attention every toy; he listens to every sound; he springs to meet his mother's caress, and smiles back in her face; and by the gratification of these and a hundred other infant activities, he stores up an unconscious experience which will by-and-by fit him for wider and fuller surroundings. In due time the infant becomes the man, with faculties mature and strong. He moves amongst men with dignity and self-respect, conscious of power to meet on every side the relations of human life. His industrial activities enable him with ease to supply the real wants of himself and those dependent on him. No base motive has drawn or driven him into marriage; and within his home his domestic instincts and his emotional nature are satisfied, whilst his intellect is fully exercised and interested, by the companionship of his wife, and in the rearing of his offspring. He responds eagerly and pleasurably to the social, civic, and political claims, that embody his share of public duty. If any worry overtakes him he deliberately lets his thoughts escape from the narrow interests of a transitory life to an environment as boundless as the Universe itself. The true principles of science taught him in his youth have been the foundation of self-culture. Astronomy, chemistry, physics, etc., have been his recreational studies; and now irritation of feeling is allayed by an intellectual range that gives a glorious sense of liberty. He sees above, beneath, beyond the world of strife and struggle. Impersonal Nature in its profound majesty, its unalterable sublimity, penetrates him; and at the promptings of a religious feeling, far removed from the abject, craven superstition of the savage, he relinquishes his passionate human emotions, and enters into the Universal Calm. When so inspired he brings his thoughts to bear on Humanity at large. He thinks of the innumerable races of mankind, and the brave battles waged against obstructive forces all along the path of human history. In every field of life's activities he sees martyrs who ungrudgingly have shed their blood for truth and progress. He feels the bond of brotherhood that links the lowest savage to the noblest of his kind, the slave to the freeman, the pauper to the prince; and whilst his heart palpitates with the tender emotion of sympathy in the joys and sorrows of mankind, his eager personal desires and personal hopes are stilled, and he lays his life down on the altar of Humanity. No selfish wish, no despicable fear, will make him swerve from duty; his whole moral nature is enkindled with religious fire. He will be true to the highest, noblest interests of mankind, and with simplicity of heart expend his strength in pure devotion to their cause.

Now, the process of advance from the infant stage through all the intermediate to the mature stage here described, is subjective education. And the vital question of this education is, how to ensure steady advance without the dropping of *valuable* stitches from the available material—in other words, how to secure the unfolding of the organism in every part of its complex structure fitted to the environment of civilization. For observe, there *are* stitches to be dropped in the process of education, as belonging to the baser social structure of uncivilized life. We may, I think, assume that all infants of the nineteenth century have germinal instincts, impulses, and sentiments, which education has to guard from springing into active vitality or to suppress.

The warlike proclivities of the ancient Briton, for instance, were useful and necessary to him; but they are intolerable and immoral in the modern Anglo-Saxon. Revenge was respectable, so long as there was no organized government in a community to protect individual interests; but it is immoral now. Jealousy and cunning were savage yet important weapons to the abject slavish wife of the Middle Ages; by their aid she kept her social position at the side of her brutal lord. They are base and utterly unworthy of the free emancipated woman, aspired to * in the present day. Tyrannic force was actively required when physical strength secured survival of the fittest. But now the passive force of independent character, accompanied by gentleness of word and action, makes the greatness of the man. The love of thrift, with the impulse to accumulate, and a blind admiration for wealth, are useful forces in the early days of

* The marriage laws are not yet in accordance with the modern sentiment; wives are not free.

a nation's industrial life. They distinctly tend to the subjugation of outward nature and the establishment of the whole community; but in our age these forces require modification. The accumulation of national wealth has ceased to promote progress and to be favourable to the welfare of all. Riches in the hands of the few does not lessen misery in the hearts and homes of the many, and it is not the creation of more wealth, so much as the distribution and right expenditure of wealth, that will claim the paramount attention of the rising generation.

The British race has passed very rapidly from a primitive state of life to a highly complex, and in many respects highly artificial, social state. It results from this that right education in the present day is necessarily a process of sifting and selection, and we educators have a fourfold duty to perform. We have to inculcate real knowledge, that is science, in order that right and true action in the world may result. We have to repress, in the young. instincts that are natural and therefore quite unblamable. We have to adequately develop other existent instincts and faculties. Lastly, we have to stimulate, and by attractive influence bring into rapid evolution, some elements of character which are destined to be progressive forces in the social life of the future, but which at this moment are lamentably deficient, simply because we belong to a nation whose transformation from barbarism to civilization has been rapid, and has occurred amidst conditions of social inequality perfectly unprecedented and hitherto unknown.

Now, why do I say we educators? I am not myself a mother, a School-Board functionary, or a certificated teacher; and my reader may be free from all conscious responsibility in relation to the young—nay, may be a bachelor, who prides himself upon a wise relinquishment of parental pleasures and duties, in consequence of the social pressure of the day. Nevertheless, we are educators, although sometimes it may be we are unconscious of the fact. The young are ceaselessly imitating our manners and customs, taking in our ideas, and moulding themselves according to the model we present. In short, each adult generation *en masse* exhales an atmosphere, moral, intellectual, and emotional, which the juvenile generation inhales, and, by the assimilative properties of its nature, forms into the noble or ignoble Humanity of the future. The atmosphere around our free, unfettered social units is a powerful influence in this formative process. Perhaps no influence is greater with masculine youth in its teens, than the example of wealthy bachelors whose ample command of money and uncontrolled liberty excite cupidity, envy, and emulation. Every adult man or woman (unless insane) has resting upon him educational responsibility. We are all, directly or indirectly, acting upon, therefore educating the young ; and it is a momentous part of the religion of the future, not yet recognized and acknowledged, that we should cease to act unintelligently in the matter, and should enter consciously and conscientiously into the enlightened fulfilment of this social duty.

I shall speak of education as divisible into four departments, although these run into one another and present no clear and distinct line of division anywhere. To the department of physical development attention has been turned of late years, and in all the great public schools for boys, and many of the high-class girls' schools, there is a marked improvement in the education of the muscles and the physical training generally. From the spontaneous activities of the nursery to the sports and recreational activities of the school playground is an effective transition ; and I have no comment to make here, except that we have hundreds of city schools without playgrounds at all, and hundreds of thousands of children (I do not mean low-class children) debarred, by the poverty of parents, from access to the expensive schools where these playgrounds are provided.

In popular language the term education points specially to mental development and the acquisition of knowledge. Our systems of education are methods by which to promote intelligence and furnish the mind with facts as rapidly as possible. When a merchant or capitalist says of his son, " He must be *well* educated," he generally means, he must be intellectually quick and sharp, ready to perceive and respond to the forces that underlie industrial life, and so keep his place in the mercantile ring. He must have the manners and habits of his social class, and so much of moral principle as to be neither above nor below the level of that class. He must not find immoral the practices of commercial life, and assume superiority by condemning them ; nor may he stoop to baser practices, and be tabooed as less than gentleman. When an aristocrat desires the best education for his son, he means University culture, embracing the physical training of the high-class British athlete, the classical acquirements, the knowledge of Greek and Latin, and of such sciences as Oxford and Cambridge afford; in short, the muscular and mental development which will make him an ornament to society in his own class.

Now, of education in this narrow sense it is curious to observe the changed conditions time has naturally brought about. At the beginning of this century the work of education was left entirely to parents, and society never interfered or inquired how the parental duty was fulfilled. George Eliot gives us in the Garth household a true picture of the period of which she writes. The mother, a noble specimen of womanhood, making pies at a well-scoured deal table, whilst also giving lessons to her youngest boy and girl, who stand opposite her at the table with books and slates before them. Mrs. Garth thought it good for children to see that a woman with her sleeves tucked up above her elbows could correct their blunders without looking, and know all about the subjunctive mood and the torrid zone. As life became more and more complex, however, parents shunted their educational duties, and schools multiplied in all directions. Education was the term applied to the machinery of school life, and the schoolmaster became the great educator of the nation. Nowadays, society steps into the parental position, and throughout the whole of the lower social stratum at least, takes each child out of its home to pass under the schoolmaster's machinery, and be drilled into a measure of book-learning. In the middle and upper social strata no such action is likely to occur, because a firm belief in the importance and efficacy of school machinery is all but universal. John Stuart Mill was educated by his father; and William Pitt, who was Prime Minister at the age of twenty-four, received the greatest part of his education at home under the close superintendence of *his* father, the Earl of Chatham.

In 1837, Isaac Taylor wrote a book upon the merits of home education, in which he lays down this proposition, " that even if schools, and large schools, were granted to be generally better adapted to the practical ends of education than private instruction, nevertheless the welfare of society, on the whole, demands the prevalence, to some considerable extent, of the other method. The school-bred man is of one sort, the home-bred man is of another; and the community has need of both; nor could any measures be much more to be deprecated, nor any tyranny of fashion more to be resisted, than such as should render a public education, from first to last, compulsory and universal."* But behold in the present day the belief in public education has become all but universal. When education is discussed, the relative merits of this or that school, Rugby over Harrow or Eton, Uppingham over Charterhouse, etc., may be brought forward, but never the relative merits of an education entirely conducted at home. An essay written upon the lines of Isaac Taylor's book would certainly be condemned as teaching old-fashioned doctrines now happily exploded.

A mixed system of home and school education, however, exists in all large towns, where middle-class day schools are of a high order. The great public boarding-schools are too expensive for some parents, and there are others who revolt from the tyranny of an ill-formed public opinion, in favour of a school-life that separates parents and children, and who therefore adopt a method which seems to amalgamate or embrace both public and private training. Many a family removes from a country house to a town house, that the children may alternate between school and home, and either reap the benefit of a double training, or, which is more likely, lose the benefit that either system taken separately would afford. In the department of intellectual education this mixed method may prove satisfactory. The lessons at school and the cramming and tutoring at home may tend to hasten mental development; but in the departments of emotional and moral education I have no hesitation in calling it an unmitigated failure. Nor is it difficult to see the cause of this. The children are poor little planets revolving daily in an irregular orbit, whilst without innate forces to hold them steady in the mazes of a complex sphere of action. Perturbations inward and outward occur, an erratic or wandering tendency of mind and body is engendered, and incoherency of thought with confusion of sentiment and feeling are the general result.

"The master has no right to interfere with me out of class," said a little fellow of ten; "he never asks what we do in play hours, and he'd better not!" And to his mother, who proposed to visit the school and be present when the Latin lesson was given: "Oh, what a shame! Other boys' mothers don't come to school. They'll laugh at me. Don't do it, mother. You shall not, you must not do it!" On the momentous question of authority what are this child's ideas? That he is bound to obey his master in school, his mother at home, and in passing from the sphere of the one authority to

* "Home Education," p. 22.

that of the other, he is as free as an Arab on a desert plain, provided the policeman, whom he always calls "Bobby," is not near enough to watch his proceedings.

An innocent trickiness is this boy's characteristic, and I take him as an extremely common sample of the shaping of character produced by middle-class day schools. The childish trickiness would be amusing, were it not that one knows it to be the germinal form of that hideous and degrading social vice, duplicity or cunning. The boy is perpetually shifting his moral standpoint, and conforming his words and manners to the conventional propriety of home, the stern autocratic discipline of school, or the unprincipled freedom of the playground circle—a promiscuous group full of incongruous instincts and no clear ideas of what is right or wrong. His native cleverness comes out in the rapid transitions, and he learns to present one side of character and conceal the other with facility and promptitude.

Now, I hold it an essential in education, that a reasonable authority, without pressing upon or limiting a child's innocent freedom, should surround him continuously, persistently, and uniformly, that he should grow up subject to authority, and with no notion whatever that he is at any time free from it. Authority is the great condition whereby his moral nature will be shaped, his civilized and civilizing instincts developed, his barbarous instincts checked, his habits formed conformably to right action; until the period is reached when developed reason within becomes a steadfast guide, and the outward authority dies down before the inward authority of the mature judgment and tender conscience of the educated human being. Observe, there is no abrupt transition here. If the outward authority is fully reasonable, it is the same in quality as the reasoning monitor within; and the child passes easily, without jar and without hazard, from the guidance of external leading-strings to the no less effective guidance of the delicate forces of rational self-control. The outward authority I speak of, however, must be consistent with itself.

Do I therefore indicate that one person alone must sustain it? By no means. But I do indicate that in the perfect education, which the (alas, still distant) future is certain to bring, all the adult members of each group in the centre of which children move, will act from moral principles, reasoned out and voluntarily adopted. Their precepts and example, therefore, will not be discordant, but harmonious, forming an atmosphere of consistent and coherent authority which the little ones will pleasurably and readily breathe. In Unitary Homes training will prove easy and natural, for scientific forethought will place around children a daily, hourly environment suited to the development of character, the evolution of noble sentiment. In the Unitary Home we shall no longer "give ourselves mighty pains to teach our children what they ought to know," whilst "what they certainly ought to feel" we foolishly leave "in the main to chance, and habit, and the evil communications of the world." * On the contrary, what children feel will be a subject of careful consideration. The whole sphere of youthful emotion will be laid open to sympathetic investigation, and become a field for the exercise of parental guidance, and judicious, responsible control.

Meantime we assuredly ought not to add to the difficulties of our transition epoch by deliberately placing children in conditions confusing to the moral nature. No amount of gain in mental development can make up for what we sacrifice in the departments of moral and emotional development; for these are of far greater import in relation to individual happiness and general well-being. The sentiments of the future generation, what they shall be, whether base and worthless, or noble, dignified, tender, and religious, rests in our hands. Are we to run all risks in respect to the emotional side of human nature in order to rear a generation of clever speculators, adept workmen, industrious slaves, and, at the head of all, belligerent statesmen—each individual bearing the distinctive characteristics of his social grade, but one and all true to the national type so well named by Mr. Matthew Arnold, the British Philistine?

As the eyes of an earnest and devoted mother fall upon this page, it may be that annoyance is the feeling she experiences. "I have," she says to herself, "a large family and a narrow income. Even if I believed in the superiority of boarding-school education (which I do not), I could not afford it for my children. Yet here is the day-school system, and I think with some reason, condemned. What shall I do?" There is, alas! *at the present moment* no available alternative without serious drawbacks. Home education, if it were possible, would, by reason of its limitations, prove defective. There are many middle-aged men and women who suffer continuously from the natural effects of a childhood and youth

* "Self-Formation," vol. i. p. 152.

"cribbed, cabined, and confined," that is to say, spent in a home that failed to exercise all their human powers, and consequently failed to launch them into life fully equipped for the complex social state around them. The complexity of human nature requires more diverse and liberal surroundings than household education presents. Strange that the various evolutionary forces of industrial and social life have culminated in thrusting children into an environment as much too wide for satisfactory training, as the former environment was too narrow. And as regards authority; parental despotism *was* apt to be rigid and somewhat stern, but now with parental control relaxed, school discipline only occasional, and our universal indiscriminate respect for the freedom of the British subject, however juvenile, the child with us is frequently a law to himself, unconscious of authority and surrounded by precepts and examples that are discordant, mutually destructive, and utterly confusing to the immature intellect.

The true path of education lies between the two extremes. Happily for the children of the future we could not, if we would, revive parental despotism; and the purely household education, which for small families or solitary children was a dreary desolateness, is now and for ever a thing of the past. But man is able, when he so wills it, to consciously evolve a new domestic system, that will not separate parents and children, but will embrace home influences and school advantages, and place the little ones in surroundings that are wide and liberal, authoritative, and yet so free as to meet all the varied requirements of complex infantile humanity, preparing for an elevated social life amidst modern civilization. The process of aggregation of families into composite domestic groups has been sufficiently explained in Chapter XVI. I must not further refer to it here. All radical reform is of slow growth. Its first step is a painful one, viz. the rousing of intelligence to a sense of miserable deficiency in things as they are, and the sharp awakening of feeling from a slumber full of pleasurable and soothing delusions. What I aim at is no agreeable task. I desire, if possible, to shake public faith in our mechanical, scholastic forms and methods of education, and to stimulate parents to look into the whole matter from a scientific standpoint.

To every devoted mother I say, "Follow your children to the day school as frequently as you can, and form some sort of relation with the schoolmaster, so that your children may at

least recognize in you and him not two authorities but one." *
What you want for your children is that their bodily and
mental powers and their best feelings should be exercised,
whilst their sentiments and habits are formed to some extent
each and every day, so that at maturity they may be full-grown
in body, mind, and character, able to control their passions,
and shape their conduct in accordance with goodness and
happiness, in themselves and all around them. Now, a child's
daily supply of vital energy and nerve force is, of course,
limited. If these are unduly expended in the activity of one
part of the organism, the other parts necessarily suffer. If a
man's nerve force is transformed into muscular energy, and
spent on ploughing all day long, he will be unable either to
think or feel at night; and if he tries to read he will probably
fall asleep. One-sided activity is bad for a man, but for a
child it is simply ruinous, seeing that what is required in the
child is *many-sided growth*, to secure which there must be
diffused nerve force and a variety of diverse activities. By our
present system too much blood supply and nerve force are
directed to one part of the brain, and expended in intellectual
processes. There is little or no surplus for the healthy nurture
and growth of the emotional and moral nature, and in many
cases an insufficient supply for physical enlargement and
support. Calculate for yourselves the number of hours spent
over lessons at school and at home, and consider the number
of books carried to and fro by the children. Do not these
represent an amount of vital activity in one department of
education, out of all proportion to the vital activity in other
equally important departments?

But besides causing children to spend far too much time
in committing to memory facts and taking in new ideas, for
the most part abstract, we stimulate this kind of cerebral
action by various artificial devices. Through rewards and
punishments, prize-giving, public exhibitions and scholarships,
we create a highly artificial, I may call it a hothouse culture
for the intellect, which is detrimental to the character, or to
the physical health, or in many cases, to both. Again, the
payment of teachers by results † is an unhealthy stimulus

* A schoolmaster remarked to me: "Twelve years ago parents came
often to hear the children at their lessons; now I never see them —they
even send the fees by the children! Can it be that they take less interest
than formerly?"
† The evils of this custom have been perceived, and it is now partially
relinquished.

under which many teachers break down, whilst straining every nerve to stimulate the children to efforts which are injurious to them, in effect producing (I quote Mr. Herbert Spencer's words) a bodily enfeeblement, which no amount of grammar and geography can compensate for.* The only healthy stimuli to growth in a human being are the natural ones. First, there is the functional pleasure. If a child has *no* pleasure in exercising the faculty of acquiring knowledge, it must be that there is something wrong. The faculty has been overstrained, or the child is ill, or the knowledge presented is unsuitable, in its essential nature, or by the form in which it is brought before the infant mind. Secondly, there is the pleasure of using the knowledge acquired, which creates a desire for acquiring more. Thirdly, there is the pleasure of winning the approbation of teacher and parent. Prize-giving brings with it two injurious sets of effects, the one negative, the other positive. On the negative side it diverts the attention from the simple pleasures that are the natural reward of effort, thus preventing the forming of an *all-important* association of ideas; whilst on the positive side it creates a craving for something *over* and *above* the natural reward of effort, and so gives to Humanity an artificial requirement. "Why do you like school?" I asked a child, and the frank reply was, "I want to get prizes." The equally frank statement of a lad of eighteen is, "I crammed at my boarding-school, beat the other fellows, and brought away prizes, and now I've forgotten all that rubbish!" Contempt for knowledge is the very common tone of mind engendered by cramming, and where a boy is clever but not bookish, he will to a certainty acquire the bad habit of veneering his mind as it were with a thin layer of information for immediate use in the gaining of prizes and admiration, but which will never be assimilated by his intellect or woven into an experience to guide and shape his future life.

* A school-board teacher tells me that at the last examination the children of her class, aged ten and eleven, committed to memory the following passages of Scripture:—Proverbs viii. from first to twenty-first verse, the fifty-third chapter of Isaiah, 304 lines from the Psalms, 24 lines from the Paraphrases. Also from the thirtieth to the eighty-second question of the Shorter Catechism of the Church of Scotland. The whole of the Sermon on the Mount, which was repeated almost by heart; various extracts from Saint Matthew's Gospel. And to the end of the twelfth chapter of II. Samuel, as well as the whole of I. Samuel, they prepared so as to be closely questioned upon it. This represents one single branch, only, of the examination these children of Standard V. undergo. Who can doubt that bodily enfeeblement is here the natural result?

In a class of fifty boys let us suppose that half of the number possess the temperament called nervous sanguine (and are therefore highly excitable), whilst the other half are nervous lymphatic. The former are thrown by the competitive system into an excited state of keen pursuit; the latter, incapable of the strain, are discouraged, and fall into languor and indifference. The state of keen pursuit is analogous to the attitude of mind in a savage, hunting for prey. It is the barbarous predatory instincts that are aroused in the child. The *prize* is before him, he sees that alone : his intelligence is narrowed, it will be receptive of nothing but what relates to the prize ; his feelings are cruel in their blank indifference to the desires of his companions; and when success crowns his efforts, the pride he feels has nothing to do with knowledge gained ; it is the innocent but quite uncivilized joy of an ancient Nimrod who has beaten his fellows in the field.

But the barbaric instincts of the chase are not confined to the bosom of the successful young Nimrod. The unsuccessful competitors are secretly suffering jealousy and envy, those passions of the uncivilized man which modern education is bound to keep latent until they die out of the race ; whilst the poor children who were never in the running, are cherishing an inward sense of self-depreciation, which, if no counteracting influence intervenes, will make them grow into cravenhearted, abject-spirited men and women, subject to the rule of any tyrant they are near, or slaves to the tyranny of senseless fashion ! My contention, then, is this, that in order to make prominent the natural rewards of intellectual effort, and to bring into play all the gentle, pleasurable emotions that are appropriate to civilized life, and the charm of which must be experienced *early* to prove effective in moulding character, we must *banish prizes* from our schools, and firmly remove every species of artificial stimulus, replacing these, however, by the natural stimuli, and never relaxing effort in this direction until every child is freely responding to these natural stimuli.

I have spoken of love between child and teacher as the *third* natural stimulus to effort, but I conceive that there are some children, perhaps many, in whose natures this is the primary, the necessary prompter to effectual effort ; and if *this* stimulus does not exist there is no liberal putting forth of the child's whole powers in their normal strength. But here, with the British instinct for the practical, my reader may accuse me of Quixotic folly. Do I demand the creation of a bond of love between the schoolmaster and each one of his fifty, or a hundred, or a thousand scholars? To do so were indeed to build my theory upon "such stuff as dreams are made of." The schoolmaster, however, with his class of forty, fifty, or even thirty boys is nothing but machinery, a part of that mechanical system which we employ to turn out clever men and women. The whole system, *in so far as it is inorganic* in its working, will be discredited and discarded when the public cease to aim at producing sharp and keen-edged tools for the warfare of a competitive age, and seek instead to produce the ripe and mellow fruit of a civilized humanity, viz. men and women full of individual character, and gifted with that kind of real culture which Mr. Matthew Arnold expounds as *the study of perfection.*

There has hitherto been a universal aiming at perfection in the methods only of *imparting selected knowledge.* Grammar, arithmetic, music, singing, drawing, are now taught scientifically, and with marked success. The improvement is apparent on every side, and in so far as these intellectual and æsthetic advantages are valuable to humanity, the present educational epoch is vastly superior to any epoch of the past. But there has been as yet no universal aiming at perfection in character; and in reference to all the finer qualities of character the present has little superiority over the past. The baser instincts of humanity are nearly as prominent as they were fifty or a hundred years ago, although, owing to the changed outward conditions, their form and action are somewhat different. The average man is no longer brutal, but he is intensely selfish. The average woman has gained in qualities necessary to her sex, but on the other hand she has lost considerably on the domestic side of her nature, and in character as a whole she is no whit superior to her grandmother, or even her great-great-grandmother. A larger proportion of women than formerly (whilst making due allowance for the vast increase in the general population) lead frivolous, worthless lives, and there has taken place the development of an entirely new and most objectionable type, viz. the Girl of the Period. Those general results force upon us this inquiry: "What are the most powerful factors in the formation of character?" An inquiry which, it seems to me, has as yet very little occupied the British mind, since written opinions on the subject are extremely difficult to find.

In the year 1837, a period when boarding-schools for boys

were becoming fashionable, Mr. Capel Lofft wrote strongly against the new custom, charging parents with adopting it simply because children are apt to be teasing and importunate, and parents do not like the responsibility. The advantages of school life and home life he thus contrasts. At school, boys advance quickly in knowledge of the auxiliary verbs, the mysteries of syntax, and the stories of gods and goddesses, "but," says he, "they could be without all that and yet live." At home, the heart, wherein are the issues of all good, develops itself from day to day. There children ripen in their affections. In short, they learn "their humanities, not in the academic sense, but in the natural and true one." Boarding-schools for girls were not then fashionable, and Mr. Lofft rejoices over the fact. "I am confident," he says, "that the reason why women generally are so much better disposed than men . . . is simply this. They are penetrated with the home spirit, they are imbued with all its influences. They live domestically and familiarly. Their memory is not fed to plethora while the heart is left to waste and perish. No daughter of mine shall ever be sent to school." *

Now, Miss Martineau, who wrote at a later period, speaks quite as strongly on the importance of a full and satisfactory home life to develop feeling, and through feeling character, in a child. She was a woman of calm reflection, able to recall and judge of the impressions of her own childhood. Of the birth of a sister which occurred when she was nine years old she thus speaks : "I doubt whether any event in my life ever exerted so strong an educational influence over me. The emotions excited in me were overwhelming for above two years. . . . I threw myself on my knees many times in a day to thank God that he permitted me to see the growth of a human being from the beginning. I leaped from my bed gaily every morning, as this thought beamed upon me with the morning light. I learnt all my lessons without missing a word for many months, that I might be worthy to watch her in the nursery during my play hours. . . . Many a time I feared that she never could possibly learn to speak. And when I thought of all the trees and plants and all the stars and all the human faces she must learn, to say nothing of lessons, I was dreadfully oppressed, and almost wished she had never been born. Then followed the relief of finding that walking came of itself, step by step ; and then talking came of itself, word by word. . . .

* "Self-Formation," vol. i. p. 42.

This taught me the lesson never since forgotten, that a way always lies open before us for all that it is necessary for us to do, however impossible and terrible it may appear beforehand . . . that nothing is to be despaired of from human powers exerted according to nature's laws."* Observe that in this child of nine the motherly instinct (I mean by that the protective love of a helpless being) threw into active movement quite a variety of human powers; the religious feelings, her muscular system, her memory were stimulated. The altruistic sentiment of anxiety concerning another's welfare and happiness was awakened, and the higher intellectual faculties, viz. the observing, the reflective, and the reasoning powers were all brought into play. Valuable experience was gained, orderly habits were promoted, and the motor force in all was the simple natural stimulus, love of a babe. Young Harriet Martineau was here " learning her humanities, not in the academic, but in the true sense," and such humanities are character, the growth of which it is one great purpose of education to achieve.

In 1867 there issued from the press a little book, called " Day-Dreams of a Schoolmaster," in which the author, from a wide experience of school life, both as scholar and teacher, gives his reflections on the educational system of the day, with the ring of an earnest, thoughtful mind, and the graceful touch of a poet. " At the best of the great public schools," says he, " the youngest children—bless the innocents !—are suckled upon grammar; the more advanced are too often fed upon dull books . . . the manuals for prose compositions are in many cases tramways to pedantry. . . . The whole system, and the elementary part most of all, is bookish, unpractical." † The mechanical, irrational method of teaching classics he humorously describes, and the implement of leather, by the dexterous application of which the schoolmaster was expected to quicken the apprehensions of such " children as might be uninfluenced by the monotonous music of his gerund-stone." Being himself a man of feeling, his nature revolted from mechanical instruction, and from the use of what he calls the "electric leather." He relinquished both, and strove to supply their place by simple personal influence, sparing neither time nor trouble to educe by every moral and intellectual means at his disposal the dormant faculties of his charges. What was his reward ? " The more I gave satisfaction to myself," he says, " the less

* " Household Education," p. 52.
† " Day-Dreams of a Schoolmaster," D'Arcy W. Thompson, p. 42.

I gave satisfaction to the majority of my so-called patrons; the guardians of my young pupils. . . . When I was indulging in a dream of appreciated toil I heard of complaints being circulated by such as were favourers of mechanism in instruction. Pupils in whose progress I had begun to take a keen interest, were from time to time removed. . . . 'They were not *grounded*,' said these waggish but unmannerly guardians: meaning all the while, 'they were not *ground*.'" It was not by a happy childhood spent in the bosom of his family that our humane schoolmaster learned *his* humanities. He speaks of his own dreary, weary, boyhood—of twelve precious years of youth dragged out within the precincts of the Grammar School of St. Edward's; of the monotonous routine; the daily committing to memory pages of unintelligible books. His mental training was "a continuous sensation of obstruction and pain"—his mental and spiritual parts were "furrowed," whilst his "growth in stature was left carelessly to his Maker." * As for his heart, it took its chance, and luckily for the boy one master had no vulgar instrument of punishment. "By his noble presence, and the unseen force of his character, he maintained order in his class." "Of that master," says the boy when grown a man, "I have an affectionate remembrance." But why is it, he meditates, that fathers with personal experiences precisely similar to mine still send their boys to school? "They forgive," he answers, "over the walnuts and the wine, the pedagogue that thrashed them to no moral or mental profit; the bully that appropriated their weekly sixpence . . . and 'depend upon it,' they say to themselves, 'if there were no virtue in birching, caning, Latin verses and Greek what-you-may-call-'ems, they would not have held their ground so long amongst a practical people like ourselves.' So Johnny is sent to the town Grammar School, and the great time-honoured gerund-stone turns as before, and will turn to the last syllable of recorded time." †

For the gerund-stone our schoolmaster would substitute an easy *viva-voce* conversational method of instruction in elementary classes, for coercion the more than hydraulic pressure of a persistent and continuous gentleness. The necessity for physical chastisement rests mainly, he says, upon two blemishes in our school system, viz. the mechanical nature of the routine, and the crowding of our class-rooms. Flogging can never instil courage into a child; it has helped to transform many a

* " Day-Dreams," p. 240. † " Ibid., p. 40.

one into a sneak. Let us discard punishment and our ridiculous scourges, and depend for stimulus wholly upon love. "Let us endeavour to make our pupils love their work. They may live—God knows—to love *us*. . . . Ah! believe me, brother mine, where two or three children are met together, unless . . . the spirit of gentleness be in the midst of them, then our Latin is but sounding brass and our Greek a tinkling cymbal." *

The schoolmaster, then, sees as clearly as did Miss Martineau and Mr. Lofft, that to feed the memory to plethora whilst the heart is left waste and desolate is no education ; that to stand outside the region of feeling and drill the children into obedience, order, mechanical proficiency, and even correct knowledge, is not education. The true method is to find our way into the region of feeling, and, keeping the children's feet steadily within the charmed circle of pleasurable sensations, to train them, step by step, to the enjoyment of useful activities, to manual dexterity, to effort in pursuit of knowledge, to manly dignity in defence of the right, to sympathetic jealousy over the rights of their fellows, to gentleness towards all mankind, to firmness in discharge of duty, to admiration for all that is noble in character, to love of truth and justice and faithful adherence to both, and to veneration of age, of experience, and of virtue.

But now this quality of character which we call veneration. Are we middle-aged men and women not struck by its almost total absence in the juvenile population around us? The average child of this generation is wide-awake, clever, and somewhat forward. He may be outwardly respectful, but inwardly he is contemptuous of his elders; and the only mentor whose spoken advice or warning he instinctively loves to obey is the juvenile whose experience is precisely as long and as broad as his own! Is the child to blame for this? Not so. Our system is to blame — the gerund-stone, the ridiculous scourges, the prize hunts, the intellectual strain, the emotional blunting, the competitive examinations, the bookish life, the mechanical school routine, and the *want* of that *domesticity* which is the sole condition whereby may be nurtured into vigorous life the finer qualities of human nature of which veneration is one.

"Domesticity!" says my reader, a delicate mother of seven boys whose exuberant youth is to her nervous system

* " Day-Dreams," p. 306.

somewhat overwhelming. " Do you suppose our school-bred children are never at home? Why, we have them in the domestic circle more than a quarter of each year. *I think* myself the holidays are far too long!" Ah! there, my friend, I perfectly agree with you; for observe, "at home in the holidays" means wild freedom for the young ones. The parent who would train them to discipline in holidays would be indeed a bold innovator upon usual custom, and to say truth, if the strain has been severe and the young noses have been kept stringently to the gerund-stone during school term, a relaxation of the whole system, even to the opposite extreme of license, may be a lesser evil than I think it. But it *is* an evil, nevertheless. The holiday season, with its indulgences, its artificial excitements, its idleness, its languor, and often its *ennui*, is as injurious to character as is the school life, with its mental strain and stiff routine. A tender instrument, of feeble capacity and small compass, but mysterious power of growth, is put into our hands. The purpose committed to our charge is to evoke from it continuous sweet strains, until its feebleness has become strength, and its quivering semi-quavers a volume of harmonious sound. Is there wisdom in straining and screwing it up to concert pitch or far beyond, and then for months leaving it alone, its music dormant, but not so its sound? It can emit barbaric notes, and in the exercise of these it attunes itself to wooden tom-toms, until all its music is degraded into harsh and jarring discords, or petty, jingling jigs, that have no rhythmical relation with the sublime harmonies of the heavenly spheres.

But now an ordinary schoolmaster interrupts to ask with practical acumen if I would shorten holidays to lengthen out the toil and drudgery of already hardworked teachers. I answer that I know no social class that more deserves our sympathy and commiseration. To lighten the toil of teachers and of taught is what I aim at. To remove from both the horrible incubus of competitive examinations, to induce the public to set aside the yoke of universal custom, and leave each conscientious teacher free to exercise his own judgment, and evolve his own method of attracting his pupils, of winning them to effort in attainment of that culture which has perfection for its end. But the conscientious teacher, when relieved of personal tension, and under no constraint whatever to overtax his own powers or those of his charges, will cease to give long holidays. He knows too well the deterioration of his work that that involves, the loss of personal influence he thereby incurs. A change in respect to holidays will be a necessary and not difficult step in the transformation of our system of education from an artificial, inorganic mechanism into the rational, natural process of training children in the path of progress.

Think of it, ye toiling pedagogues! think of the changed conditions of your life, when parents demand no strain and will not welcome youths whose arms are filled with prizes and their memories with mere book-lore; but who uphold you in the simple task of awakening general intelligence, of healthfully developing the body, mind, and feelings, of inspiring to useful activities, of inculcating the uniformity of the laws of nature, and guiding to conduct in harmony with such natural laws as regulate the life, the health, the happiness of individuals singly, and of individuals combined to form a civilized society.

If we view this whole matter in the abstract, and judge it upon rational principles, what ought we to expect from children nurtured in an artificial social atmosphere and subjected to violent action and reaction; to perturbation from exertion that is extreme, to idleness that is extreme; from mental cramming, to mental starvation; from a school life that has the excitement of a racecourse, to a home life stagnant by contrast; from routine that is mechanical, to freedom that is license, and so on? By the laws of nature which we profess to teach our children, dare we expect that they will grow up calm, self-sustained human beings, with minds that are the abode of peace, and lives that move in rhythmical harmony with the gentle pulsations of universal happy life, the labour of each day relieved by the constant return of tranquil pleasures and heartfelt delights? I answer, no—a thousand times no. A weird-like spirit of unrest, an eager selfishness and greed, a joyless joy, a soulless mirth, these are the natural consequents of the antecedents we have looked at, and these are what we see. The poet Heine, if he lived again, would but repeat his sad strain; alas for the British "ease that is ill at ease, that prudish insipidity, that varnished rudeness, in short the whole unrefreshing life of those wooden butterflies that hover about in the drawing-rooms of the West End." * John Bull is the coldest friend. He works " day and night to replenish his exchequer, in order to balance accounts with his extrava-

* " English Fragments," Heinrich Heine, p. 16.

gance."* He is a mechanical formality, an open display of egotism. Wherever he may be, it is his own comfort, his own immediate, personal comfort, that is the greatest object of all his wishes and endeavours. "Should you wish to make an intimate friend of him, you must pay your addresses as to a woman; and if at last you succeed in gaining his friendship, you soon find it not worth the trouble."† " For Heaven's sake send no poet to London. That stern earnestness in all things, that colossal uniformity, that mechanical motion, that irksomeness of joy itself, that inexorable London, stifles phantasy and rends the heart!" ‡

My brave compatriots!—and by this term I address myself specially to women, for my sex in this generation is extremely brave—we fight for our rights, we step fearlessly into the arena of public life and enter the lists with men in so far as they will permit us; we speak boldly of subjects on which our grandmothers kept silence; shall we be daunted by the magnitude of any task to which Nature calls us? She calls us loudly in this age to scientific reforms; she exposes to our eyes the hidden errors of the past; she makes us keenly feel the imperfection of human character within us and around us; she shows us unconscious evolution everywhere entering upon a conscious stage, in which man, in the van of circumstance, will shape the course for future humanity. Let us respond to her call by acting on the knowledge, that human character does not depend on inherited nature alone, nor on religious instruction alone, but on the *whole surroundings* of childhood from its infancy. Let us recognize that educational reform does not imply the establishment of school-boards merely, the improvement of text-books merely, or even the higher education of women. It implies a change of system; and the task before us is to initiate a new departure, and evolve a system by which children will have a calm and tranquil youth, free from all artificial excitement, a constant supervision, authority that will guide them without coercion, training in the emotional and moral departments of education as well as the mental, in short such surroundings as will teach them their *humanities*, not in the academic, but in the real sense.

* " English Fragments," Heinrich Heine, p. 9.
† Ibid., p. 25. ‡ Ibid., p. 8.

CHAPTER XX.

THE TREATMENT OF EVIL-DOERS.

"The spirit of Howard is on its pilgrimage; and barbarous as is still our treatment of the guilty, better days are in prospect."—HARRIET MARTINEAU.

LET us suppose that all babes are born with perfect normal organizations, that is with sound physical constitutions, and such latent intellectual and moral powers, as will, when developed, place them on a level with the highest type of existent humanity. In such a case education, beginning in infancy and rationally and systematically pursued in the nursery, the schoolroom, the home, could perfectly well dispense with every form of punishment, and permit the very idea of punishment to become as completely a thing of the past as legal torture, at this moment, is a thing of the past.

The essence of punishment, as it existed in former ages, is vindictiveness; and a race that has outgrown the moral standard in which revenge is right and proper, and on the contrary regards revenge as thoroughly base and despicable, must logically put punishment out of court. Universal gentleness in the treatment of the vicious as well as the virtuous, the guilty equally with the innocent, is the only principle of action that accords with our modern philosophy of life and the vital moral sentiments of modern human nature. If punishment, therefore, is to be temporarily retained, it will cease to have any emotional connection with punishers, whether these be the friends of a murdered man, the pursuers in a case of libel, the schoolmaster with his rod, or the nurse placing an infant in the corner. The connection will be entirely between the punished and society in general. In the relations between these two alone have to be found the necessity and justification of punishment.

If we go down into the slums of a great city, we may be surrounded by children who freely strike out in self-protection, and in spontaneous aggressiveness; but it might prove difficult to decide whether nature or nurture created the blow, or whether both these forces combined made up the antecedent cause. Such children from infancy breathe the very atmosphere of savage assault. Not gentleness, but tyranny and force are the ruling conditions of their young life, and such nurture bears fruit in revenge, and other barbarous moods of mind. In the rich man's nursery may be found a babe who slaps the naughty chair that gave his head a blow. And here again the impulse may proceed from nature, or nurture; for an ignorant nurse will teach a child the foolish barbarous action. But nevertheless, our civilization has produced a high mental type in embryo; and nature speaks strongly in thousands of children precisely as it did in that child, who, at five years old, asked the meaning of the expression, "hanging a murderer," listened to the explanation with earnest attention and wonder, and then said eagerly, "Will hanging this man make the other man alive again?" and upon being informed that unhappily it would not, exclaimed with a strong feeling of commiseration, "Then why kill him? for when he is dead we can never make him good again."* The normal nineteenth-century intelligence requires proof that punishment protects society from evil-doers, or reforms the criminal.

Now, the real fundamental thorough-going factors in the work of protecting society and reforming criminals are *Nurture* and *Nature;* but unhappily, as yet these two have not combined for the production of a perfected Humanity. Nurture, in the mass of the lower classes and a large proportion of the upper, distinctly fosters evil propensities; and Nature, in her uniform action, which we term the organic laws of heredity, brings children into the world who are simply the counterparts of their mentally and morally defective, in other words vicious, parents. There are year by year born in our midst unhappy babes predestined to a life of crime. The class is not a large one, as compared with our population, but it exists; and these babes, if they live, will develop into thieves, or murderers, or criminals of some sort, as surely as the male infant will become a man and the female babe a woman. What would a perfect Nurture (if we possessed it) accomplish for these babes? Nothing effective. The power of nurture is limited. It can

* Hill, on "Crime," p. 173.

direct the forces of nature, but it cannot alter the intrinsic quality of the raw material which nature provides. If the raw material is the criminal type of brain, the culture which aims at perfection will fail utterly to reform it into the type of virtuous man.

But though nurture should fail, may not punishment accomplish this miracle of reform? We have the evidence of Prison Governors, and Prison Inspectors—men of observation, reflection, and earnestness, to the contrary. Mr. Chesterton, Governor for twenty-five years of the Cold Bath Fields House of Correction, writes: "I have been forced into the conviction that there is little hopeful expectation of the rescue of habitual thieves, and systematic evil-doers. My twenty-five years of observation have not encouraged me to rely with assurance on their corrigibility."* Mr. Frederick Hill, late Inspector of Prisons, writes: "Nothing has been more clearly shown in the course of my inquiries, than that crime is, to a considerable extent, hereditary; crime appearing, in this respect, greatly to resemble pauperism; which, according to the evidence collected by the Poor Law Commissioners, often proceeds from father to son in a long line of succession."† Mr. J. B. Thomson, Resident Surgeon of the Perth Prison, writes: "The great corollary from the whole study (of prison life) is that crime is intractable in the highest degree. . . . The facts press strongly on my mind the conviction that crime in general is a moral disease of a chronic and congenital nature, intractable in the extreme, because transmitted from generation to generation." Mr. George Combe, in his "Moral Philosophy," speaking of prison discipline in the United States, writes: "I have put the question solemnly to the keepers of prisons, whether they believed in the possibility of reforming all offenders; and from those whose minds were most humane and penetrating I have received the answer that they did not, and that experience had convinced them that some criminals are incorrigible by any human means hitherto discovered. These incorrigibles, when pointed out to me, were always found to have defective organizations. . . . They are morally idiotic; and justice, as well as humanity, dictates that they should be treated as moral patients. They labour under great natural mental defects, . . . to punish them for actions proceeding from these natural defects is no more just, or beneficial to society, than it would be to punish

* "Revelations of Prison Life," by G. L. Chesterton, p. 129.
† Hill, on "Crime," p. 55.

men for having crooked spines or club feet." * I could refer to many other authorities on this point were it necessary and did space permit.

Now, when a rational society accepts these facts, and regards the class of born criminals as moral idiots, it will not only cease to punish from anger or revenge, but it will cease also to attempt impossible reform ; and in reference to crime—simple self-protection, present and future, will form the basis of its corrective action. The change which has already occurred in the treatment of the insane (our mental idiots), pretty well represents the change which is certain to occur in our treatment of the criminal class—our *moral idiots.*

Fifty years ago we maltreated lunatics. We bound them hand and foot, we punished them for their congenital defects, we shunned and hated them ; and because they were victims of pitiful disease, we made them also victims of unnecessary and cruel sufferings. At the present moment lunatics are not enemies, and not even disturbers of the public peace. They are simply patients to be tended with kindness and regarded with compassion. It is true that we err in our method of treatment. We build huge asylums, and gather them together in unwieldy groups, depriving many who are perfectly harmless of a liberty which they would not abuse. But our principle is right, although our practice is defective. We profess, publicly and privately now, to secure for curable lunatics the best medical treatment, and for the incurable, the greatest personal comfort possible.† We shall learn in time to regard, in a similar manner, our *moral idiots.* Hatred of them and fear of them will die out. Punishment will seem to us absurd. Prisons will become asylums, in which restorative treatment for the curable, and the greatest comfort possible for the incurable, will be provided.

The modern scientific treatment of the insane is grounded, Dr. Maudsley tells us, on the principle of removing the various conditions which appear to have acted as causes of the disease. Let us see if this principle is applied also in our modern treatment of curable criminals. By curable criminals I mean that large class of individuals who (without congenital or structural defect) commit petty thefts and other misdemeanours in consequence of—bad training and ignorance—drunkenness

* "Moral Philosophy," George Combe, p. 306.
† I refer my reader to the last chapter of Dr. Maudsley's work on "The Physiology and Pathology of the Mind."

and other excesses—poverty and destitution. Within our prisons, undoubtedly, these causes of crime are to some extent removed. Drunkenness and disorder are not permitted there; whilst the necessaries of life, and a small measure of training and discipline, are supplied. Nevertheless, the methods pursued are wholly and radically unscientific. Warders are chosen without regard to their personal fitness for the office. There is very little discriminating treatment of prisoners. Convicts are mingled together without reference to character or age; official inspection is practically useless, and government reports are unreliable. Mr. Francis Peek alleges that English prisons are engaged in the manufacture of hardened villains out of reclaimable criminals!* In short, the facts brought forward from various quarters distinctly prove, that at the present time our prisons neither deter from crime, nor reform criminals.

Apart, however, from prison defects, our criminal system is at fault. Mr. Hill, late Inspector of Prisons, tells us: "The average period of imprisonment in England is about fifty days, and in Scotland, about forty. The utter insufficiency of such periods to produce a permanent good effect on the character of those imprisoned, must be apparent to every one."† The hatred and anger roused in the bosom of his more fortunate fellow-creatures by a criminal, varies according to the magnitude of his crime; and as our laws have come to us from a semi-barbarous race, whose impulse was to act from revenge and retaliation, the leading principle of the criminal law of Britain is to award punishment for different offences in supposed *proportion* to their *magnitude*. Now, the carrying out of this principle is incompatible with the civilized view—that the amendment of the criminal is the primary object of the punishment. It is clearly impossible for a judge to foresee the length of period required for the cure of each criminal's moral disease; and in effect we apply our remedies quite irrespective of any rational diagnosis of the disease! Hence there results—that in Scotland, in one year, no fewer than "six hundred and ninety persons were committed to prison who had been in confinement at least ten times before. Of these three hundred and ninety-three had been in prison at least twenty times before, and twenty-three at least fifty times." ‡

These figures show clearly how unscientific is our system.

---

* "Official Optimism," *Contemporary Review*, July, 1884.
† Hill, on "Crime," p. 183.   ‡ Ibid., p. 28.

We do not remove the conditions that act as causes of crime. We punish, and yet let loose again, offenders no better prepared than before to withstand the temptations of a life of social liberty. It is true that, during the last sixty years, many important reforms have been carried out in British jails. Cruelty and oppression are no longer practised as in former ages. Cleanliness and sanitary conditions are carefully secured ; and as regards juvenile offenders the establishment of reformatories was a stride of progress in the right direction. But notwithstanding all these improvements, there is at this moment an immense expenditure of public funds, and of conscientious effort on the part of prison managers, officers, and chaplains, utterly futile because the *system* is *wrong*—the criminal law of Britain is based upon a wrong principle. " No one," says Mr. Hill, " thinks of sending a madman to a lunatic asylum for a certain number of days, weeks, or months. We content ourselves with carefully ascertaining that he is unfit to be at large, and that those in whose hands we are about to place him act under due inspection, and have the knowledge and skill which afford the best hope for his cure ; that they will be kind to him, and inflict no more pain than is necessary for his secure custody, and the removal of his malady; and we leave it for *them* to determine *when* he can safely be liberated." *

On these two lines must run our action towards criminals as well as towards the insane. If a man is unfit morally to be at large we must narrow the conditions of his life, but make that life as natural, easy, and enjoyable, within the restraints, as is compatible with his steady growth in industrious habits, pure sentiments, and kindly feelings ; and we must on no account restore him to liberty, until there is conviction in the minds of those who watch his daily conduct that he will no longer abuse that liberty. Otherwise we stultify our own efforts for reform, and sin against our poor moral patient.

And here I must call the attention of my reader to the subject of restraints in general. When I express privately my belief, that individual character results from inherited nature and surrounding conditions—that therefore it is irrational to blame severely and punish sharply a naughty child, or a turbulent criminal—I am often met with the grave assurance that anger is righteous, and that we shall inevitably bring suffering on the innocent, if we refuse to whip the disobedient child, to punish with severity the hardened criminal, and to hang the

* Hill, on " Crime," p. 151.

murderer. Now, what *I* advocate is not *laissez-faire*. It is the systematic, rational treatment of evil-doers—from the troublesome infant, and the juvenile pickpocket, to the burglar, the fraudulent bankrupt, the felon, the traitor, the murderer, and the born-criminal class. It is, in short, the science of necessary and beneficial social restraints, which must be applied in the nursery, the schoolroom, the prison, with universal gentleness, and for the threefold purpose of: first, directing and improving imperfect character; second, reforming abnormal, or temporarily diseased character; and third, protecting society from the corrupt infusion of morally insane character.

In the nursery the scientific forming of character requires the utmost skill. The task clearly belongs to the educated female adults of each generation; and, that cultured mothers should shunt this duty and surround their children with an inferior class of persons, incapable of training scientifically, is a shameful error in a scientific age. There are thousands of mothers of this generation who may well feel reproved by the action of that nine-years old child of the last generation, who threw herself on her knees many times in a day to thank God that He permitted her to see the growth of a human being from the beginning.* Wilful disobedience, obstinate persistence in wrong-doing, or any other naughtiness should never be overlooked, even in a babe of only a year old. And yet it is of the utmost importance that no child should at any time be startled or exposed to a nerve shock, by being addressed in an angry tone, or by receiving a hasty blow. How, then, ought we to act? In the nursery stands, we shall suppose, a baby prison-house. It is a goodly sized circular basket, weighted so that it cannot be overturned, and softly lined and cushioned. Baby creeps to the fire. He is gently removed. He creeps there again and again. Nurse lifts him quietly and calmly; places him in the basket, and gives him toys. There he remains, until the impulse to disobey has worn itself out; the attraction of the fire has been forgotten. The child of two flings his ball in baby's face, and though conscious of the wrong-doing, persists in the amusement. He is firmly placed within the basket, where he lies down to kick and scream till he is tired, or contrite. When these children pass out of the nursery their nerves are healthy and strong. They know no craven fear, for gentle kindness has formed the moral atmosphere they have breathed. They are trained to docility and prompt

* "Household Education," by Harriet Martineau, p. 52.

obedience, and understand perfectly the simple principle, that —if they abuse liberty, their liberty will be abridged ; and they are sensitive in a high degree to affection, for love has surrounded them, and from the very dawn of consciousness formed the *one* stimulus to painful effort, and to successful effort the natural and abundant reward.

In the schoolroom the teacher's labours are not burdensome, since the children have already received in the nursery a systematic, carefully considered training. They obey readily, and meet their teacher with confidence and respect. He interests them in every lesson ; and never overstrains their mental powers. His checks and restraints are principally emotional. If playfulness intrudes, and the serious work of the class is impeded by some little urchin's fun, the master checks it without scolding. By the use of Mr. E. T. Craig's register,* he shows that child *his* opinion of his conduct, and withdraws all signs of personal favour and approval, till the culprit proves by earnest endeavour that he is correcting the fault. The education given is not mechanical and not indispensably allied with books. It aims first at imparting a knowledge of the most useful facts and principles of life ; second at perfecting character ; third at giving more or less of æsthetic culture. The study of human nature and of right conduct in social life is a part of the daily course ; and the children become perfectly true critics of character, to the extent of their childish capacity. They know which feelings in themselves and others are civilized and which are barbarous, and therefore to be restrained and checked. When they have done wrong the teacher assumes that it arises from ignorance of what is right, or from weakness of self-control ; and every

* Mr. E. T. Craig, formerly director of Lady Noel Byron's Agricultural School at Ealing Grove, invented an instrument by which to register good or bad conduct ; and in the use of it he was able to set aside prizes, place-taking, and every kind of reward and punishment. In the year 1841 this instrument was described by the inventor to a large and influential audience at a meeting held in Edinburgh, and reported in the *Scotsman*. " Mr. Craig showed the method of working the instrument ; exhibited diagrams of the method of registering the conduct ; and related interesting cases illustrative of the moral effects of the use of this characterograph." In the *Co-operative News* for April, 28th, 1883, the subject is thus referred to : " The moral results from this instrument must be very great, for it appears to supersede all rewards and punishments. It also renders it unnecessary for the master to express anger or irritation of feeling (the worst example a teacher can set to his pupils)." Some teachers might find the method transitional and therefore useful.

encouragement is bestowed to strengthen the powers of self-control.

At the age of sixteen or seventeen the intellect and moral powers of the average human being are sufficiently developed to control the propensities; therefore adults resign authority, and to the young people themselves, not singly but in their corporate capacity, is entrusted the regulation of conduct, and discipline of turbulent elements within their circle. They associate in proper form for this purpose. They elect conduct committees, from amongst their members, invariably choosing those of high moral type. These committees deal with each case of disobedience to rules, insubordination, infringement of the rights of others, and so on, as it arises; and thus youthful public opinion is enlisted and exercised in defence of the virtuous, peace-loving members of society, and in protection of the whole community. This method secures the happiest results—for antagonism between old and young entirely disappears; and at one and twenty the youthful generation has already acquired useful knowledge of important and practical principles in the science of sociology.

When this system of training is universal; and the children of every social class have partaken of its benefits; *then*, but not till then, will criminals stand out convicted of *moral insanity*. A perfect *nurture* has failed to make them worthy social units; therefore *nature* must have rendered them incapable of social life. Their organism is defective. Either the intellectual and moral powers are intrinsically inferior, or the lower propensities are out of all proportion strong, and the ill-balanced human being is unfitted for civilization, *i.e.* an associated life the essential condition of which is, that intellectual and moral forces should everywhere dominate, and restrain the animal or lower propensities.

How, then, is society to treat this morally deformed class? Scientific restraints it *must* impose for its own protection. Will these be the scourge, or forty or fifty days' imprisonment; the dungeon, the treadmill, the hangman's halter? Clearly they will not. The treatment must be of an entirely different nature; and the enlightened action of an advanced society must become analogous to the ignorant action of an earnest Church in the Middle Ages, with precisely opposite or contrary results. Let me here explain by quoting from Mr. Francis Galton. "The long period of the Dark Ages under which Europe has lain is due, I believe in a very considerable degree, to the celibacy enjoined by religious orders on their votaries. Whenever a man or woman was possessed of a gentle nature that fitted him or her to deeds of charity, to meditation, to literature, or to art, the social condition of the time was such that they had no refuge elsewhere than in the bosom of the Church. But the Church chose to preach and exact celibacy. The consequence was that these gentle natures had no continuance, and thus by a policy so singularly unwise and suicidal that I am hardly able to speak of it without impatience, the Church brutalized the breed of our forefathers. She acted precisely as if she had aimed at selecting the rudest portion of the community to be alone the parents of future generations. She practised the arts which breeders would use who aimed at creating ferocious, currish, and stupid natures. No wonder that club law prevailed for centuries over Europe; the wonder rather is that enough good remained in the veins of Europeans to enable their race to rise to its present very moderate level of natural morality."* The policy of the Church in the Middle Ages will be pursued by society in the twentieth century; but in the reversed mode. It will gather poor criminals into its bosom, and secure for them a safe and happy refuge; exacting, however, celibacy. The racial blood *shall not* be poisoned by moral disease. The guardians of social life in the present dare not be careless of the happiness of coming generations, therefore the criminal is forcibly restrained from perpetuating his vicious breed. Now mark the result. Not the gentle natures (as in the case of the Church) but the criminal natures will have no continuance. The type will disappear: whilst evenly balanced natures, the gentle, the noble, the intellectual, will become parents of future generations; and the purified blood and unmixed good in the veins of the British will enable their race to rise *far above* its present moderate level of natural morality.

To promote the contentment of congenital criminals within their prison-home, where they are detained for life, an alternative to celibacy might be offered, viz. a surgical operation † rendering the male sex incapable of reproduction. Were this course *voluntarily* chosen, sterile males might then be permitted the society of females without danger to posterity, and since

* "Hereditary Genius," F. Galton, p. 356.
† The surgical treatment indicated is not the operation ordinarily performed upon some domestic animals; this, applied to human beings, would be morally and physically injurious. The reader is referred to the *British Medical Journal* for May 2nd, 1874, p. 586.

fuller social life tends to make all human beings happier, these convicts would become more manageable, and coercive restraints would cease to be indispensable. Although, as I have already stated, cruelty is no longer rampant in British prisons, the "desperate gang" as they are called are commonly, though not universally, treated with stringent, hard repression, and often subjected to bullying. A member of the Howard Association who has devoted much time to prison visitation says, in reference to this point: "A man of brutal instincts may do well even under bullying." This proposition I feel convinced is false, and I think we ought to condemn the mode of treatment that called it forth; whilst I heartily concur with the further statement made by this prison visitor: " But a man of nervous temperament, who has been in a respectable position socially, cannot put up quietly with the coarse despotism of an average warder, and so he finally becomes mentally deranged— a mere lunatic, ready to fly at those who control him. Penal servitude is not meant to *manufacture violence*, as I am afraid it sometimes does." *

Rough handling and brutal words are wholly inadmissible whether directed to first offenders or to the desperate gang. The principle of abstract justice condemns them; and they are simply suicidal in view of society's proper aim—the reformation of criminals. The folly of this false policy has in all civilized countries been recognized by a few enlightened prison governors, and success has invariably attended their attempts to institute a discipline of decreased restraints, and increased self-dependence. When, for instance, Mr. Obermair was appointed Governor of the Munich State Prison, " he found from six hundred to seven hundred prisoners in the jail in the worst state of insubordination, and whose excesses he was told defied the harshest or most stringent discipline. The prisoners were chained together. The guard consisted of about one hundred soldiers who did duty not only at the gates and round the walls, but also in the passages, and even in the workshops and dormitories; and strangest of all protections against the possibility of an outbreak, twenty to thirty large savage dogs of the bloodhound breed were let loose at night in the passages and courts to keep their watch and ward . . . the place was a perfect Pandemonium, comprising within the limits of a few acres the worst passions, the most slavish vices, and

* "Official Optimism," Francis Peek, *Contemporary Review*, July, 1884.

the most heartless tyranny." Mr. Obermair gradually relaxed this harsh system. The dogs and nearly all the guards were dispensed with; and the prisoners were treated with such consideration as to gain their confidence. In 1852, Mr. Baillie Cochrane visited the place, and his account is as follows: — The prison gates were "wide open, without any sentinel at the door, and a guard of only twenty men idling away their time in a guard-room off the entrance hall. . . . None of the doors were provided with bolts and bars; the only security was an ordinary lock, and as in most of the rooms the key was not turned, there was no obstacle to the men walking into the passage. . . . Over each workshop some of the prisoners with the best characters were appointed overseers, and Mr. Obermair assured me that if a prisoner transgressed a regulation, his companions generally told him, '*es ist verboten*' (it is forbidden); and it rarely happened that he did not yield to the opinion of his fellow-prisoners. . . . Within the prison walls every description of work is carried on; . . . each prisoner by occupation and industry maintains himself; the surplus of his earnings being given him on his emancipation, avoids his being parted with in a state of destitution." *

Since poverty and destitution, physical disease and inherited alcoholism, ignorance and the degraded nurture that tends to crush the humanities and develop the brutal propensities of man, are *all* causes of crime, it follows that the scientific treatment of crime must embrace the thorough eradication of these various causes; and the only available, comprehensive and effective (although slow) methods to adopt have been pointed out in my chapters on Poverty, on Heredity, on our Domestic System, and on Education. Within reformatories and prisons, the scientific training essential in the forming and reforming of character (and which, at present, in the homes and schools of our lower classes is nowhere to be found) may be partially supplied; and the class of criminals who have no defective mental and moral structure will recover self-respect, imbibe virtuous desires, and attain to habits of industry and self-reliance. To these liberty must ultimately be accorded; but the State will not neglect its duty of affording aid in establishing a life of self-support, and effective guardianship against the many dangers that beset a criminal at the critical period of liberation.

* This account is extracted from Mr. Herbert Spencer's "Essay on Prison Ethics."

And now one word regarding the expense of prison discipline, prison training, prison industries, and the economic theory —that the cost of the criminal should be reimbursed to the State by the criminal himself—a principle which results in the adoption of the iniquitous practice of leasing convicts. This system is pursued in the Southern States of America and elsewhere, and I entreat my reader to refer to the *Century Magazine* for February, 1884, and read there an article upon the subject, which clearly and forcibly sets forth the inhuman cruelties, the grave injustice, the indefensible errors, that naturally flow from this altogether mistaken method of dealing with criminal life. The theory of self-support in prisons is unscientific. Economy is of course desirable, but the prison holds an important place in the investments of public money for the improvement of public morals and the securing of public safety; and to sacrifice these aims to any mere money consideration is both grossly immoral, and foolishly impolitic.

From five of the largest prisons in the United States, with an aggregate population of 5300 convicts, there escaped during twelve months only one prisoner. In all the State prisons of the country, not kept according to the lease system, with a population of 18,400 there escaped in one year only sixty-three. Whilst in Tennessee alone under the lease system no fewer than forty-nine convicts out of a population of 630 escaped in one year. In Texas, by the official report for 1881 and 1882, there occurred under the lease system 397 escapes, and but 74 recaptures. In the previous years of 1879 and 1880, 366 convicts had escaped, and of these 123 were recaptured. "Now, in the interest of the Texas taxpayer," says Mr. Cable, "from whom the lease system is supposed to lift an intolerable burden, as well as for society at large, it would be well to know what were the favourite crimes of these 366 escaped felons (since unreformed criminals generally repeat the same crimes again and again), what moral and material mischief 123 of them did before they were recaptured, and what the record will be of the 243 remaining at large, when the terms they should have served have expired. These facts are not given; we get only as it were a faint whiff of the mischief in the item of $6900, expended in apprehending 100 of them."* Mr. Cable clearly proves his statement, that it is a fatal and inhuman policy, to act upon the theory that the community

* "The Convict Lease System in the South," *Century Magazine* for February, 1884, p. 594.

should not be put to any expense for the reduction of crime, and the reformation of criminals. The principles of justice require that the expenditure of public funds should be wisely regulated and conscientiously administered; but an enlightened society will as little grudge the expenses of a scientific and pathological treatment of its poor *moral patients*, as our humane society of the present day grudges the best medical treatment for its *physical* and *mental pauper patients*.

Expenses in the building and outside adorning of prisons, however, may well, I think, be curtailed. I heartily concur with Mr. Hill when he thus writes : " A gaol with a stately and imposing exterior has a mischievous tendency to give importance to criminals and dignity to crime, which the poor, but honest, man is likely to regard as a kind of injustice towards himself. I cannot therefore but hope that the fashion which led to the erection of such gaols as that at Reading and the New City Prison at Islington will soon pass away, and that we shall rid ourselves of that strange kind of vanity which causes us to make a parade of moral deformity." * There is no injustice and no impropriety in the productive employment of prisoners. Lunatic patients of the public are partially self-supporting, and that the public should maintain in idleness its moral patients is wholly unnecessary; but the labour exacted must be strictly proportional to the individual capacity, for observe, whilst its indirect result is economical, its direct aim and purpose is medical and educational. It is intended primarily to promote the physical health, the mental comfort, and the moral improvement of the patient.

Educating ignorant criminals, reforming corrigible criminals, and restraining from crime incurable criminals, are duties of the State to be stringently, faithfully, and gently discharged, until the glorious period is reached when science, having been for generations persistently applied to all the various causes of crime, and having removed these causes—by restraint of the too rapid increase of population, by careful attention to the laws of heredity, by the scientific training of each individual member of the community, and by well-ordered domestic and social life—the criminal nature will have become extinct, and crime itself be simply historical—a thing to be studied with interest as an extirpated social disease.

* Hill, on " Crime," p. 297. This applies equally to our magnificent lunatic asylums. Why should we make a parade of mental deformity ?

CHAPTER XXI.

THE LAND.

"In every country which enjoys the European system of civilization, the right of property has ever been in a state of evolution, always tending to give a greater degree of independence to the individual owner; in other words, the evolution has always worked in favour of individual egotism. Who can say that the evolution is now complete, or that we have yet realized the highest ideal system in the disposition of our property? A progressive evolution is for every society one of the conditions of existence. The right of proprietorship cannot, therefore, remain stationary."

"The future will place civilized societies in an inexorable dilemma: justice or death."—CHARLES LETOURNEAU.

THE subject we have now to consider is the relation of private landowning to society, civilization, and progress; and the question before us is, What changes in the conditions of landholding are necessary for the future well-being of man? In its simple aspect the relation between land and labour on the one hand, and human life on the other, is very clear and definite. All classes of land animals (and man is one of these) are in various ways dependent for subsistence upon the produce of the land; and when man emerges from savagery another element, namely, labour, enters into the conditions of his life, and becomes along with land essential to his existence.

Of food fitted for the nourishment of man, land, in its uncultivated state, produces very little, save some wild fruits and edible roots, and many wild animals which he may eat, if, in the process of hunting them, he himself is not eaten. But land placed under the additional forces of man's physical energies and intelligent control, produces a variety of objects, comprising plants, animals, and minerals useful to him—a wealth of raw material, which, submitted to the further elaboration of his efforts and genius, help to create for him a civilized life. In other words, this raw material supplies man with food, clothing, shelter; with the comforts and luxuries that his developed nature demands, and with all that is necessary for the existence of literature, science, and art.

Passing over the nomad and purely pastoral forms of primitive associated life, let us glance at the agricultural stage, a form in which the labourers on the land are manifestly the all-important social units. They sow, till, and reap the fields, and tend domestic animals, whose skins and wool are made into coarse garments by other members of the group. The latter are dependent for the raw material of their industry, as well as their food supply, upon the agriculturists who form, if I may so speak, the foundation-stones of the simple social structure. Community in land is the prevailing system, and annual division amongst the families for purposes of cultivation; whilst weapons, fishing-boats, tools, and such-like movables are the only individual property.

Gradually, however, a change takes place: communal possession of land gives way to individual possession; and force in one form or other is the sole cause of this change. External aggression initiates militant activity; and in the process of chronic resistance to invasion, and direct aggression upon others, there is produced the class inequalities which distinguish a militant type of associated life, and along with these a complete individualization of landownership. Land has now become private property in the hands of bold and crafty individuals of the social group, who compel the cultivation of the soil by landless men of their own race, and by prisoners of war, spared on condition that they perform hard labour. The institution of slavery is thus established, and becomes a leading factor in the promotion of civilization. The lords of the soil are free to spend their energies in warlike activities whilst carefully protecting their serfs and slaves, the produce of whose labour is secured by this dominant class, and more and more appropriated to its own peculiar benefit. Simple garments, simple dwellings, and the mere necessaries of life do not suffice for its requirements, and slowly there uprises a new form of labour and a large class of workers upon raw materials, producing, not necessaries, but a variety of commodities to gratify the desires of pomp-loving, barbaric chiefs. These workers, whose labour is all absorbed by the chiefs, are fed from the produce of the land. And now let us see how this is done. The chiefs, whilst exacting hard labour from their slaves and serfs, keep them to a bare living; but the return of the labour at all times exceeds the bare living, and when the harvest is good exceeds it very considerably. The whole of the surplus, which is really rent, belongs to the chiefs, and here is their source of power; with it they can support a greater or less number of landless men engaged in ministering to them in all manner of ways and means. The latter group themselves around the castles of the chiefs—filled with military retainers —and thus we have the beginnings of towns.

As the surplus produce of the land increases under better conditions of cultivation, the inhabitants of towns increase in number. Markets and stores are instituted, and a commercial system is introduced. Barter, that clumsy method of trade exchange, speedily gives way to the use of money; and the growth of the whole social organism, in expansion and complexity, moves on apace. Meanwhile, slavery disappears; but workers on the land (the all-important social units) are poor as before. Starving competitors for work drive no hard bargain with masters in full possession of the soil—masters without whose leave they cannot grow the simple fruits and grain necessary for a meagre living. For food enough to live they readily give the labour of their whole lives; and since nature bestows an abundant return, there is plenty of surplus produce for landowners to absorb and employ as they choose. Into the towns they send it; and there it gives constant stimulus to material and mental progress, and creates new departures in social life.

Class inequalities amongst the workers there slowly appear; the mentally stronger dominate the mentally weaker, and organization of labour takes place. Directors and controllers of labour, in the production of commodities required by the chiefs or dominant class, demand a greater return for their work than a meagre living; and out of the surplus produce of the land they secure from the chiefs what forms the foundation of wealth in a new social class, namely landless capitalists. These men, possessing brain-power and money-power, become supreme factors in altering social conditions. They promote manufacture and commerce by a method precisely similar to the landholders' promotion of the cultivation of the soil; they press down the labourers to a bare living, and take, in the shape of profit, as much of the remuneration of the joint labour as competition permits, which they apply to the satisfaction of their personal desires and the carrying out of their schemes of manufacturing and commercial enterprise. Entering the field of luxurious living, they emulate landholders in the purchase of valuable commodities, and so stimulate general trade.

Meanwhile, mental activity grows by exercise, and in the intercourse of urban life intellectual power is rapidly evolved. Education is initiated, aptitude and skill are highly prized and rewarded. Invention profoundly modifies the primitive modes of production, and genius aspires to understand and govern the forces of nature. One direction taken by mental activity eventuates in a social force of primary importance—I mean the Church, or religious organization. Many of the finest minds of early ages have been allied with the priesthood; and the Church's desires for stately temples, gorgeous shrines, and decorative worship, have enormously aided the outward development of architecture, sculpture, painting, and music, and the inward growth of æsthetic capacity. But priests and all whose labour is absorbed in the construction of temples and the requirements of religious worship must be fed on the produce of the land. The priesthood is a privileged class, freed from industrial labour and military activity, and maintained in leisure by rents, tithes, or the voluntary offerings of the people. This leisure has made it possible for some within this large class to advance civilization by devotion to literature, history, philosophy, and art.

The purely militant stage of social life is gradually outlived; the society becomes of an industrial, peaceful type. Nations now exist possessing enormous wealth in the form of material commodities; wealth in the form of intellectual endowments, and the educational institutions that promote knowledge; wealth in the form of ornament, all that embellishes existence and makes beautiful the surroundings of human life; wealth in the form of social feelings with their attractive charm in refinement of mind and manner; and all this wealth has come into existence through the natural action of evolutionary forces—an action creating step by step a *system* of social interdependence and regulation. The prominent features of that system are: first, private property in land; second, great social inequality; third, poverty of manual labourers; fourth, a large town population and a small or minimum peasant population. Its less prominent but no less decisive feature is the complete social subjugation of the weak by the strong.

The pillars of the system throughout the whole process of its growth have been labourers on the land; and these have scarcely at all partaken of national wealth. The products of their labour—the food which they reaped from the earth—was the motor force which vitalized and energized the whole evolving social organism. Food was the ruling power which decided the extent and growth of economic life, as well as the form of its development. But observe, these food-producers have never been determinating factors of the destination and consumption of the surplus food ; and here lies the key to a great social problem—the essential point of an argument, which the student of social science is bound to comprehend. The holders of the land, and these as a rule have not been workers upon it, were all along initial directors of the destination of its produce ; and up to the present day, landlords, including the proprietors of coal, iron, and other mines, capitalist employers of labour, the State churches, and the hereditary rulers, are at the very fountain-head of modern civilization. Through their action, caused mainly by egoistic forces, such as selfishness, rapacity, greed, ostentation, tyranny, pride, agriculturists have been kept diligent and limited in number ; operatives have been kept at maximum toil ; and the resources of the land have been developed and improved, until land and labour together suffice to support an enormous mass of individuals of entirely distinctive character, activities, social position, and social worth, but all alike in one particular—they are daily, hourly, consumers of the produce of the land.

A good harvest that is general over the world has this natural tendency—to send activity like an electric current through the economic system. A bad harvest, if it is universal, will cause universal depression : not agriculture alone, but manufacture, commerce, science, art, education, literature, recreation, all will suffer, for on the production of food depends the buying power, not of nations individually, but of the whole trading world. Modern countries, it is true, are not maintained from their own food resources alone. From every quarter of the globe comes food for our teeming multitudes. It is true, also, that the machinery of exchange, *i.e.* money and our vast credit system, entering into the phenomena, are apt to confuse the mind that seeks to grasp the relations of those phenomena. Nevertheless, under close investigation, this fact clearly discloses itself—the relation of Landlordism to civilization is not accidental, it is *causal* and *necessary*.

This connection of land with the general industry and life of a people becomes obscure through the differentiation, or growth in complexity of modern society. If we look back only a century or two in the history of our own country, landed estates were small, whilst resident landlords or yeomen were numerous, and so were the agricultural labourers. At the present time, both these classes are, in proportion to our population, extremely small. The general trend of both food and labour has been into the towns. The invention of machinery and the application of science to land cultivation have lowered the amount of peasant labour required; but since it has not lowered the rural birth-rate, our surplus peasants have persistently drifted to the centres of manufacturing industry, and by competing for work with the operative class (whose birth-rate does not diminish). they keep wages low, and facilitate the enriching of capitalist employers of labour. The last mentioned class, observe, applies wealth to the further evolution of material civilization, the extension of the production of objects that minister to the comforts, luxuries, and refinements of life. Capitalists are incited to this by personal desires, selfish ambition, and tendency to mercantile speculation.

I believe it is from want of attention to evolution that political economy gives unsatisfactory explanations, and becomes less fixed and more disputed—sure characteristics of error. It may be looked upon as in, what the Comtists call, the metaphysical stage of thought; it requires to be advanced to the positive stage of science—not dependent upon abstractions, but facts; not deduction, but induction. Political economists usually hold that capital results from parsimony. In the words of J. S. Mill, "All capital, with a trifling exception, was originally the result of saving." * Deduction from insufficient premises must surely have led to this conception of a fundamental point. It is really a question of *fact*, and the fact is very different. Not individual saving, but *social seizure* is the origin of capital. The landlord does not in any sense save the produce of his land; he could not want to save what it would be useless to keep, for he could not consume it at any time. The essential quality of saving, or abstinence, *i.e.* the personal will, is absent here. The true origin of capital lies in the surplus seized from producers and sent into the towns; it originates through the selfish quality of rapacity, not the respectable virtue of prudence, as capitalists themselves would have us believe. The amount of capital in circulation does not depend on virtues of abstinence or prudence, on failure

* " Principles of Political Economy," J. S. Mill, bk. i. chap. v. § 4.

or success of mercantile men, but on the amount of food produced.

While wealthy capitalists have increased in number, landlords have diminished, partly by the action of antiquated land laws, and partly through class pride and ambition, leading to the extension of landed properties. Estates are no longer small, but large. Out of a population of about thirty-five millions, seven thousand persons only, dukes, marquises, earls, barons, and rich commoners, own no less than four-fifths of the soil of the whole United Kingdom. Whilst country population, then, has dwindled, town population has increased, and is still rapidly increasing; and this has been a marked characteristic feature of all advancing civilizations. With us, workers in towns are now out of all proportion more numerous than workers on the land, yet the latter are far more essential to the former than the former to the latter. Without the action of landlords in wresting the surplus produce of agricultural labour, and directing it to the towns, under the influence of purely personal motives, town industry would speedily stagnate and town workers languish into death. Town production must always be dependent upon rural production, and in town and country the operatives and peasants, who united form one huge class engaged in manual labour, resemble Sindbad the Sailor, on whose shoulders rides the Old Man of the Sea. It is they who maintain our leisured classes. The proletariat carries on its back all the rich and their innumerable dependents, three hundred thousand soldiers, an immense navy, a number of criminals, a million of paupers, a number of State pensioners, her Majesty and the Royal Family, all Government officials, ecclesiastics, and a vast host of unproductive consumers throughout society.

It is possible to compute in a rough way how much labour of the poor is absorbed through a rich man's private expenditure of £5000 a year. I have mentioned in Chapter XIV. that this means the absorption of the labour of seventy-five persons for a whole year in ministering directly or indirectly to the comfort, luxury, and enjoyment of the one rich man. Female labour is of still lower economic value (by reason of its quantitative relations to demand), so that one single dress worn a few evenings by a fashionable lady, may cost as much as the average earnings of a working-woman for a whole year. Truly, this civilization that we value and are bound to respect, is based upon a prodigious waste of human labour and degradation of human lives.

Slavery of the many for the comfort and enjoyment of the few, is all that man, as yet, has attained to in the carving out of his destiny; and when Mr. Henry George, in his arraignment of the whole modern system of economic life, cries out, "All over the world the beauty and the glory and the grace of civilization rests on human lives crushed into misery and distortion," no earnest, enlightened person will deny his indictment, but, on the contrary, all such will admit that these words are full of sober truth and right reason. The truth concerning the matter is this: class inequality and the superior position of the rich over the poor has no justification in ethics; whilst it has a distinct evolutional justification. Under perfect ethical conditions every child born into a society would necessarily have equal chances of life and comforts and luxuries with every other child. At the present day one British babe may be born to an income of £100,000 a year, and another to no income at all, but to a perpetual struggle for the barest existence, and a pauper's grave at last. This is surely the height of social injustice; and the system that permits it is ethically odious. But the question is not really one of ethics at all; it is a question of possibilities, in a problem relating to development from a lower to a higher state. This development has taken place, and whether we approve its process or not, we are bound to accept it as the natural and inevitable result of all prior conditions; and to recognize that conditions have not yet sufficiently altered, to make a much better system possible.

Mr. Henry George and other revolutionary land reformers are eager for immediate change in the structure of society; but the fitness of the units for a superior social state is a consideration they entirely overlook. To my mind they resemble devoted physicians applying the most superficial remedies to a disease, imbedded in vital organs, which they guilelessly omit to diagnose. To break down the system of land-owning or to abolish private property in land; to sweep away landed gentry; to establish peasant proprietors or government tenantry; to institute co-operation or communal land-holding; to dispossess the rich and by revolution create social equality;—these are the remedies for social misery believed in by the above physicians. Three fallacies are implicated in this belief, and must be dealt with here. The two first fallacies are positive;— first, the impression that landlordism is the main cause of poverty; second, that class inequality is an unmitigated evil.

The facts, as I have shown, are directly opposed to these doctrines. Landlordism and social inequality have been powerful factors in the creation and support of civilization; and it has yet to be shown that civilization would endure were these factors cut down and destroyed. The third fallacy is a negative one;—it lies in the leaving out of the complex problem, a law of life which in reality controls its determination. The law is this : all living beings tend to increase at a more rapid rate than the means of their subsistence. This tendency controls the destinies of the lower animals, of all the races of uncivilized man, and as yet of civilized man himself. Until man learns to control it, the dominating cause of human misery is left untouched. Population presses on subsistence year by year. Were we to revolutionize society, and by freeing the land from the present system, cause it to produce double what it does now, population would soon rise equal to it, and happiness be as far off as ever.

Better circumstances invariably act upon the working classes of England as a stimulus to the birth-rate. In proof of this assertion I point to the following facts. The Registrar-General, in his report for the year 1876, writes as follows:—
" The state of trade and national industry is strikingly exhibited in the fluctuations of the marriage-rate of the last nine years. . . . The period of commercial distress which began about the middle of 1866 and continued during five years . . . influenced the marriage-rates of these years which were 17·5, 16·5, 16·1, 15·9, 16·1, and 16·7 (in the 1000) respectively. In 1872 and 1873 the working classes became excited under the rapid advance of wages and diminution of the hours of labour, and the marriage-rates rose to 17·5 and 17·6 respectively." In his report for 1881, the Registrar-General again accentuates this important point : " The marriage-rate," he writes, " reflects with much accuracy the condition of public welfare." And further on : " The birth-rate was at its maximum in 1876 and fell uninterruptedly from that date year by year, in natural accordance with the corresponding decline in the marriage-rate." These years represented another period of commercial depression. We have here, then, incontrovertible proof of the national tendency. The mass of our people increase their numbers so soon as they are more comfortable, and the marriage-rate for each year may be called the pulse or indicator of the nation's economic health. Its fluctuations coincide with the upward and downward

movements of commercial activity. To the theory of Malthus concerning population, objection is made by Mr. Henry George and others, on the ground of its repulsiveness. But for ugly facts, ugly theories are requisite as explanation; and the odious law of population accords with the wretchedness which it explains.

Society has during the last fifty years enormously increased its machinery of free education and philanthropic instilment for the elevation of the masses; but nowhere is a scientific knowledge of the law of population, with its far-reaching results, carefully instilled. Under all the mercantile and ornamental instruction freely bestowed, the essential character and ordinary habits of the social units remain unchanged; and an easier outward condition of life is still, and will long be responded to by, not a higher standard of living, but an accelerated marriage-rate and birth-rate. To destroy the richer classes, or even to render them straitened in means, would be to create depression in every branch of manufacture, trade, science, and art, that ministers to the amenities and refinements of civilized life. The supply of wealth from these sources would necessarily sink to accord with a lessened demand; for the working classes, suddenly enriched, would assuredly make no efficient demand. It is refinement alone that demands refinement, culture that demands culture, education that demands education, and if we disinherit the educated, cultured, and refined classes—a small number as compared with the unenlightened and as yet non-elevated masses—we have nothing better rationally to expect, than a general abasement of our national civilization.

It is true that the partial civilization we possess is by no means to be defended at every point. It is grossly imperfect. If the gold of true refinement is in it, false glitter and base alloy are in it as well. Our upper classes, made effeminate by ill-spent leisure, and often inflated by tyranny and pride, demand all the artificial pleasures of a voluptuous and inane life. Civilization is not by them directed to the highest and noblest ends. Nevertheless, if the control of human labour passes out of their hands into those of the lower classes, retrogression, not progress, will result. There is no escape in logic from the cruel conclusion, that the Old Man of the Sea must for the present remain where he is. If he is suddenly thrown off from the neck of Sindbad, civilization will adjust itself to a lower level. It will lose much of its precious gold;

whilst permanent happiness will be no nearer, but on the contrary will be further off than before.

The scientific path of reform is an entirely different one from that pointed to by Mr. George and his friends. Inner and outer improvement must march forward together. Adjustment of outward conditions must be carried on comformably with advance in the social units; and this advance is dependent upon the transmutation of lower into higher instincts under heredity and training, the acquisition of knowledge in regard to facts of life, and the growth of inclination to subject individual action to conditions of general happiness. Changes in this direction are in actual progress now. The outcry against injustice, the aspirations for liberty, ascending from the lower classes prove this fact; and a still greater proof lies in the rapid growth of social feeling and decay of anti-social emotions.

The evolutional justification, which I have given, of landlordism and class inequalities, with all that has been stated in defence of the present disposition of property relates to the general system only; and by no means implies that improvement is not called for at once. On the contrary, improvement, I believe, is urgently required, for as individual progress in mental and moral elevation proceeds, the defects of the system are more and more felt. Imperfections that had no special significance in the past are pressing heavily on society now; and speedy removal of these ought to take place. The various questions of practical land law reform it is not my purpose, however, to enter upon. I can only state my opinion that we are already prepared for, the abolition of the right to *settle* and *entail* land; the abolition of the law of primogeniture; the facilitation of the transfer of land; the imposition of a graduated land-tax pressing most heavily on large estates (in order to lessen the extremes of social distinction); protection of tenants by law; and, in short, a gradual movement towards equalization in society. A perfect ethical code belongs to a future age; but although not realizable now, it indicates the true path of advance. Social justice must ultimately be reached by progressive evolution; and since private property in the soil of a country constitutes an injurious monopoly of that which is necessary to the existence of all mankind, it is clearly a social injustice; and society will ultimately therefore assume possession of the soil, and institute a system of nationalization of the land.

CHAPTER XXII.

SOCIALISM *VERSUS* INDIVIDUALISM.

" It is only quite recently that there has come into existence anything like a truly *positive* philosophy, *i.e.* a philosophy of *action*. The intellectual power of enlightened man has at length become sufficient to grasp the problems of social life. . . . Nothing remains to be done but to apply the established canons of science to these higher fields of activity. Here there is still competition. Here the weaker still go to the wall. Here the strong are still the fittest to survive. Here Nature still practises her costly selection which always involves the destruction of the defenceless. The demand is for still further reduction of competition, still greater interference with the operations of natural forces, still more complete control of the laws of nature, and still more absolute supremacy of the psychic over the natural method of evolution."—LESTER F. WARD.

CIVILIZATION is a general term bound to become wider and fuller in meaning as Humanity approaches an ideal state of social life. To the popular mind the word suggests railways, telegraphs, steam navigation, machinery, and all the other striking features of our modern industrial society, whilst a thinker might define civilization as embracing the whole results, subjective and objective, that have flowed from association in the life and labour of man. These results, however, are not confined to the production of an innumerable variety of material objects that minister to his comfort and enjoyment. They extend to a sphere of delicate immaterial relations including mental and emotional, in one word, psychic, forces. The qualities of man's nature have become manifold, the range of his thought immeasurably increased since the days of his savagery. New developments of feeling and sentiment have become social forces of enormous power and diverse signification. Sociality has given depth, height, strength, variety, delicate purity, to the relations of man with his fellow-men, and has continuously and persistently altered his immaterial environment, for the stores of his accumulated knowledge have been added to in every age, and consequently the young intellect of each new generation has unfolded in a richer and wider mental atmosphere.

Now, in my last chapter I had to deal mainly with material civilization. I slightly sketched the order of its progress. The method I may thus describe. Man has made himself acquainted with the natural laws of matter by means of patient observation and tentative experiment; and then has so handled matter as to direct its forces into channels carrying benefit to himself. Thus he has constituted himself (not the creator), but the complete controller of such natural laws of the material universe as lie within the reach of his physical activities and mental comprehension. By this method and no other has our whole advance along the line of material civilization been hitherto accomplished; and all further extension of the comforts and amenities of social and economic life is certain to be obtained through persistence in this satisfactory and available course.

Throughout the domain of non-material civilization, however, man has not constituted himself controller of natural laws, although he has subjugated many vital forces in the orders of life beneath him. Complicated vegetable and animal forces, for instance (all subject to natural law), are enlisted in his service and prove submissive to his dominion. On every side prevailing forms of vegetable life, cereals, fruit trees, plants and flowers of infinitely varied tints, bear witness to the art and skill of man; whilst the animal kingdom, despotically ruled by mysterious biological laws, has yet provided him with faithful servants of his will, and happy partakers of his enjoyment in a life which, to dogs and horses, is wholly artificial. In the order of his own social life man's position is entirely different. If we take an objective view of society, what it presents to us is an infinite variety of complicated movements produced by natural forces which are pursuing a wild unbridled course. That course is the path of least resistance, for no psychic force has intervened to adapt the course to any ultimate aim or purpose, or to harmonize those lines of least resistance with the line of permanent and universal advantage to man. "He has made the winds, the waters, fire, steam, and electricity, do his bidding. All nature, both animate and inanimate, has been reduced to his service. . . . One class of natural forces still remains the play of chance; and from it, instead of aid, he is constantly receiving the most serious checks. This field is that of society itself, these unreclaimed forces are the social forces, of whose nature man seems to possess no knowledge, whose very existence he persistently ignores; and which he consequently is powerless to control."*

In the field of society the supreme forces broadly recognized are religious and political. Government on the one hand restrains and dominates such social forces as conflict with general happiness; and popular religion, on the other, rules in the emotional and moral sphere, and guides the spiritual destinies of mankind. Naturally, then, the gradual awakening of the general mind to the miseries of our imperfect social state has been accompanied by a proportional increase in religious fervour, and in faith in legislative measures of reform. But, in individual minds here and there, a new conception has arisen, namely this—that control of social laws of nature and subjugation of the social forces may be attained to through a *scientific method;* and man may rightly aim at becoming master of himself in the field of bewildering chaotic confusion, which we call nineteenth-century civilization, as effectively as he is master in the domain of vegetable life, and master of the lives and destinies of all domestic animals. This conception is based upon the theory that, first, all the complicated movements of industrial and social life—the spontaneous actions of a freeborn race, the conflicting desires of palpitating hearts, the noblest aspirations of enlightened men, the sweetest hopes of religious women—all these and many more are natural *forces* obeying natural laws; and that, second, these laws may be studied and understood, and instead of the order which at present prevails and is replete with misery, a new order may be substituted, for man may apply inventive genius to the natural social forces, and under the new knowledge of the laws that regulate them, divert these forces into fresh channels of happy activity and so initiate decisive change.

This bold conception is conspicuous in the present day upon the horizon of intellectual thought, and its outcome in the cut-and-dried schemes of State Socialism, the Democratic Federation, and Christian Socialism, must here be slightly glanced at and appraised. But first in reference to Government and the Church—political agitation is one of the most prominent features of the present age. I have already shown, however (Chapter XIV.), that legislation scarcely at all affects

* "Dynamic Sociology," Lester F. Ward, vol. i. p. 35.

our material and moral progress. External government has been and is a necessity of social life, yet the institution *as it at present exists* contains no element of progression, gives out no impulse of real advance. Mr. Ward has aptly pointed to the analogy between political events and meteorological phenomena. The latter comprise clouds, winds, tornadoes, cyclones, thunderstorms, followed by universal calm and general sunshine; yet "a retrospective view shows that the corresponding seasons of different years do not materially differ, and the state of the world, meteorologically considered, is the same now that it was a century or ten centuries ago." * To vast cosmical processes, and not to surface turmoil, is due the great secular changes in the history of the globe. And similarly, political events excite men's wildest passions. They elevate to brilliant hopes and then reaction brings despair. But all the time, "human progress depends upon deeper laws (than those involved in politics), and would not have been arrested had there never lived a Wellington, or a Washington," † or, I may add, a William Ewart Gladstone.

Of the progressive element in the great religious systems of the world I shall speak in my next chapter. It is sufficient here to point to the existence of Christian Socialism as proof of the continuity of the profoundly religious spirit from age to age. Individuals of strong conservative tendencies may call our modern new-born spiritual force utopian, and feel only jarred and otherwise uninfluenced by it, whilst nevertheless that force is eagerly imbibed and joyously assimilated by innumerable minds that, hitherto, had been nourished upon ancient dogma, and are hereditarily endowed with a strong conscience and a deep religious fervour. To these minds there is no vital dissonance between the old spiritual life and the new. They adapt themselves to fresh conditions, adopt new forms of expression, and manifest religion in their daily life precisely as before. To them Utopianism, or, what is correctly termed "Meliorism, is the great spiritual motive-power of the world, and is rapidly rising to high-pressure. Stronger to-day than ever before, it is likewise assured against the aimless waste in which it formerly frittered itself away. . . . With Utopianism for its propeller and Science at the helm, Humanity cannot but speed forward to a new heaven and a new earth." ‡

* "Dynamic Sociology," Lester F. Ward, vol. ii. p. 245.
† Ibid., p. 244.
‡ Norman Britton, in *Progress*, September, 1884.

It was in Germany amongst the Teutons, a race of peculiarly solid and rational mental quality, that the great modern reformation of religious thought began. From Germany came the impulse to movement—the force which broke up and ultimately destroyed the profound stillness of an intellectually glacial epoch, when individual minds were held fast locked in the mechanical embrace of a rigid and stern ecclesiasticism. And it is perhaps somewhat significant that in Germany, amongst the Teutons of the present day, a reformation in social thought has arisen and is rapidly extending its influence beyond the land of its birth. "All socialism of to-day, whether found in Paris or Berlin, in New York or Vienna, in Chicago or Frankfort-on-the-Main, is through and through German. . . . One of its leading characteristics is its thoroughly scientific spirit. Sentimentalism is banished, and a foundation sought in hard, relentless laws, resulting necessarily from the physiological, psychological, and social constitution of man and his physical environment." *

If sentimentalism, however, is banished, the sentiment of the movement commands respect. One of its strongest opponents thus writes : "The pretence that Socialism springs in the breasts of its advocates from a low ideal of life, is the very reverse of the truth. When the history of Socialism comes to be written without passion or prejudice, many a man, now regarded as an ogre, will be recognized as a devoted lover of mankind. The moral springs of Socialism are not envy and greed. On the contrary, a noble impatience of the misery of their fellow-men, and the injustice under which they suffer, supplies the motive-power of ninety-nine hundredths of the socialistic propaganda. In our contest with Socialists," continues this opponent, " there needs enter no element of bitterness. Our political ideal is the very negation of theirs, . . . but our difference is not moral or religious in the ultimate sense, but economical and political. . . . Feeling of any kind other than *desire to arrive at truth* should not be imported into it." † Observe, this utterance is typical of the changed mental attitude of all earnest opponents in the work of social reform. Evolution in sentiment and the growth of sociality have rendered impossible to advanced social units the brutal intellectual wrestling and passionate word-assaults of the past;

* "French and German Socialism in Modern Times," R. T. Ely, p. 156.
† " D.," in *National Reformer*, November 5th, 1882.

and so far as the *leaders of thought* are concerned, the new war of opinion upon which we are undoubtedly entering, viz. a contest between Collectivism and Individualism, may be, and we anticipate will be, fought out zealously to its noble end, truth, upon the elevated platform of calm reason, with a constant evidence of the cool head and the warm, that is, social and humane, heart.

It is not with the many defects of educational systems and modern domestic life that crude Socialism deals. It takes no note of the utter inefficiency of our present social surroundings to help forward the transmutation of lower into higher instincts, which is the primary condition of real advance in immaterial civilization. The economical and political systems of modern life are what this Socialism deals with, and the relations that exist between capital and labour are what it attacks and broadly condemns. The constant pressure upon rural labour exerted by landlords and its beneficial effects in evolution have been fully explained in my last chapter. The pressure on labour, which formulated Socialism specially points to, is that arising from competition and the possession of capital by a minority. Organization (in the form of factory employment and division of labour) has taken place in the field of production by individual action of capitalists in their own interests, and this organization, which is really socialism, *i.e.* concerted action for social ends, has had the effect of enormously promoting the increase of national wealth—the necessaries as well as the luxuries of life—and so benefiting society as a whole. But in the field of distribution, organization has very little beneficial effect. There is, it is true, an increasing tendency to union amongst capitalists in the interests of their own individual class; but this union is monopoly, not socialism, and acts injuriously upon society as a whole. The great social force that holds sway in the field of distribution is competition. Manufacturers compete with one another for purchasers in the market composed of distributors, and merchants compete with and undersell one another in a chronic struggle for the custom and patronage of the general public. The outcome of this strife and antagonism is, on the one hand, tyrannic oppression of labour, and on the other, adulteration of the products of labour. Its general result is the subserving of society's wants in an exceedingly clumsy, unsatisfactory manner, accompanied by an increase of poverty and a degradation of public morals.

From the whole field of economic life, say the State

Socialists, competition must be banished and organization made to take its place. On the method by which this shall be done Socialists differ, but all are alike in this particular—they seek to impose on society an economic system regardless of the history of the past; that is, without due consideration of the point, whether evolution forces have prepared the social units in habits and quality of mind, for the new system they propose. The system proposed is, that the State should own the land, all instruments of production, and capital; that the State should organize all labour and direct the distribution of all produce; and that throughout society social equality should be established and maintained.

Now, the monstrous evils pointed out by Socialists are, for the *most part*, lamentably true, even to Mr. Hyndman's somewhat exaggerated picture of adulteration—"a fair representation (of the age) would be a keg of bosh butter, a bale of shoddy cloth, and a wooden ham." * But the whole argument of my last chapter concerning landlordism applies equally to our competitive mercantile system, and proves—first, its evolutional justification; and, second, that for us at present any much better system is simply impossible. Without bloodshed no sudden change could be accomplished, and the strength and superior quality of the executive required to maintain the new order would prove an insurmountable difficulty. Again, the habits of the people, rendering them unfit for a more associated life, and their inability to restrain the birth-rate would, under conditions of social equality and liberty, speedily cause poverty to reappear. The new order of things, if temporarily established, would inevitably break up in anarchy and confusion.

The renovated system proposed by State Socialists is but "the baseless fabric of a vision" ("such stuff as dreams are made of"); nevertheless, and to my reader the statement may appear paradoxical, whilst condemning as impracticable the schemes of the Socialists, I assert, that they are right in their primary principle—interference, and right in their ultimate aim —social justice. This proposition may be stated in terms of social science, thus: first, the possibilities of evolution extend to inventive interference with social nature, that is, the subjugation of social forces by psychic force; second, human evolution advances towards a perfected social state determined by pure ethics. No other state but that of complete equality in the comforts of life could satisfy all mankind; and that this

* Debate on Socialism between Mr. Bradlaugh and Mr. Hyndman.

end will be attained follows from an important psychic law which may be termed "the law of the elimination of evil," i.e. *under the spur of pain, discomfort, and injustice, it is impossible that man's endeavours should cease until every preventible evil of human life is overcome.**

It is of fundamental importance that the student of evolution should not overlook this law of mind. It obviously and naturally follows from the simple law of life pointed out by Mr. Herbert Spencer, which develops in low organisms, and is continued up to the highest species. "At the very outset," says Mr. Spencer, "life is maintained by persistence in acts which conduce to it, and desistance from acts which impede it; and whenever sentiency makes its appearance as an accompaniment, its forms must be such that in the one case the produced feeling is of a kind that will be sought — pleasure, and in the other case is of a kind that will be shunned — pain."†

The psychic law, which I have termed the law of the elimination of evil, is simply a development on intellect of this early biological law of conscious life, and increases in strength as intellect expands, and knowledge extends in society. When applied theoretically to society, in accordance with the knowledge already possessed of man and his environment, the power of this psychic law in anticipating future changes must be immense. It will mainly realize for sociology what Mr. Buckle speaks of as "the final object of every scientific inquiry, namely, . . . the power of foretelling the future." By its aid and that of knowledge, the social philosopher, we may hope, will be able to predict the general characteristics of man's future state as confidently as the astronomer predicts the eclipses of the sun and moon. Towards the ideal of Socialism society must slowly move, but crude socialism in method has gone astray, and real socialism is yet in an early stage. Of the term —socialism—the only definition wide enough to be scientifically correct is this—*concerted action for social ends*. This definition makes all of us *socialists*, for no one, not even the strongest

---

* The importance of this law is to be found in its applications. Its recognition disposes at once of the Comtist schemes of social organization. It shows that the projects of Professor Beesley, Dr. Kaines, Dr. Congreve, Mr. Frederic Harrison, for class organization of labour are delusive and unpractical ; for, while directors or employers of labour are to have larger incomes and more comforts than labourers, the less fortunate majority will feel the unfairness of their position, and continue the agitation for social justice until all special class privileges come to an end.

† "Data of Ethics," II. Spencer, p. 79.

individualist in theory, will deny that for social ends concerted action is sometimes necessary. It is the scope of the action, and the method to be employed, on which professed Socialists and Individualists disagree.

Socialism is of two kinds, coercive and voluntary; and in the process of its evolution, the coercive has always preceded the voluntary. In slavery and militancy it first becomes prominent. Under these forms of associated life fear develops the quality of obedience in the growing mind of barbarous man. The scourge and the club, visible to the eye, prompt action in the direction required, without appeal to an undeveloped imagination. In feudal times progress is observable. Direct coercive restraints are somewhat relaxed, for the power of obedience is stronger, whilst a few primitive abstract ideas concerning self-interest come into play. In modern industrial life coercive restraints are of a wholly different order. They belong to the system, and not to arbitrary, personal authority. They appeal to self-interest by mental concepts and vivid imaginations of possibilities of evil, and prudence creates action for social ends.

The motives to socialism of a purely voluntary nature are intellectually, emotionally and morally superior. The mind trained to obedience by despotism and authority, and to prudence, by self-interest under social restraints, becomes fitted to instigate action of an ego-altruistic nature. Ambition, patriotism, and love of approbation, become motives to union for public ends, and a popular government such as our own is made possible. The ego-altruistic motive that comes into play in the voluntary productive associations of workmen is of a different and twofold nature; it is prudence directed towards self, and public spirit directed towards the group; and forms a social force, in the germinal stage, of extreme value. The purest form of socialism depends for existence and development on this force allied with sympathy—a delicate appreciation of the rights of others, and an imperious conscience, or sense of duty—the whole directed by far-reaching intellect to the reconstruction of society on scientific methods. The sympathy that enters into this socialism springs from an entirely different source; not economic, but family life, is its origin. But it is not necessary for me here to trace its evolution, I have only to point out one consideration in relation to it. The break up of the family system shown in Chapter XV. is an interruption to the development of sympathy; and the more or less rapid growth of the true socialism will largely depend on advance in the reconstruction of family life, on the broader basis.

A closely associated, yet broad, domestic life will give increased capacity for concerted action. Communism in industrial life will be by superior social units voluntarily assumed as the system most in accordance with their sympathetic feelings, and higher moral nature. From small beginnings communism will surely increase. Communistic groups scattered over the country will grow in number, trading as companies in the general community under the terms of the old system. With multiplication of these groups, alliance between them will take place for mutual benefit, and the state of aggregation will give conditions allowing a new form of mercantile life to arise, *i.e.* a system not governed by supply and demand, but by knowledge of wants in the groups, and knowledge of their individual capacities for the production of the commodities required. The absence of personal ambition, and predominance of public spirit, within the groups, will stimulate invention for general benefit ; and the new system will be methodized and made scientific throughout. Its superiority in subserving society, over the old competitive system, will be evidenced, and the latter will be gradually superseded and die out.

The present system of industrial life is the natural product of unconscious evolution, full of inherent imperfections, and *necessarily* creating a state of inequality and social injustice. The future system will be the product of inventive genius consciously and carefully adapted to a society become capable of conforming to it. The law of "survival of the fittest" will assert itself in relation to economic life. It is thus that the dreams of the old Utopians and the conceptions of the modern Socialists will become true. The psychic method of evolution permeating the whole field of human life—*eugenic*, domestic, industrial, and social; and dominating heredity, training. education. production, distribution, and legislation, will gather up all natural forces at present pursuing a "wild, unbridled course in society," and control them to the production of general happiness.

To return to the present aspect of society ; the socialism that manifests itself in co-operative unions for production and distribution is in one sense a failure. It does not improve economic society because itself subject to competition. No permanent rise of wages can be secured by industrial co-operation alone, in a society in which the law of population is uncontrolled. Nevertheless, its existence is strongly beneficial; it improves the social units, awakens intelligence, strengthens the ego-altruistic motives; creates sympathy and public spirit, and fosters the virtues of civilization. Industrial societies of this order are valuable agencies for the real elevation of the masses and ought to be encouraged.

Government, local and central throughout society, is another form of socialism, *i.e.* concerted action for social ends; and it changes continuously in adaptation to change within the society itself. In its genesis, "government, like most other human institutions, has been the product of egoistic attempts on the part of *crafty* individuals to meet and supply a popular demand. . . . It is the keen-sighted few who perceive the wants of the many, devise means to supply them, and anticipate rich rewards from the befriended and grateful community." *
We see, then, that the socialism of government in its beginnings was of a low order dependent upon selfishness, and in its action it was wholly coercive. Later, in consequence of the growing intelligence and resistance of the people, governments change in character, and become more popular. The governing classes no longer dare to oppress the people, and find it their interest to be just.

The socialism of popular governments is clearly of a higher order, showing public spirit and better concerted action for social ends. The functions fulfilled by government are maintenance of order, protection from aggression, and public accommodation. It will be seen that as society improves, the need for the two first functions will diminish, whilst the last will increase in importance. The socialism manifested will become of a still higher order, and the spread of the knowledge of social science will cause the aims of government to be still better attained. A marked feature of the popular governments of the present day is the absence of unanimity of opinion on first principles. It will be evident to the student of social science that this state of things cannot continue. Sooner or later false theories of society and government will give place to true. A better selection of legislators will be made by a more educated people; not social position will decide the choice, but scientific knowledge of society, and high personal ability. Government by party will disappear—superseded by deliberative assemblies certain to prove progressive agencies in evolution.

* "Dynamic Sociology," Lester F. Ward, vol. i. p. 585.

Returning once more to our present state of society, socialism is showing itself in two ways—first, in philanthropic action all over the country, which, stronger to-day than at any previous period, is the outcome of the highest motives of mankind; and second, in the efforts of the people to alter the political and economical system of society by the action of the State in the direction of justice and equality. This agitation is a very significant fact: it represents a great social force certain to gather strength with the growth of knowledge, public spirit, and sensitiveness to personal rights. The hunger of the people for social justice can never be appeased by any economical system that the individualist can offer, wherein wealth will flow to the clever and the lucky, the cunning and the greedy, and be handed down by inheritance and bequest from generation to generation. In the competitive system, even when modified by extensive co-operation, controlled reproduction, and nationalization of the land, a satisfactory social equality in the comforts of life is a natural impossibility; for differences in the acquisition of wealth must follow from the natural differences in the qualities of the individuals competing. The prudent and wise will secure more of these comforts, than will those who are less highly gifted, although just as desirous of happiness. Similarly where unlimited private property—to keep or sell, to give or bequeath—is permitted in society, there social justice cannot be; the principle of equal rights must always be infringed. It is only in a scientifically organized system of communism, that equal, or almost equal, happiness can be given to all, irrespective of natural endowments, and only in such a system can the purest and highest ethics be practised and obeyed. Communism—the enlargement of the family system to national boundaries—presents the true goal to which the modern popular social agitations tend, for in anything short of this, strict justice and equality are not to be found, and the conditions of discontent remain.

I have said the agitation against present social arrangements is certain to increase; it remains to consider in what direction action will take place. There are two, and only two, general directions of popular reform: first, the *revolutionary*—the driving straight at the established system with the intention of overthrowing it and setting up some new order in its place; second, the *legislative*—the aiming to improve the present system, by modification or destruction of its *worst* features, its *extremes* of injustice and inequality. Whether coercive social revolution is or is not to take place in any country, will greatly depend on the spread of education and of true views of society among the people of the country. In these days, more than at any previous time, ignorance of social forces among men of public spirit and strong emotion is becoming a danger of the first magnitude; any one may understand this who thinks deeply of the moral causes of the recent dynamite conspiracies to destroy public and private property in English towns. I have elsewhere pointed out how retrograde and futile for the increasing of human happiness is the method of revolution. It is the *direct* method of human emotion with intellect unenlightened on natural evolution and causation; it seeks blindly to overthrow what is really the highest product of evolution for the time being, with the impotent intention of putting in its place some system which may be ethically better, but which cannot organize individuals ethically unfitted for a superior system. The method of legislation and gradual improvement is, we may hope, the only method likely to be used at any time by the social agitators of, at least, the countries of England, France, and the United States of America, so long, at any rate, as Socialists in these countries are unmolested respecting their opinions. If they are in error they can be instructed, but force resolves no questions, whilst it invites dynamite in return. The causes for public discontent are not trifling. That one man should have £300,000 per annum at the same time that six thousand of his countrymen, each as hard-working, or more so than himself, get no more than this sum divided among them for the year; that many men should toil ten hours a day for pittances that barely keep them from starvation—this is a state of things which we may be sure a people, growing in knowledge and morality, will not much longer permit to exist. The individualist of the "Liberty and Property Defence League" may cry out about the sacredness of the liberty to contract, amass, and bequeath, and the rights of property, but *laissez-faire* on these questions will not satisfy a hungry people conscious of a state of injustice.

It is usual to consider a man spending an income over personal gratifications, as in a *passive* condition towards society, or if *active* as one who benefits society by "giving employment." Also it is considered that the difference between rich and poor is a state of *inequality merely*, with which the working people are not directly concerned and of which they ought no more to complain, than invalids should complain that other people are healthy. But what is the truth ? The truth is that the rich owe everything to the poor ; that they are a parasitic class dependent entirely for support on the labour of others ; that the money they spend represents a power (socially permitted) to absorb the labour of others, to exact lifelong services from others, giving in return not their own services, but a little of the power (money) which they themselves exercise. If we were to place a rich man with all his money—paper and metal—on an uninhabited but fertile island, he would be reduced to his natural stature. In order to live he would no longer be absorbing the whole labour of scores of fellow-men, but only of one man, that is, of himself; he would no longer be a parasite but a working man earning his own living. It is obvious that (apart from the law of population) if the working people had only themselves to support and no rich class, with its enormous absorption of labour besides, they would have more of leisure and comforts, and would not have to toil so hard. There can be no ethical justification for the absorption by *one* of the labour of *many;* it is an unfair, that is, an immoral state. The ethical unit in labour and exchange is: for the labour of one, the product of one, or what amounts to one, in return.* Nevertheless, full evolutional justification for the present state of things there certainly is. It is better that some should be rich and civilized than that all should be miserable and savage. So long as general ignorance, low morality, and an uncontrolled birth-rate, which are the main supports of a rich class, are present, so long is it well to have inequalities and injustices in wealth-holding.

But although the rich, at present and for some time to come, can by no means be dispensed with, the *very rich may.* The rich, as a class, are necessary to civilization, but the legislative method of improvement, which is the true direction of the force of popular agitation, is—to reduce the *worst* conditions of unfairness; and this may be done to the benefit, not the injury, of society; for the evils of excessive wealth more than counterbalance its good. Excessive wealth naturally promotes in the whole class of the rich, emulative, selfish extravagance, deteriorative of human nature, and therefore it is injurious to society at large. The individualist, who looks upon property as something passive in society, something

* Subject to some modification relative to children, the aged, the weak, the diseased, and the criminals.

which can be handed down in any amount with perfect justice from father to son, is manifestly in error. Social expediency, not social justice, is the support of private property, and of the privilege of absorbing the labour of others without adequate return. Society will almost certainly before long think it expedient to cut down the largest fortunes by heavy cumulative taxation; and to prohibit the inheritance of property above a certain value (high at first), as Mr. J. S. Mill suggested.

The prevailing ideas of justice in society have arisen naturally in the past among the strong and the privileged few, and have been readily accepted by the docile and oppressed many. The foundation of these ideas has always been *power*, not *equality*. It is considered perfectly just that the clever and good man should have (if he can get it) a higher reward for his labour than the ignorant and stupid man. Of course the origin of this idea is nothing more than the fact that in a state of competition the clever are *able* to secure more of the good things of life than the stupid; and it is the clever and not the stupid who *form public opinion*. That these ideas, and the facts that have given rise to them, have played a very important and useful part in human evolution in the past is quite evident; but they are not likely to continue indefinitely into the conscious evolution of the future : a new conception of justice, having its test in social equality, is the growing idea of the masses (who are now rising into power), and before these masses the ideas of privilege will be found unfit to survive.

How very clumsily and unfairly the old system has worked is apparent on the least inspection. The rewards of life have gone almost entirely to low orders of cleverness—not to speak of cases of still worse forms. The best and most usefully clever men have had in the past, as they generally have now, only poor rewards for their labour, and often neglect or bad treatment from society. This genetic, or natural method of evolution, in which the most worthy are *supposed* to come to the top and the bad to be crushed dead at the bottom, is a very *imperfect, cruel, clumsy* system, which humanity will tolerate no longer than is necessary. The good and clever will have to learn : first, a lesson of equality with their inferiors in capacity ; second, that the improvement of humanity must be continued in the future by an enlightened socialism and not by the blind forces of competition.

But if the clever and good have to learn this lesson of equality, how much more the rich and lucky ; people who from no merit of their own, but by mere accident of birth and circumstance, have been enabled to absorb the labour of others! Justice is certainly not to be found in leaving these persons without interference, to live the life of parasites; rather will it be found in taking from them as much of their unjust privilege of social power, as may be deemed wise, whilst keeping always in view the well-being of society at large. Individualists of the *laissez-faire* school deprecate interference with the social system to the extent here indicated, but I have not been able to find that they have any reasons except those of expediency for limiting the action of the State at all. That legislative interference with social arrangements has been usually unwise and had often better not have been tried, is ground for their very useful protests, but this does not touch the *principle* of State control over its members, in any way. The ignorance of rulers and ruled, and the infancy of social science, are sufficient reasons for past failures in dealing with little known social forces; in short, nothing different could have been expected.

State regulation, I believe to be wholly a matter of *power* and *expediency*; it has never been, and I think it never will be affected by any other considerations whatever. In a general state of ignorance it will be marked by injurious action and failures, but we may hope that in the future, as knowledge of the real increases, wrong action will cease, and the organization of the State will be marked throughout by wise adaptation to human needs in view of the greatest happiness possible. As a matter of fact State interference to any and every extent is beyond argument as a principle; it represents *force*, the force of the ruling class, or the force of the majority. If intellectual conceptions are erroneous, its regulations will be unwise; if they are scientific, its regulations will be useful; but regulations of one order or another it will certainly make. Hence the vast importance, in this age of rising republics, of instructing the people, high and low, upon the true nature of society, and the best courses to pursue in furtherance of human happiness, thus aiding the upward evolution of humanity. If the independence and power of the masses grow out of accord with their real knowledge of things, disastrous and bloody revolutions become possible; with abortive attempts of ignorant socialists to change the social order in a way that is impossible as *stable* change, resulting in awful misery and in the degradation of all classes.

When government loses its evil characteristics and becomes

an enlightened and progressive agency, State education of the people will be directed to new ends. Its aim will be to impress such knowledge on the rising generations as will not only prepare them for social life, but instruct them how to preserve and increase happiness in the world, and avert misery from all; it will, in short, instruct them in the science of society and true meliorism; in the best methods for repressing anti-social feelings; in the formation of noble ideals of conduct; and in the religion which, for the establishment of general happiness, must dominate each individual life, and the whole order of society. This, indeed, must be the aim of government in relation to the young. The exact means to be adopted will differ in each age. Where parents are superior much will be left in their hands, but inferior parents cannot be permitted to train up children in inferior ways, making them not only unfit but injurious members of society, members who will perpetuate a bad state, and defeat the object of scientific socialism. I believe a time will come when government, acting on its right of force and expediency, will take up and sequestrate a degraded minority in the population, the small class of individuals who from defective natures and bad circumstances are unable to live in society without propagating their kind, or lessening general happiness; these will be kindly dealt with and left as much at liberty as is consistent with public safety, but the object will be to put an end to their race and class.

Let us now observe the difference of opinion at the present moment upon the policy of State aid in two important matters, viz. the feeding of starving board-school children, and the erecting of wholesome dwellings for the poor. Individuals are usually slow to recognize that government interference is merely a matter of power and expediency; and in their search for ruling principles, they either discard logic entirely, or form rules of expediency for limiting State action, and regard these as moral principles. Some people object to State education; others agree to this policy, but object to have starving school children fed by the State. But every policy carries some bad results, and the choice is really between a number of evils. The right aim of the governing people is to choose that policy which tends to the greatest good, and a good or bad choice is made according to knowledge and discernment. As State feeding would undoubtedly tend to lower wages and to increase parental neglect and the number of destitute children (Chapter VI. page 77), I am of opinion that the policy productive of least suffering would be that of leaving the feeding of starving school children to voluntary agencies. Apart from expediency, there are people professing moral objections to the position of being forced to help their fellow-creatures. "Why," it is asked, "should a man without children be taxed to support or educate the children of others? Is it not unjust that the earnings of the prudent should be taken to save the improvident from the natural consequences of their folly?" I answer: First, the rewards of life depend more upon the economical state of society than upon efforts and merits; a man's income is what he receives from society; the amount of it is determined by forces over which he has no control, and in which justice has no place. A clever physician may be able to demand the fee of a guinea a visit; but if another of equal reputation enters his neighbourhood and charges half a guinea, the first will probably have to lower his fee, or lose his income; and even if he lowers his fee the sum of the two incomes of these physicians sharing the patients between them will be less than the amount of the single income originally derived from the same source. If employers do not think it unjust to lower wages, neither should they deem it unjust were government to lower their incomes to the amount that wage-earners receive. It is not justice but competition that rules incomes; a man's gains are what society permits him to *seize*, whether it be little or much, whether he be working hard or not at all. Now, if society gives arbitrarily, why may not the State equally take away arbitrarily? It is simply a question of expediency; and within unmoral conditions the appeal to justice is illogical and out of the question. Outside the unmoral conditions, what justice adjudicates is that all men should be socially equal in respect of liberty and command of comforts. If we take the average amount of all incomes as representing the fair sum for each worker to receive, we find that the reward of life obtained by a large number of people rises considerably higher than this sum; these then gain what we may call an *unearned increment*, larger or smaller according to circumstances; whilst another still greater number of people are unable to obtain this average amount, although they may work hard and well all their lives. It is surely in the highest degree just and reasonable that the more fortunate class should be required to give up some of this unearned increment, to be applied in succouring poverty and educating poor children. But second, the improvident and immoral are *naturally deficient* in some good qualities, and the highest religion demands that they should be kindly treated and if possible saved from the torture which nature would bring upon them. We are happily learning that there are ways of diminishing evil and the number of evil-doers, without our having recourse to nature's methods. Wherever no evil is likely to accrue it is incumbent on a humane people, by the agency of government, to equalize social conditions, and to require from the fortunate the means whereby the unfortunate may be saved from suffering. In reference to the miserable accommodation for the poor provided in large towns by the action of natural competition, what is the best policy to pursue at the present moment is a difficult question to answer. Building them dwellings by the State is justifiable only by invention of a system under which such action would certainly prove beneficial to the poor, and I have pointed in Chapter XVII. to the provision respecting increase of population in the Peabody Dwellings which must have important salutary social effects. It is more the state of public ignorance preventing the acceptance of a scientific scheme, than the existence of real difficulties, that here stands in the way of successful action. I believe that in the order of conscious evolution, the State is certain to become more and more philanthropic in action, for the sufficient reason that the members of the State will become more humane and public-spirited. But between voluntary and State agency there will always have to be decided which should undertake any particular case. Each agency has its peculiar merits and demerits, and each case its peculiar conditions. Science and experience, therefore, must decide in each case how the evils may be best overcome.

The only true and possible way to social justice and equality, that is, communism in social life, and socialism in economical organization, is by successive stages ; first, associated domestic life ; second, communistic industrial life superadded ; third, federation of communes, and a State-organized system embracing entire social life. The want of fitness in individuals precludes any other course but this. Adaptation has to be gradually evolved in the units to permit of the formation of a superior social system. Particular aggregation of units must precede general aggregation and State organization. The theory of crude socialism that general reorganization of society may take place irrespective of particular organization and aggregation in society, is an idea only possible to minds having no conception of the real facts and forces of social evolution. To the evolutionist it seems certain that there must first arise, amid much difficulty and many failures—voluntary communes, and that then State socialism will develop, out of the voluntary federation of these communes. All this may and must take place with perfect peacefulness in general society, without conflict with the present system, which will itself improve under the various meliorating influences of increased knowledge, and better habits of life.

From personal and class despotisms, through despotisms of ruling popular majorities, we advance to the final despotism of reason and science. Individual liberty must always be conditioned; hitherto it has been painfully restricted in many directions of usefulness and harmlessness, and left unrestricted in ways of hurtfulness. It is necessary that the conditions of liberty should be consciously and collectively made consistent with natural causation and the happiness of society. The despotism of science—the true positivism—is a despotism to be hailed with joy, to be united with our most cherished ideals, and to be laboured for, with our most earnest endeavours; for under it liberty has its widest possible bounds. No longer trammelled by fashion and convention, by ignorance and arbitrary rule, it gives free expansion to all variety of noble individualities, and to all action which is innocent of evil. A scientific socialism must set the final, true bounds, to human liberty; knowledge and reason declare what these shall be, and society collectively enforces them. They are entirely determined by the actual nature of the social organism with its environment, and the real conditions of general happiness.

The gradual enlightenment of mankind must eventuate in the positive philosophy—*the philosophy of action*, the true regulator of society throughout its length and breadth. The psychic method of evolution with its persistent gentleness and teleological control of natural forces, must finally supersede, in human society, the genetic evolution of mindless, cruel competition—the survival of the strong and the destruction of the weak. The era of science will be the utopian golden age; the beautiful Psyche (the mind) now wandering aimlessly through the world will, in the evolution of happiness, be sought out and cherished; and be raised to the Olympus of a new heaven above the sweetness of a new earth, there to be joined to Love and reign evermore in all true hearts as the genius of Socialism and the guardian of Individuality.

CHAPTER XXIII.

RELIGION.

" Oh, priests who mourn that reverence is dead !
Man quits a fading faith, and asks instead
   A worship great and true.
I know that there was once a church where men
Caught glimpses of the gods believed in then :
I dream that there shall be such church again.
   O dream, come true, come true.

    *   *   *   *   *

" So all intolerable wrong shall fade,
No brother shall a brother's rights invade,
   But all shall champion all ;
Then men shall bear, with an unconquered will
And iron heart, the inevitable ill,
O'er pain, wrong, passion, death, victorious still
   And calm though suns should fall."
                                                       W. M. W. COLL.

To understand human nature, by careful study of its highest and best, as well as its lowest and most imperfect specimens, is an essential condition of progress. Mr. Froude calls forth our gratitude by the frank confidence and unreserved honesty with which he has boldly led us into the privacies of Carlyle's inner life, and laid bare before us the secrets of his soul. Many of his worshippers no doubt mourn the desecration of their idol, and the destruction of its shrine. But Carlyle was essentially heroic, and martyrdom in the cause of truth is not a *rôle* from which he would have shrunk. It is no ignoble mission to fulfil, that of lying after death upon the world's dissecting table, if, in that passive act, a clue is found to some bewildering problem, and the way made shorter to the goal of all true spirits such as Carlyle's, viz. human happiness. Vivid portraiture was a pre-eminent gift of Carlyle's, and his introspective tendency, aided by the friendly touches of Mr. Froude's pencil, have placed before us the entire portrait of a being of noble aspirations, of delicate sensibilities, of robust conscience, of broad intellect, throughout the whole range of personal thought and feeling.

This psychological study is valuable material for the comprehension of Humanity and the solution of the great problem which lies still before us ;—what is vital religion ? That Carlyle's nature was profoundly religious will be, I think, freely admitted by all students of his outer history and inner life, whether they are Christians or Atheists, Pantheists or Theists, Positivists, Secularists, or Agnostics. His books, his public actions, his private letters, and the introspective scrutiny and self-criticism which is found in his journal, all witness to the truth of Mr. Froude's assertion, "He was a man indeed in whom was no guile." Sincerity was the basis of his spirituality, and a rigid, stoical determination to do the right, so far as he knew it, was the backbone of his moral nature. The visible life, wrought out by these factors amidst surrounding conditions, was "a model of simplicity and uprightness, which few will ever equal and none will excel." In Mr. Froude's judgment the result redounds doubly to his honour for reasons relative to his inherited instincts and his chronological position. Nature had made him, says Mr. Froude, "weak, passionate, complaining, dyspeptic in body and sensitive in spirit, lonely, irritable, and morbid." Time had placed him in an age of theological disintegration. The creed of his fathers was to his acute intellect fundamentally imperfect or false. To his heart, therefore, it could administer no solid hope or comfort to sustain him on his way through life. "The inherited creed had crumbled down, and he had to form a belief for himself by lonely meditation." * This he never succeeded in accomplishing. He gazed out reverently into the vast Universe, and peered patiently into every crevice of his own secret thought. He questioned the Eternities, the Immensities, the Infinitudes, the Silences—from whence no whisper came to him of the God he sought. Slowly and sadly his honest soul adapted itself to the inevitable ; and the confession of his utter failure to formulate a theological creed was publicly made in his Rectorial Address. To his conscientious mind the occasion was one of solemn obligation to make his mark upon the rising generation. He pressed upon the students the importance of religion, but he told them that theology *was not essential.*

* "Thomas Carlyle. A History of his Life in London," vol. ii. p. 458, by James Anthony Froude.

Admitting, then, that Carlyle had no theological creed, let us see if his religion proved "a solid hope and comfort to sustain him on his way through life." At the end of a career like his, which, take it all in all, was really noble, with work successfully performed and action made conformable to moral principle through many weary struggles, a religious soul was surely entitled to rest serenely in the lap of spiritual comfort, and fall asleep in the arms of a divine peace. But the aged Carlyle, with all his goodness, had no permanent enjoyment of satisfaction of heart, and a soul at ease. Theologically creed-less, his religious creed was like a leaky vessel that bore him on the sea of life with the waves ever surging at his feet and threatening his destruction. A fretting conscience and the strain upon his nerves left by work, when work had become no longer possible, made life a burden to him. He sat under the shadow of his "general canopy of sadness and regret," or called up memories that stung him into painful consciousness of pungent remorse. Now, why was this? Was peace denied him because, Prophet as he had felt himself to be at the beginning of his career, he had made no fresh discovery of God, no new revelation proportionate to the world's emergencies hovering on the brink of chaotic thought? Not so; for although his massive intellect, grasping the intricate phenomena of modern life, derived no comfort from the sight, and he turned from it, as "a pestilent congregation of vapours," his star-gazing had long since ceased. Theology was not essential. *That* question was at rest; and the sting of his conscience had no relation with the speculative side of his intellect.

The grief of his old age was mundane, not supernatural, not cosmical. The subject-matter of his remorse was small offences in his daily life. Matters belonging to the commonplace, the practical, precisely what in the days of his ardent, enthusiastic youth he would have deemed insignificant, if not contemptible. In old age, with the wisdom of a long experience made during hours and years of solemn thought and earnest effort, how different was his intellectual view and the conviction of his moral judgment. "Two plain precepts there are," he calmly and confidently asserts :—" Dost thou intend a kindness to thy beloved one? Do it straightway, whilst the fateful future is not yet here. Has thy heart's friend carelessly or cruelly stabbed into thy heart? Oh, forgive him. Think how when thou art dead he will punish himself." In old age, when his own life-happiness was for ever marred, Carlyle made an important discovery. Religion may be too lofty and grand. It may soar to the skies and spurn the ground. But if it misconstrues humanity, misapprehends the delicacies of domestic relations, misappreciates the love and tenderness of fellow-beings, and misdirects the emotional forces of a man's nature, it is what? false? nay, not false, not a sham, an hypocrisy, but —a failure! In the hour of bereavement its afflatus is gone. It cannot sing a man's sorrows into the painless stillness of death, because it has not kept his conscience clear, his heart at peace with all mankind.

When manfully fighting with the pen for justice and truth in his lonely study, and bravely exposing the world's shams and delusions, Carlyle's religion braced his nerves. When meditating under the starry heavens, when moralizing on man's poor lot, and comparing his paltry idols, his joss-house and bambino, with the dumb Eternities and Immensities—religion dominated his mind. But in the privacy of his home, at his domestic hearth, in the closest and most precious relation of his human life, the "nature of the beast" was uncontrolled by his religion; and earnest as he always was for the general happiness and goodness of mankind, he yet never sought to comprehend the nature of the individual who was nearest and dearest to him, nor studied to promote her happiness and goodness in each step of the life-journey which they made together. His religion lay outside of this boundary and left all the forces within it unreclaimed.

Now Mr. Froude, in his character of biographer and friend, rather than public teacher or scientist, hints that Carlyle's mind latterly was morbid; that his failures of conduct compared with other men's were scarcely immoderate, and his remorse was, to say the least, excessive. Comparison with other men's failures in duty, however, is of course wholly irrelevant. And to Carlyle's profoundly religious nature these bitter waves of keen remorse were simply the natural, inevitable accompaniment of his intellectual awakening to some new, stern facts of life. "Omnipotence," he writes in his journal, "has developed in me these pieties, these reverences, and infinite affections." Alas! it was only in old age that he knew the direction that these forces should have taken, the channels in which they should have run—in his old age, when for the recasting of his own life, the renewing of its sacred fire, the reawakening of its sweetest music, the revival of its buried joy, the knowledge had come too late. Carlyle's keen intellect did *not* exaggerate the misdirection that his religion had taken, or the miscarriage to happiness that it unwittingly had brought about.

That we may rightly understand this misdirection and perceive the important lesson it conveys, let us recall a graphic picture of husband and wife given by Mr. Froude—Mrs. Carlyle stretched on a bed of suffering; her jaw dropped, her mouth open from nerves and muscles injured by a fall. Her husband leans on the mantelpiece, gazing at her, happy no doubt that the accident had spared her precious life. " Jane, ye had better shut your mouth," says he. She tried to say she could not. But again comes his exhortation—" Ye'll find yourself in a more compact and pious frame of mind, if ye shut your mouth. Ye ought to be thankful the accident was no worse." " Thankful," exclaims the suffering, worried, and exasperated wife; " thankful for what? for having been thrown down in the street when I had gone on an errand of charity? for being disabled, crushed, made to suffer in this way? I am not thankful, and will not say that I am." He leaves her, saying he is sorry to see her so rebellious, and then writes to his brother, " She speaks little to me, and does not accept me as a sick-nurse, which, truly, I had never any talent to be"!!* The reader's indignation at a husband's want of sympathy, the impulse to smile at the blindness of a wise philosopher, all quickly pass into sympathetic emotion with the anguish of regret, the keenness of the aged Carlyle's remorse, when his piety took a new and unquestionably sound direction, and he saw clearly, when too late, what duty had required of him.

Now, turning from the story of Carlyle's secret mind, its restless searchings after God and a new creed, its negative conclusions, its stoical pursuance of duty, its discovery of the mistaken path, its remorse and sadness, and half despair, let us glance at the general mind of the epoch in which we live, in so far at least as light is thrown upon it by the public utterances of the foremost writers of the day. Behold! we find the same questions that perplexed and disturbed Carlyle, the same doubts and scruples as to duty, the same waves of religious aspiration and religious fears that he had to contend with surging around us now in all directions, and shaking, at least, if not destroying, the foundations of our inherited creeds, and poisoning the sources of our springs of inward joy. It is precisely as though Carlyle's prophetic mind first catches and

* " Thomas Carlyle. A History of his Life in London," vol. ii. p. 272.

reflects, as in a mirror, a little cloud on the horizon destined to grow and expand, until it casts its shadow over the whole heavens, obstructs the sunlight, and commands the observation of all mankind.

In answer to the question, "What is religion?" Mr. Herbert Spencer holds that religion is concerned with that which lies beyond the sphere of sense.* The primitive human mind had no religious idea and no religious sentiment whatever. These ideas and sentiments have been evolved; and the progressive stages of their evolution may be distinctly traced from the savage's dream which suggested his belief in ghosts, accompanied by a strong fear of the supernatural, to the pantheon of pagan deities, and down to the one awe-inspiring Jehovah of the Hebrews, and the tenderly attractive Christian God of Love. This development of religious consciousness is not destined to cease. According to Mr. Spencer it will go on in a definite direction, becoming more and more logical, and less and less grossly material, until the mental conception is indistinct—a vague overmastering consciousness of the Unknowable accompanied by a feeling of awe.

But this, objects Mr. Frederic Harrison, is the mere ghost of religion, not its substance; and we had "better bury religion at once than let its ghost walk uneasy in our dreams." Mr. Harrison holds that Mr. Spencer is mistaken; that it is theology not religion that his philosophy necessarily destroys, (and theology is, to Mr. Harrison as to Carlyle, non-essential). Religion remains though theology, which deals with the assumed authors or controllers of nature, we already are, or certainly shall be, compelled to give up. "Religion" (happily for us) "is not a thing of star-gazing and staring, but of life and action." It has to do with the natural emotions of the human heart, not such emotions as wonder, a vague sense of immensity, an unsatisfied yearning after infinity. These are of the nature of disease, "a metaphysical disease of the age." "The roots and fibres of religion are to be found in love, awe, sympathy, gratitude, consciousness of inferiority and of dependence, community of will, acceptance of control, manifestation of purpose, reverence for majesty, goodness, creative energy and life. Where these things are not, religion is not." But these things require in the object that calls them forth, "some sort of vital quality;" and whilst the religions of the past have been

* *Nineteenth Century Magazine*, January, 1884. "Religion: a Retrospect and Prospect," H. Spencer.

anthropomorphic, the religion of the future will be "frankly anthropic." "Humanity is the grandest object of reverence within the region of the real." Humanity is the power which controls the life of the individual, and is fitted to call forth his religious feelings and to influence his conduct. "Shall we cling to a religion of Spiritism, when philosophy is whittling away spirit to nothing? Or shall we accept a religion of Realism, where all the great traditions and functions of religion are retained unbroken?" * So far Mr. Harrison. Mr. Spencer, in reply, attacks with his invincible logic the anthropic god of the Positivist's worship. If the Unknowable is no fit object to satisfy man's religious sentiments and guide his conduct, still less will this divine Humanity prove the god we seek. Think: what is humanity composed of? If we exclude (though why we should do so I know not) criminals, paupers, beggars, Humanity is composed of men and women of extremely commonplace intellectual calibre, whose most useful deeds result, for the most part, from no higher motive than personal ambition, pursued with absolute disregard of human welfare. The Positivist can speak of holy Humanity. "Had I to choose an epithet, I think 'holy' is about the last which would occur to me."† And why should we feel grateful to this non-holy Humanity? You call it a Power; but after all Humanity is but the transitory product of "that great stream of Creative Power unlimited in Space or in Time," which, because essentially unknowable to man, you refuse to accept as the basis of religion. Humanity is only a bubble thrown up by that creative stream, a dull leaden-hued thing, though here and there some noble specimen of man or woman shows us the beginning of the bright iridescence which will one day characterize the bubble, when entire modification of human nature has taken place through the action of moral agencies, which—Mr. Spencer does not at this point go on to indicate.

Meanwhile British minds that are neither speculative, nor strongly emotional, but eminently practical, are equally exercised upon this momentous subject of religion. To one of these Mr. Spencer's philosophical labours appear similar to what Isaiah described in the manufacture of idols; Mr. Spencer "works his words about this way and that, he accounts with part for ghosts and dreams, and the residue

* *Nineteenth Century Magazine*, March, 1884. "The Ghost of Religion." Frederic Harrison.
† Ibid., July, 1884. "Retrogressive Religion," H. Spencer.

thereof he maketh a god, and saith, 'Aha, I am wise, I have seen the truth.'" "To me," says Sir James Stephen, "the whole theory is a castle in the air, uninhabitable and destitute of foundations."* Turning, then, to Mr. Harrison's theory, he asks, "How can a man worship an indefinite number of dead people, . . . many of whose characters were exceedingly faulty, besides which the facts as to their lives are most imperfectly known? How can he in any way combine these people into a single object of thought?" And again, passing from the intellectual conception of Mr. Harrison's god Humanity, to the emotional state described as religious, Sir James remarks, this "language about awe and gratitude to humanity represents nothing at all, except a yearning after some object of affection, like a childless woman's love for a lap-dog." And in reference to the practical outcome of Mr. Harrison's Positivism, Sir James asks, " Has it the smallest prospect of being able to 'govern men and societies'?" What will it "do with the vast mass of indifferent and worldly people? It can neither hang them nor damn them," whilst Christianity, he infers, certainly did both when engaged in the work of establishing itself. That such new religions are nought is the conclusion of this practical mind ; but finally, Sir James asks, why religion is wanted? What would the use of a new religion be if we possessed the materials whereof to form it? "We can get on very well," says he, without religion ; "for though the view of life which science is opening to us gives us nothing to worship, it gives us an infinite number of things to enjoy. . . . Love, friendship, ambition, science, literature, art, politics, commerce, professions, trades, and a thousand other matters will go on equally well, as far as I can see, whether there is or is not a God or a future state ; and a man who cannot occupy every working moment of a long life with some or other of these things must be either very unfortunate in regard of his health or circumstances, or else must be a poor creature."

To these public utterances my readers will doubtlessly add many private doubts and questionings of similar character. I have heard the calm assertion made : " My personal requirements are exclusive of religion. They are for the most part supplied by material objects that minister to man's worldly comfort and well-being." I have also heard expressed this mercantile view : " We are thought to stand in need of a new

---

* *Nineteenth Century Magazine*, June, 1884. "The Unknowable and the Unknown," Hon. Mr. Justice Stephen.

religion; but until philosophers and thinkers have settled the matter amongst themselves, and are prepared with a religion of real solid substance, I mean to hold on to Christianity. In active life we may afford to go creedless; but in the hour of death, I fancy, men want the comforts of religion, and Christianity has served our forefathers well, and will in all probability last out my time."

Now, observe, there is one point on which the whole of these minds exercised upon this grave subject of religion appear to think alike. All religionists, from St. Paul and the Christian Fathers, down to Mr. Spencer, Mr. Harrison and Sir James Stephen (in his attitude of willingness to part with religion altogether), treat it as consisting essentially of mental conceptions and beliefs. The essential point to them in religion is its objectivity—the object-matter, which contemplated in the subject-mind, arouses religious feelings and prompts to pious actions. This object-matter, observe, has altered in every stage of man's evolution from savagery to civilization, whether along the line of spiritism—from ghosts and souls of ancestors, to the God of the Trinitarian or the Unitarian, and down to the Unknowable Eternal Energy, from which all things proceed; or along the line of Realism—from the grossest fetich, to the Sun, the Moon, the Earth, and down to that Humanity which Comtism proclaims the grandest object of reverence within the region of the real. The object-matter, I say, has continually altered; and in each age religionists have asserted the perfection and all-sufficiency of their own object of worship, whilst condemning as imperfect, and therefore loudly rejecting, the object of worship of others. To this reverenced object-matter and the varying doctrines concerning it, we are, I think, justified in applying the term theology—a word which signifies a leading abstract intellectual conception, a temporarily fixed idea; and allied with that, a body of doctrines to be believed and duties to be practised. These doctrines and duties have altered with the altering ideas; but *one* characteristic has been common to all theologies, viz. *an emotional bearing on life*, and here is the distinctive element, the essential feature to which, I think, should be applied the term *religion*.

The study of the long series of alterations in theological conceptions, with their appended dogmas and practices, discloses a certain order of change, namely this: the later form has always been superior to the earlier form which it superseded, in respect of this *emotional bearing on life*. The newer

system of thought has proved better calculated to rouse and
develop those emotions that are most powerful in prompting
actions conducive to human welfare. Immaterial civilization,—
the power of mankind to unite and to create an associated life,
tending towards peace, order, and progress, in short, human
happiness,—has been enormously promoted by this factor which
I conceive to be religion ; and I define it as not theology and
not morality, but an emotional tendency which in the past has
led man to apply theological conceptions to human life, and to
shape individual conduct in accordance with the result. The
result, that is, the ideal of life, has altered, and the morality
or conduct ensuing has altered; but religion has persisted
throughout all the changes ; and in its ultimate analysis, what
it subjectively reveals is a capacity for dealing with abstract
ideas and a conscience or duty spirit, which compels the realization in actual life of *ideals of conduct*.*

Religion, then, has by no means passed away; on the contrary, it is stronger, more profound, more complex in its relations, and more intensive in its vitality than ever before. The
days of its weakness, as well as of its grossness, are to be
found, not in the present age, but in the history of the past.
The childhood of man was the childhood of religion, and the
efforts of the savage to frame ideals of life, and shape his
conduct thereby, are as feeble as his theological conceptions
are crude and foolish.

Putting aside barbaric religion, let us think for a moment
of Hebraism, with its awful Jehovah, a wrathful, revengeful
God. In the emotional bearing of this theological conception
upon life, fear predominates, and Hebraism gives to sacrifice
and the propitiation of this fear-inspiring God, an important
place in the conduct of life. The later form takes up sacrifice
into the theological conception, and by expanding the latter, so
as to embrace propitiation and substitution, lowers the emotion
of fear and makes to predominate the emotion of love. All

---

* This view of religion accords precisely with the theory of the deriva-
tion of the word from an Aryan root, *lag, to regard;* the opposite of *to
neglect*. In Latin, *religens* is the converse of *negligens*. *Religion*, which is
applied to such diverse forms of thought and action, does not denote particular objects of regard, allied with particular forms of conduct; but only
the connection in the mind between some abstract object or objects of
strong regard, and the conduct following from this regard, the connection
being emotional in respect to its energy or motive, and intellectual in
respect to its comprehension of relations, and determination of conduct—
a complex condition of mind due to evolution in man, and not possible in
less developed beings.

the tenderer, gentler feelings of the mind are aroused in contemplation of the noble, selfless Jesus, and the Christian rule of conduct becomes such as to promote the development of the benign side of human nature, and of all the social qualities that ennoble civilized men. In its approach to perfection religion becomes more and more powerful in the direction above indicated, and when theological conceptions have given way to scientific considerations of man and nature that are true, it is difficult to estimate how great the strength of this factor may prove in promoting man's rapid and happy advance.

Instead of saying this age can do without religion, it were truer far to say, this age makes so great a claim upon religion, so incisive a demand upon its virile strength, that although maturity, not senility, characterizes it, yet religion staggeringly recedes for the moment:—it must breathe deeply, it must gather up its forces ere it can rise triumphantly to the new occasion. The age, undoubtedly, has changed its ideals. But why? Not so much from the action of any negative force, such as scepticism, still less from wilful opposition to the profound pieties and reverences of religious man. It is science that has mainly done this thing. Science has destroyed the conceptions of all earlier ages by proving them to be imperfect or false, and it is precisely because man's reverence is not dead, that he quits the fading faith of the past, and asks instead :—

"A worship great and true."

At the present day religious minds that understand the "solid methods by which truth is separated from mere opinion, and science winnowed out of philosophy,"* and who confess and yield to the authoritative character of truth, are called to a "higher heroism than that compassed by the martyrs of theological religion. What is asked now is a devotion unspeakably more self-denying. To work quietly, unremembered and it may be unnoticed for an end which we shall never see—this is the demand made from the bulk of its adherents by the religion of a scientific age, that religion which sooner or later will subjugate all older forms and conquer the world." † It is not the inward comforts of an easy-going faith, not the childish rewards and punishments of an anthropomorphic God, that the religion of an advanced civilization can offer. It has no dreams of heavenly personal delights, whilst brethren of

* "Natural Religion," Professor Seeley.
† ".D.," in *National Reformer*.

mankind are weltering in sin or withering in despair, and it has no fears of torturing hell-fire. Such forces as these the noblest men no longer need to prompt them to duties which it were utterly base to repudiate or neglect. Religion has to shake itself free from artificial, groundless, and imaginary fears, begotten of the old dogmas.

Theological religions have used, and done good service with, fear-forces. Ideal terrors supplementing real terrors have immensely aided man's development in power of self-control. They have prepared the mind through exercise, through heredity and natural selection, to follow ideals of life, and guide conduct by abstract ideas. But at this moment this terror, like anger, jealousy, and other emotional survivals of the past, is wholly evil in its action on the best of mankind. Advanced religion must discard it, for this Religion deals with the real, and the future of its history has to unfold itself within the sphere of the actual, whilst conforming its ideals to advancing truth.

Now, I have said that everywhere religion is present with us, it is not a vanishing quantity, it is deeply implanted in, it rests broadly over humanity; nevertheless this age's religion as a *cultus* has yet to be born. Its "substance," as Professor Seeley puts it, " is broken up and distributed under other names or under no name." * That substance must be integrated, the scattered forces must come to focus, it must rehabilitate itself in preparation for the arduous task, the sacred toil, that it alone may compass and fulfil. Religion, *as a cultus* adapted to the age, is bound to possess—first, its object-matter, that is, its clear, intellectual conceptions; second, its body of doctrine; third, its definite principles of practice. The first consists of true ideas of the universe, society, and man, of man's relations to nature and society, and of the conditions of life and happiness; †

* " Natural Religion," Professor Seeley.

† George Eliot was strongly aware of the religious importance of science; she wrote long ago of "that invariability of sequence which is acknowledged to be the basis of physical science, but which is still perversely ignored in our social organization, our ethics, and our religion. It is this invariability of sequence which can alone give value to experience, and render education in the true sense possible. The divine yea and nay, the seal of prohibition and of sanction are effectively impressed on human deeds and aspirations, not by means of Greek and Hebrew, but by that inexorable law of consequences, whose evidence is confirmed instead of weakened as the ages advance; and human duty is comprised in the earnest study of this law and patient obedience to its teaching."—From the *Westminster Review*, January, 1851.

with noble ideals of human character and conduct, personal, domestic, and social. The second consists of the necessitarian or causational doctrines of philosophy which, in disclosing man as the direct outcome of individual heredity and general environment, destroy all logical ground of hatred and blame, and justify a social principle of universal love. The third consists of principles of practice based upon scientific meliorism, and therefore rendering it certain that the morality or social action of the *cultus* will promote the highest interests of all mankind— in short, such ethics as will tend universally to " make things in general better without making things in particular worse, and things in particular better without making things in general worse." * But better! what is better? I will add—better is simply more in consonance with the conditions of human happiness, that is, with such a state of society, of public opinion, of law, of private thought, feeling, and action, as will let happiness come to the individual and stay with him undisturbed, without diminishing by one iota the similar happiness of others.

But if a *cultus* such as this is urgently required, how comes it that we have it not? Why does the *cultus* towards which all theological religions and every form of supernatural and natural worship tended, and whose advent they pointed to and gave promise of, through all the dark ages of man's gropings after truth, tarry to appear? For this reason: advanced thinkers have never fully recognized that theological religions have their true place in evolution. Until the advent of science they were the great developers of the higher powers of the mind; and after the advent of science they continued to be the greatest factors in the formation of conscience. Advanced thinkers not giving sufficient weight to these facts have, in denying the dogmas and ideals of these religions, rejected their method of action. The method, however, was true. It was adapted to the essential nature of evolving Humanity, and if rapid progress is to be attained, we can no more dispense with that method now, than in former ages of the world's history.

Such modern movements of reform as possess ideals that are true, and principles that are scientific, are pursuing a different method. They act almost wholly through the intellect upon the emotions, whilst the great reform movements of the past adopted the reverse process. Now, observe, the ordinary human nature of this age is weak in intellectual ability, but strong in

* Miss Bevington, in *Progress*, 1884.

emotional power. We have masses of intelligent men and women in whom logic, as applied to the abstract, is a force of such minimum quantity as to be practically *nil*; whilst conscience or the sentiment of duty is a force of maximum quantity and enormous practical power for good or for evil. Allied to false intellectual conceptions and dogmas, this force often works mischief, in society, that is lamentable in its disastrous results. How is it to be dissevered from the false conceptions and dogmas and firmly linked to the true? By logic? Assuredly not; but by the very method which has evolved conscience—the method of religion, and according to the order of change obtaining in all transitions from earlier to later forms. *The new form of religion must be superior to all theological forms in rousing and developing those emotions that are most powerful in prompting actions conducive to human welfare.*

Direct and earnest appeal, then, to the best emotions of mankind; touching pictures of suffering humanity, of the real griefs, the intolerable wrongs that, whether they touch the outward hem of our own garments or not, ought to be borne in upon our souls (for are we not all brethren of mankind?), attractive pictures of the happy future of Humanity—that glorious future which scientifically *may exist*, and religiously *ought to exist*, and assuredly *will exist* if we fulfil *our* part in advancing the highest interests of man—these are the teachings of the modern *cultus*; and by such lessons that *cultus* will prove its oneness with the great religions of the past, and take its place as inheritor of their sacred joys, and fulfiller of their sacred work.

The theological religions of the past have evolved conscience; the religious organization of the future must direct conscience as well as carry on its further evolution. It must gather up all the tender, generous, dutiful feelings of the deeply religious nature of man, and whilst allying these to truth, enlist them in the cause of progress. There must be no misdirected religious energy, as in the case of poor Carlyle, who, wasting his youthful fervour in star-gazing, sinks in old age into feeble self-reproach for duties unfulfilled. There are no duties insignificant in the Religion of the Real; its disciples are taught to bring religion to bear upon *every relation of life*, whether individual, domestic, or social; and to permeate every sphere with religious emotion, whilst every action is guided by scientific thought.

Now, as regards the actual state of things, in every town

throughout the country large enough to be a centre of intellectual activity, there are many individuals in whom the mental transition is already complete. They are untrammelled by theological dogmas and ideals ; but silence in respect to these matters characterizes them ; and in action they tacitly carry out the doctrine of *laissez-faire*. Meanwhile, on every side they are surrounded by the type of mind different from their own—a type religiously strong, intellectually weak—men and women whose feelings are enslaved and whose conduct is ruled by conceptions and ideals they instinctively feel to be false. Within the shelter of a Church, where they now only catch *glimpses* of the gods they once firmly believed in, they are struggling towards truth, in their most rational moments rejecting the old dogmas, in their emotional moments clinging to them still. Have the latter no claim upon the former class of minds in their transitional state ? Should these not bring to those all the aid that the advanced position affords ? To give no aid in intellectual effort is surely to fall short of the true ideal of man's highest duty. To unite in groups, however small, to combine with one another, and begin a work of reconstruction, is what the age requires of us. No aggressive action is called for, and no abuse or interference with the institutions that exist—but to take up a position at once bold and free ; to embody modern conceptions and ideals in language that is adequate ; to adorn them with all the attractions that are really theirs ; to point to them emotionally ; to give to religious faith, that is falling away from the old abstract ideas, the stimulus and power to lay hold upon the new ; in a word, to create a *cultus* adapted to the present age.

Standing as we do at the beginning of an epoch of Conscious Evolution, the vista before Religion is indeed vast and magnificent, its work enormous. It has to reconcile the heart, and put into the hand of man the golden thread of science which will guide him in the labyrinth of life. It has to disclose all the grovelling instincts of humanity, and show the scientific method of their sure destruction. It has to nurse the nobler instincts, and to promote the rapid birth of the germs of a perfected Humanity. It has to make its way into every social channel, and reclaim the forces springing from unselfish emotion, that through mistaken aims and means are only producing social disorder. It has to call man's attention, engrossed in the taming of the Titans—that is, the subjugation of the material forces of *external* nature—to the infinitely subtler and more important task of procuring health, physical, spiritual, and social, for the world at large. It has to reveal the new heaven and the new earth, and to inspire mankind with the faith, the energy, the indomitable will, to move in that direction.

"Let us be worthy of the opportunity which is ours. Future generations may be happier than we are in many respects, but for the heroic pleasures—the pleasures of devotion to an ideal, which shall not be realized till the grass has long grown over our graves—there never was, and perhaps never will be, a time equal to that in which we live."*

* " D.," in *National Reformer*.

If religious minds are prompted to immediate action on the lines of Scientific Meliorism, I should be glad to know of it; and may be communicated with through the following address:—George A. Gaskell, Esq., 6, Chester Street, Bradford, Yorkshire.

CHAPTER XXIV.

SCIENTIFIC MELIORISM.

"*Meliorism* implies the improvement of the social condition through cold calculation, through the adoption of indirect means. It is not content merely to alleviate present suffering, it aims to create conditions under which no suffering can exist."—LESTER F. WARD.

"Man has already furthered evolution very considerably, half unconsciously, and for his own personal advantages, but he has not yet risen to the conviction that it is his religious duty to do so deliberately and systematically."—FRANCIS GALTON.

GENERAL happiness cannot co-exist with the evils of our present social state. These evils are poverty, a lifelong struggle for existence, the birth and survival of individuals weak and unfit, disease, premature death, enforced celibacy, late marriage, drunkenness, disorganization of family life, severity and harshness in training children, widespread prostitution, war, competition causing survival of militant instincts, social injustice, inequality of comforts and luxuries, tyranny, crime, barbarous treatment of criminals, disrespect of natural function and consequent injury to health, conventional folly, social repression of innocent enjoyment, subjection of women, legal restraint of exuberant life, social and religious bigotry, the feebleness of religious guidance and confusion of religious thought.

Partial views of man and his social life have hitherto prevailed and given birth to specifics of all kinds for the cure of the diseases of society; and these, in the growing tenderness of Humanity, have been eagerly adopted and promptly applied, to prove, not only fruitless of good effect, but in many cases fruitful of bad and disastrous effects.

Scientific Meliorism deals with society *as a whole* and *throughout all its parts;* and one great object of this book will be unachieved if I fail to impress upon my reader the high im-

portance of a full comprehension of each and all of the groups
or classes of social phenomena and their inter-relations. Viewing society as a whole we realize that there are *no remedial
specifics in the case :* that general happiness will be attained only
by a process of evolution; and that the process is one of continual readjustment of multitudinous relations, or unceasing
adaptation of individual human life to a social environment, and
of social environment to individual human life. The evolution
of social environment proceeds towards the highest ethical
state which implies a system of society of perfect justice and
equality. But the realization of this state requires a perfected
Humanity, hence the path of progress is in the gradual
improvement of individuals—the creation of a superior race
whose spontaneous impulses will construct and support an
improved and improving social system.

Unconscious evolution has carried us forward from savagery
through many transitions to a state of civilization which,
although grossly imperfect and partial, contains within it *a new
element of advance,* capable of intensifying the action of blind
evolutionary forces and immensely increasing their momentum.
Not universally, but here and there throughout society, the
power of reason in Humanity has become strong, and, aided
by a scientific knowledge of man and the conditions of his
life, this psychic force—the conscious element in evolution,
the power of design—may initiate a fresh departure. Reason
must be directed, however, to the invention of an effective
policy of Meliorism which so combines the practical methods
of reform, as that each will add strength to all and the result
prove a powerful factor of change in the society on which it is
brought to bear. Scientific ideas of the universe and the
nature of man, and an accurate acquaintance with the facts of
our national history, are essential to the creation of such a
policy.

The British nation is growing out of militancy, but the
strife of competition throughout the whole sphere of its industrial life gives free play to the passions of militancy—rapacity,
intrigue, antagonism, cunning, tyranny—and permeates society
with the warlike spirit. Advance in morals is the sure step to
a better and happier future; but how is this advance to be
secured? Man's moral nature is dependent upon heredity,
upon training, upon environment. His goodness is conditioned by his life, and life at present, in modern countries, is
unfavourable to a high moral state. The policy of Meliorism

has to embrace and combine rational breeding, rational training, and a new order of life in which sympathy and cooperation will take the place of individual competition, and happiness, not wealth, be the clear aim of man.

Now, the conscious element in evolution is as yet far too weak to alter *general society* much for the better; and the State is wholly unfit to reorganize society. Conscious evolution, however, if *socially localised*, i.e. centred in groups of human beings within general society, becomes at once capable of effective action. Whilst Meliorism must present ideally an "organization of all facts, forces, and phenomena into an orderly and connected system,"* its practice during a transition epoch will primarily be carried out in individual groups, by clusters of select human beings whose high standard of intellectual and moral attainment fits them for union in domestic life under conditions of voluntary socialism.

The disintegration of the ancient family group, the unfitness of an archaic domestic system to achieve the great ends of *rational training* and *the initiation of habits of rational breeding*, is the central source of our widespread social corruption. Consequently a reconstruction of family life upon modern principles and the formation of a domestic system adequate to the above ends is pre-eminently a feature of Scientific Meliorism. In this channel must flow the *main* remedial force of the new policy; although beyond its boundaries there are minor channels in which tributary streams of Scientific Meliorism will permeate society and aid its gradual reformation.

In the lower social strata the relinquishment and discouragement of all such patronage of the poor as undermines their self-dependence, or tends to their rapid multiplication and produces deterioration of race, is immediately required. Parental responsibility should be strongly inculcated and upheld. Public teaching should be given on all natural laws affecting society, and especially the laws of health, of increase, of heredity; and, under conditions respectful to human dignity, Malthusian doctrine and knowledge of Neo-Malthusian art should be carefully instilled.

In the higher social strata advanced Religion has to accept and meet the new position. It must reorganize and re-direct the religious forces of Humanity, and guide the conscience of man to the final path of Scientific Meliorism. The welfare of future generations must be unselfishly promoted; the purest,

* "Dynamic Sociology," L. F. Ward, vol. i. p. 9.

noblest feelings of man aroused and enlisted in the cause of progress through *union* for social reconstruction, for scientific education, for gentle training of the young, for associated domestic life, for the facilitation of happy marriage, and for the welfare of all mankind, whether good or bad, clever or dull, fortunate or unfortunate.

Experiments in living have to be preceded by union for the thorough investigation and discussion of the various evils of our present system of social, economic and domestic life. Unanimity of opinion must be reached upon the principles of social and sexual morality to be strictly observed by individuals, co-operating in the establishment of an Associated Home or the carrying out of a practical experiment in domestic life, according to the new order or system. The ends to be attained in Unitary Homes are economy by means of joint labour and joint expense, to the relief (in great measure) of monetary anxieties and domestic worry; stability of social position, *i.e.* the certainty to an individual that his home will not dissolve or break up independently of his will; the utility and recognition of all personal labour, with the banishment of individual idleness and its accompanying misery, *ennui;* social intercourse and enjoyment relieved of all conventional tyranny and senseless *etiquette:* freedom for friendship between the sexes and such conditions of domestic union as will promote development of mental capacity and altruistic sentiment in each; early marriage without disregard of social responsibility, and based upon mutual knowledge of character, habits, and tastes; a fitting refuge for old age, rendering impossible the premature destruction of valuable social forces which *age alone* can supply, and securing the material, intellectual, and emotional surroundings necessary for comfort and happiness up to the last moment of life.

Further, it is in Unitary Homes that sexual conduct and parentage, with its far-reaching results for good or evil, must be controlled and directed towards *racial regeneration*, and that a thoroughly progressive educational machinery must be devised and acted upon.

The scientific study of man's nature gives sexual passion an honourable position in life; whilst its indulgence, under present conditions, is a perpetual source of personal degradation and crime, social immorality, misery and disease, of reckless increase of population, of constant pressure of labourers on work—of life on the means of subsistence. On the conscience of each adult generation it rests as an imperative social duty so to influence the young generation as that this great passion shall subserve physical and social health and elevate not degrade the race. A due activity in growing organs strengthens organic function; therefore with early marriage and freedom to young love, checked only by scientific knowledge of the laws of health, propagation at the *age of maturity* is bound to put forth vitality of maximum quality. In conscious evolution sexual functions cease to be regarded as essentially allied with propagation. They are regarded, however, as properly subject in youth to parental and social control; and that control acts as a perpetual restraint upon licentious, dissolute tendencies, and a shield to the young love that seeks personal happiness consistent with domestic purity.

No less potent is the action of control in another direction. Physiology of sex and the laws of inheritance are carefully studied by guardians of domestic peace who, rejecting ordinary custom and habit, accept as their guide science which points to philoprogenitiveness. or love of offspring, as the best motor force in reproduction. Were this force the antecedent cause of parentage throughout the nation, disease and premature death would be undermined and gradually subside. "Sympathetic Selection aud Indiscriminate Survival" gives way before that "Intelligent Selection and Birth of the Fit" which is a fundamental condition of social well-being—the master-spring to a rapid evolution of general happiness.

The transition, however, from our present state of confused sentiment, illogical thought, and disastrous action in the field of *eugenics*, to clearness of purpose and consistency of life, must necessarily be a work of extreme delicacy and patient endeavour. Its achievement requires the nuclei of Unitary Homes. Its nurture must take place in the bosom of a superior domestic life. The process, in short, implies an alteration in Humanity itself, to be brought about only by such preparatory alteration in outward conditions of life, as will set up and bring into constant play the interaction of new social forces. Individualism in domestic life is at the present moment a great impediment to progress. It vitiates the action of true socialism outside domestic life, and gives us misshapen units, unfit for a better social system scientifically directed to the attainment of general happiness by banishment of tyranny, despotism, self-will, pride, and every anti-social emotion, and establishment of the perfect justice and equality essential to the highest ethical state.

Socialism—concerted action for social ends—*must* lay hold of the *family* and fashion it anew, adapting it to the formation of a superior material of human life composed of individuals whose enjoyments lie chiefly in sympathy, and whose spontaneous impulses are towards a life of elevated socialism. Not only is our present domestic system so wanting in point of delicate adaptativeness to human nature as to be wholly incapable of dealing with sex relations and adjusting them to stirpiculture and the sterilization of diseased organisms; not only is it so feeble as to be impotent in the regulation of the conduct of masculine youth outside its boundary, but it is destitute also of the elements required in the organizing of a Progressive Educational System.

Home education has during the last century almost disappeared in the disintegration of family life. Nevertheless the strong forces of aggregation in society, which under diverse conditions of industrial and conventional life group mankind in masses, have moulded schools to massive proportions. The youth of the nation, cut off in a great measure from the varied relations of domestic and general society, that are calculated, observe, to teach mankind " humanities not in the academic but in the real sense," is congregated in universities and large public schools for superior culture, and in vulgar day schools for culture of a less exalted order. In the former young men and maidens are separated. Domesticity, that important quality of human nature on which depends the consolidation of society, is disregarded; whilst to the development of unanimity of social aims, community of interests, affinity of tastes, harmony of habits and sympathy of sentiment between the sexes, no attention is paid during the plastic period of life when individual character is in process of determination. In day schools boys and girls are often associated, but under such conditions of mechanical routine, invisible nerve tension, visible cramming, conflicting and alternating authorities, irregular and erratic forces of moral control, as to make these schools hotbeds of evil, fostering every anti-social instinct of man.

Co-ordination in the life of the young is the demand of Scientific Meliorism. The nursery, school, and playground must be harmonized, and the entire juvenile orbit, within and without the home, governed by intellectual and moral forces of fixed congruity.

In the future, when we may hope, a public sentiment will exist " in favour of scientific education as strong as it has ever existed in favour of religious education," * and religious forces will be in accordance with that sentiment, a school system truly progressive and independent of home may become possible. But the fact that at the present moment the above sentiment exists in select circles and is by no means a strong public force, commanding recognition, clearly demonstrates that the invention of educational machinery of scientific structure and progressive nature will follow not precede the organization of a new domestic system. Domestic love is bound to be the great motor of the machinery. The springs and checks of true education can be found only within the circumference of *home*. Culture and book-learning have to be displaced from their dominant position, keen competition and emulation to be subdued, cramming, prize-giving, and every artificial method of reward and punishment renounced.

The primary object and aim of scientific general education is not culture, but the adaptation of individual character and habits to the prosecution and enjoyment of a social life. This implies the development and growth of a variety of sympathetic emotions, and the repression of all anti-social feeling, leading to the gradual extinction of, first, the passions of militancy—tyranny, dominancy, fierce aggression, antagonism; and second, the predatory instincts—rapacity, intrigue, cunning, selfish greed. It implies a persistent teaching from babyhood to manhood in right conduct, *i.e.* such domestic and social habits as tend to general happiness in an ethically organized State. It implies the imparting of knowledge of the *real* in every department of life, and the careful instilling of noble ideals of conduct in every relation of life. Lastly, it implies the embellishment of existence by culture and the practice of every exquisite, ennobling art.

These ends will not be attained by book-learning or direct teaching alone. The *environment* of the children from morning till night has to aid the educational process, and must be scientifically adjusted to this purpose. By environment is not meant simply matter and space—airy nurseries, sunny schoolrooms, and so forth—but also the inner and outer mental environment in which the infant powers will unfold themselves and grow. No barbarous tale-books are permitted to create in childish minds a fascinating environment, the false colours of which distort images, and transform to virtue vindictiveness and other hideous forms of social vice. Purity of thought in

* "Dynamic Sociology," L. F. Ward, vol. i. p. 26.

all matters of sex is secured by early and careful instruction on the subject; and no risk is run of a child obtaining knowledge of the phenomena of sex from an accidental or impure source. Children are not placed under the influence or authority of adults of an inferior order. Refinement of mind and manners they easily and naturally acquire by contact with refined human nature in adults whose presence is constantly with them.

There is in man a group of emotions of comparatively recent origin requiring scientific treatment of the utmost delicacy and precision. On its further development depends in a very special manner the rapid evolution of an ethical, industrial, and commercial system. The group is threefold—egoistic, altruistic, moral. It comprises a sense of personal rights, a sympathetic jealousy for the rights of others, an intellectual and moral sentiment of justice, or equivalence of liberty and social comforts for all mankind. The first element is perceptible throughout society and fast becoming hereditary in the British race. The second is (on a proportionate estimate) extremely rare. It must be fostered, strengthened, assiduously created in the nursery, schoolroom, and domestic circle by a system of training whose characteristic is extreme gentleness. The tender shoots of sympathetic jealousy are incapable of growth in an environment of harsh sounds or brutal force. The authority that begets antagonism has no place in the education of Scientific Meliorism.

As the young emerge from childhood the responsibilities of life become aids in education and immensely develop the above emotions. Discipline of conduct within their own order appertains to the young; whilst society within and without the domestic circle demands the thorough *regulation of young life*. Conduct clubs and combinations for a variety of social ends, both sexes taking part, arise amongst the young; and these promote in the highest degree the healthy growth of such intelligence, virtuous emotions, and habits in the individual, as are indispensable to ethical socialism. The method adopted is a just and intelligent *criticism* to which the youthful mind has previously been trained.

The moral sense of social justice for all mankind has at present little or no existence in the upper ranks of society, and the conditions of life render it extremely unlikely that the sentiment will there spontaneously spring up and be transmitted by inheritance. An obvious inference from the facts is that reconstruction is necessarily a growth from below. Rooted in social strata far beneath the upper crust, it will spread slowly upwards and outwards, till the whole of society is involved, and reorganization is complete. Pride of birth, pride of wealth, habits of domination and luxury, are antagonistic to a public spirit that seeks some method of gradually equalizing the labour of life and its rewards, and undermining class distinctions. And pride of acquisition in the fortunate capitalist class is no less strongly opposed to the principle, that reward should not be proportioned to personal capacity, that mental labour has no title to superior distinction, but, that equal useful exertion *ethically* requires *equality* of reward.

In the lower middle class lie the elements that by segregation may form into *republican homes*, capable of by-and-by aggregating into the solid foundation of a pure and elevated Republican Society. Education in large Unitary Homes, where mixture of ages, from the white-haired centenarian to the infant in arms, creates all manner of tender ties, where gentleness and love are the main stimuli in training, where authority is exercised consistently and reasonably, and replaced at maturity by reason and self-control, must eventuate in the production of a superior moral and intellectual type.

This Humanity, rich in sympathy and palpitating with ethical sentiment, will live in larger groups and unite in industry. Its instincts all opposed to strife, antagonism, tyranny, greed, and every hideous accompaniment of *competition*, will prompt it to establish *co-operation* at every available point.

Federation of communes will be an easy step of advance ; and since *co-operation*, as compared with *competition*, is a social force of immensely superior power in the promotion of general happiness, the "law of the elimination of evil" will have its natural course and issue in the universal and voluntary adoption of *economic socialism*.

The order of social evolution, computed roughly, is as follows : In the *first stage*, social equality exists; it is an epoch of savagery. In the *second stage*, differentiation issuing in class distinctions takes place ; the birth of social inequality and injustice arising *naturally* through exercise of superior brute force and cunning. Civilization has here its genesis ; and coercion, tyranny, robbery, injustice, avarice, love of power, inequality, are prime conditions of civilization and the creation of a superior race. Individuals who are inferior, then whole classes socially weak, are compelled by forces, individual and social, to minister to the wants of the strong and superior. Civilization nurtured by inequality and injustice develops in the superior classes of society and slowly spreads downwards. In the *third stage*, reaction occurs, prompted by *civilization itself!* Justice and liberty develop in the lower or inferior social classes and spread very slowly upwards without destroying a civilization, become inherent in the superior type or race of man. The *fourth stage* is one of readjustment in which civilization becomes general, and there is a gradual return to *social equality*. Ultimately society will have no class distinctions, no idlers or parasites, no poor, and no coercive government. Voluntary co-operation, or concerted action for social ends, is a self-regulating, self-controlling force which, when fully developed in the new domestic and industrial systems, is able to dominate Society throughout its length and breadth.

The path of Scientific Meliorism in its main features has now been placed before my reader. Outside the policy that will cause its direct action to become a great factor of social change, however, there are sundry courses of less direct action it is bound to pursue. These bear relation to, first, pauperism and patronage of the poor; second, the proletariat; third, the criminal classes; fourth, the position of woman; fifth, the young; sixth, conventionalism; seventh, political action; eighth, theology.

In the first relation the action of Scientific Meliorism is to carefully discriminate between benevolence that is beneficial and benevolence that is mischievous in its results on social well-being. Whilst exercising the former it gives no support to charities that hurt the independence of the poor, or relieve them of parental responsibility. It discountenances and opposes the social force of *sympathetic selection* which results in *survival of the unfit*. It seeks to initiate and press forward the counteracting social force of *intelligent selection*, which brings about the *birth of the fit*.

In the second relation, Scientific Meliorism strenuously supports co-operation in industry (whether in the field of production or distribution) as a means of improving human nature, and preparing it for general socialism.

In the third relation, Scientific Meliorism strives to enlighten public opinion upon the nature of crime and the philosophical principles of its treatment. It elaborates a new policy in which vindictiveness, the essence of punishment, has no existence;

but gentleness towards all evil-doers issues in, first, the effectual protection of society; second, the reform of corrigible criminals; third, the gradual extinction of crime. It urges upon Government a cautious, deliberate adoption of this policy.

In the fourth relation the action of Scientific Meliorism is to promote the enfranchisement of women, and at every point aid the movement of advance to the position of social equality of sex.

In the fifth and sixth relations, Scientific Meliorism inculcates by admonition and example, and especially among the young, a return to simplicity of manners, habits, and dress. It repudiates conventional etiquette, and opposes the tyranny of fashion. It promotes association of the sexes in youth, under conditions of adult control, whether the union be that of marriage, of friendship, or of simple intercourse and companionship. It discountenances and takes no part in the excitements of an artificial, frivolous society, but it creates and fosters the invigorating excitements of useful labour, alternating with unconstrained and "tranquil delights."

In the seventh relation, Scientific Meliorism agitates for alteration of the marriage laws, the laws of inheritance, and the land laws. Equality of sex is required as the basis of the marriage law, accompanied by the condition of easy divorce, in order to facilitate (under ethical restraints) the dissolution of false ties in favour of the true. The laws affecting children require adaptation to the ethics of social justice and sex equality. Laxity must give place to strictness in respect of parentage, and childbirth be recognized as an event bearing directly upon the interests of the general public. Hence modification here entails the recognition of illegitimate children and the counteracting of the vicious tendency to shirk parental duty and social responsibility. The land laws and laws of inheritance must be adjusted to a levelling process—a policy of paring down large estates and diminishing the massive proportions of private property so slowly as to create no individual suffering or social confusion. Such legislative measures being directed, however, to Land Nationalization as their final aim.

In the eighth relation, Scientific Meliorism frankly, deliberately relinquishes supernaturalism, and in the sphere of the real sets itself to the reconstruction of a religious *cultus*. It discards all theological ideals and dogmas, all selfish rewards and terrors. Religion in a scientific age is the tendency to form and follow ideals of life. It unhesitatingly embraces ideals that are true and beneficial, and becomes the *religion of the real*. The new *cultus* has its inspiration in the scientific doctrine of universal love and kindness and the evolution vista of universal happiness. It denies that the smallest duty of life is insignificant. It enlists the conscience of man. Pointing backwards it thus speaks : " The Nile overflowed and rushed onwards. The Egyptian could not choose the overflow, but he chose to work and make channels for the fructifying waters, and Egypt became the land of corn. Shall man, whose soul is set in the royalty of discernment and resolve, deny his rank, and say, I am an onlooker, ask no choice or purpose of me ? That is the blasphemy of this time. The divine principle of our race is action, choice, resolved memory. Let us contradict the blasphemy and help to will our own better future, and the better future of the world."*

The world stands in need of *heroes—heroes of peace*. " Let the spirit of sublime achievement move in the great among our people, and the work will begin." Let us invoke and act upon the noble sentiment, " I am lord of this moment's change, and will charge it with my soul."†

From the too early grave of an earnest follower of the religion of the real there comes a voice that may fitly close this imperfect exposition of conscious evolution reclaiming the diverse forces of society and directing them to the production of general happiness by means of Scientific Meliorism. " Our interest, it seems to me, lies with so much of the past as may serve to guide our actions in the present, and to intensify our pious allegiance to the fathers who have gone before us, and the brethren with us; and our interest lies with so much of the future as we may hope will be appreciably affected by our good actions now. Beyond that we do not know, and ought not to care. Do I seem to say, Let us eat and drink, for to-morrow we die ? Far from it. On the contrary, I say, Let us take hands and help, for this day we are alive together." ‡

---

* " Daniel Deronda," book v. p. 357.
† Ibid., book vi. p. 402.   ‡ Professor Clifford.

INDEX.

A

Acquisitiveness, 122-129, 267, 268, 305, 331, 332, 379-388, 433
Æsthetic emotion, 121, 124, 126, 232, 233, 267, 268, 305
Affection as motive, 6, 121, 355, 360, 431, 432
Aged, social position of the, 262, 265, 270-272, 296, 428, 433
Agriculturists, 35, 42, 250, 263-266, 379-382
Amelioration, popular methods of. *See* Philanthropy, popular.
Animals, rights of, 195-197
Appetite of eating and drinking, 169, 172, 176
Artificial birth-control, 95, 97, 102, 118, 174-176, 302, 304, 321, 334-342, 373, 427
Asceticism, 103, 169, 252, 303, 304, 310, 373
Associated or Unitary Homes, 237-242, 253-257, 278-296, 302, 351, 352, 398, 427-429, 433
Authority in training the young, 350, 352, 363, 430-432

B

Betting, 105
Birth-rate in England, 87, 88, 386
—— in France, 175, 339, 341
Blasphemy, prosecutions for, 211-214
Boys, training of, 202-208, 268, 269, 347, 349, 357-360

C

Capital, 43, 249, 254, 267, 383, 384, 394; origin of, 383
Carlyle, Thomas, 2, 103, 115, 264, 265, 409-413, 422
Celibacy, 110, 111, 138, 150, 173, 221-224, 235, 269, 270, 295, 322, 333-336, 339, 373, 425
Characterograph, the, 371
Charitable schemes, 9, 37, 53, 68-70, 75-80, 99, 101
Charity, 65, 67, 68, 97, 117
Checks upon population, 89, 95, 102, 118, 174-176, 264, 298, 302, 334, 339-342, 386, 395, 405, 427
Chesterfield, Lord, 139, 140, 176
Children, fastidious, 172
——, independence of, 58, 59, 99, 269, 283, 300, 360, 361
——, training of very young, 21-25, 48, 170, 171, 198-202, 294, 344, 346, 358, 365, 370, 371
Christian teaching, 2, 53, 74, 99, 192, 311, 312, 373, 418, 419
Church, the, 381, 391-393, 405
Civilization, 37, 64, 122, 129, 132, 195, 201, 234, 251, 324, 336, 337, 378-390, 394, 399, 427-434; definition of, 389, 390, 404, 405
Class feeling, 211, 213, 214, 293, 294
Cock-fighting, 244, 245

2 F 3

Combe, Abram, 237
Combe, George, 248
Commercial rapacity, 18, 34, 380, 382, 394
Communism, 283, 285, 398, 400, 407, 408, 433
Competition in industrial life, 31, 35, 42, 46, 245, 246, 252, 380, 394, 395, 398, 426
Conduct clubs, 208, 372, 432
Conventional propriety, 139, 150-155, 176, 180, 208, 215, 220, 221, 225-235, 256, 282, 428
Convict lease system, 376
Co-operation, 64, 119, 210, 239, 247, 285, 289, 290, 398, 399
Cotton corners, 17, 46
Criminals, congenital, 365-367, 372, 373, 377
Criticism by the Perfectionists, 208-210
Culture, 4, 60, 63, 107, 123, 124, 160-163, 233, 251, 267, 268, 343, 381, 387, 430, 431
Custom or fashion, 214-218, 225-235, 256, 313, 428

D

Dancing, 152-154, 227, 238, 280
Darwin, Charles, 89, 130, 337-342
Democracy, 248, 399, 433
Derby Day, 104-106, 221
Dinner-parties, 220, 226, 228-230, 235
Disease, cost of, 227, 228
———, inherited, 328-332, 335, 339
———, mental and moral, 331, 332, 335-367
Distributors, 45, 68, 380, 394
Divorce, 311-321 ; in different countries, 317-319
Domestic economy, 289, 290, 428
——— servitude, decay of, 287-289
Dress, 216-218, 231-234
Drunkenness, 8, 49, 52, 53, 55, 86, 106, 126, 301, 329, 367, 368, 375, 425

E

Education, definition of, 343, 346, 431
——— by the State, 405, 406
Education of children, 21-25, 48, 59, 99, 114, 144, 149, 160, 161, 170, 171, 181, 182, 198-208, 269, 294, 295, 344, 346-363, 365, 370, 371, 430-432
———, higher, of women, 160, 233, 295, 307, 309, 310
——— in anti-social feelings, 114, 128, 171, 182, 198, 199, 245, 347, 355
——— in barbarity, 22-25, 46, 48, 86, 105, 171, 201, 245, 359, 365
——— in gentleness, 26, 49, 114, 135, 140, 200, 254, 283, 291, 346, 353-360, 370, 371, 393, 431
——— in luxury, 66, 107, 149, 154, 172, 267, 268, 387
——— in self-control, 200, 204-210, 219, 283, 285, 290, 336, 350, 372, 432
——— in subservience, 97, 359
——— in vice, 99, 105, 106, 110, 152, 155
———, mixing of the sexes in, 153, 307-310, 430
———, physical, 347, 353, 362
Elevation of the people, 63, 64, 80, 117, 285, 293, 387, 388, 407
Elimination of evil, law of, 396, 433
Eliot, George, 3, 108, 180, 186-188, 348
Emigration, 67, 92, 93, 115, 116, 163, 269, 275, 300, 341, 342
Emotional development, 121, 130, 135, 137, 138, 140, 158, 167, 181, 193, 344, 420-422, 432
Emulation, 204, 277, 309, 353-355, 360
Equality, social, of sex, 112, 150, 155, 162, 165, 187, 188, 221, 233, 299, 300, 316, 435
Equivalence in labour rewards, 257, 258, 385, 401, 402, 406, 433
Ethical unit in labour and exchange, 402
Etiquette and fashion, 235, 279, 281, 282, 428, 435
Eugenics, or stirpiculture, 102, 118, 307, 315, 320, 321, 325, 332-342, 373, 405, 429, 430
Euthanasia, 118
Evils of man and society, 49, 425

INDEX. 439

Evolution, the doctrine of, 27, 81, 82, 134, 215
——, conscious, 49, 117, 118, 120, 170, 195, 196, 242, 247, 278, 303, 321, 326, 338-341, 388, 391, 392, 395-400
——, unconscious, 117, 118, 248, 251, 259, 288, 326, 338, 398, 426
—— in marriage, 317, 321
—— of civilization, 378-382, 387-391, 397-400, 407, 408, 433, 434
—— of government, 399, 404
—— of ideas of justice, 403, 406, 434
—— of landholding, 378-383, 388
—— of socialism, 397, 398, 400, 403, 404, 407, 408
Excitement, 146, 159, 220, 221, 274, 294
Excursion parties, 50-52, 104, 271
Experiments in living, 230, 236, 240-242, 278, 286, 289, 290, 293, 309, 310, 322, 398, 427, 428
Extravagance of rich and poor, 60, 66, 106, 220, 270, 384

F

Factory labour, 34, 35
Familistère of M. Godin at Guise, 253-256, 302
"Fanny Dover," 145, 146, 158, 189, 227, 289
Fashion and etiquette, 235, 279, 281, 282, 428, 435
—— or custom, 214-218, 225-235, 256, 313, 428
Fastidious children, 172
Fellenberg, Emanuel Von, 204, 205
France, birth-rate in, 175, 339, 341
"Freethinker" prosecution for blasphemy, 212-214
Free trade, 246
French Revolution, 166, 169, 192
Friendship, 222, 296, 428

G

Gentleness, 14, 41, 111, 195, 283, 345, 355, 359, 360, 364, 370, 374, 393, 431, 432

George Sand, 148
"Girl of the period," 132, 141, 142, 144, 147, 159, 356
Godwin, William, 240
Goodness as an aim of life, 7, 40
Government, 247, 248, 259, 391, 392, 399, 404, 434
Greg, W. Rathbone, 90, 95, 96, 117, 173, 303, 338, 341
Grief, restoration from, 189, 344

H

Habits, good, 281, 350, 432
Happiness defined, 54, 93, 344
—— the primary object of life, 1, 7, 39, 40, 103, 116, 119, 143, 160, 252, 285, 291, 311, 324, 333, 420, 421, 425
Harmony community, 36, 238
Heine, the poet, 215, 362, 363
History, the teaching of, 171, 181, 182
Home education, 348-352, 430
——, the British, 41, 58-60, 66, 84, 220-230, 261-276, 298
Homes in America, 263, 264
Hydropathic establishments, 292

I

Ideal of education, 344, 350-353, 363, 370-372, 430-432
—— of home life, 290, 291, 295, 296, 304
—— of socialism, 396, 400, 407, 408, 433
Ideals of marriage, 298, 304, 305, 311, 313, 334, 344
—— of social life, 234, 277, 278, 281, 284, 286, 398, 400, 407, 408, 428-433
Immigration, legislation against, 175, 342
Individual liberty. See Liberty of the individual.
Indolence, 107, 340, 428
Industrial epoch, 34, 42, 125, 127, 277, 346, 380-382
Infant labour, 34, 35, 99, 100
Insanity, increase of, 325

Insurance companies, 61, 62
Intellectual development, 115, 120, 193, 344, 381
Intellectual force. *See* Psychic force.

J

Jealousy, sexual, 129–140, 182, 190, 191, 241, 259, 345
——, sympathetic, 194, 195, 198, 201, 212, 217, 281, 360, 432
Justice, 138, 168, 195, 197, 200, 213, 218, 258, 259, 311, 326, 334, 367, 385, 388, 395–404, 426, 432
——, evolution of, 403, 406, 434
Justification, evolutional and ethical, 385, 388, 395, 402

L

Labour, organization of, 239, 254, 289, 379, 380, 394, 395, 398, 407, 408, 433
—— rewards, equivalence in, 257, 258, 385, 401, 402, 406, 433
Labourer, the British, 33, 43, 60, 83, 84, 94, 380, 382, 384
Land law reform, 388, 435
Landlordism an important factor of civilization, 129, 251, 379–382, 386, 387, 394
—— an unjust state, 250, 251, 382, 384, 385, 388
Land, the problem of the, 129, 249–251, 378–388, 435
Law of the elimination of evil, 396, 433
Laws of nature, 195, 196, 390
Legislation against immigration, 175, 342
Liberty of the individual, 195, 198, 206, 214–218, 230, 279, 280, 293, 334, 371, 405–408
Licentiousness, 106, 164, 175, 177, 178, 187, 221, 303, 304, 308, 310, 322, 341, 429
Life, definition of, 330, 331
Lodgings, solitary, 224, 295, 321, 322
Love at first sight, 305, 306

Love, the emotion of, 135, 179–192, 197, 296, 305, 309, 311, 334, 335
Lunatics, 367, 369, 377

M

Machinery, social effect of, 33, 34, 35
Malthus, Rev. T. R., Essay on Population, 89, 95, 333, 337
Marriage, 110, 116, 138, 142, 150, 151, 157, 160, 174, 183, 191, 241, 242, 288, 297–322, 333, 334, 425, 428, 429, 435
——, early, 174, 241, 300–302, 304–311, 319, 321, 428
—— laws, 299, 314–320, 344, 435
—— of the prudent, 89, 95, 303, 326, 333
——, preparation for, 156–163, 179–192, 241, 265, 296, 307–309, 322, 428, 429
—— in different countries, 314–320
Martineau, Harriet, 82, 83, 148, 149, 187, 223, 304, 357, 370
Maternity charities, 69, 70, 75, 101, 302
Matrimonial agencies, 321
Medical science and teaching, 11, 55, 98, 324–332
Meliorism, *Preface*, 392, 405, 421, 424–436
Mendicancy, 70, 71, 79
Men, young. *See* Young men.
Mental suffering, 324, 331, 425
Middle class, 65–67, 104, 110, 138, 220–235, 241, 266–271, 294, 298, 347
Militancy, British, 15, 33, 41, 344, 379, 426
Milton, John, on divorce, 311–313, 319
Moral deterioration, 12, 19, 31, 46, 83, 97, 110, 128, 301, 332, 374
—— teaching, 101, 114, 161, 193, 198, 281, 321, 322, 334, 351, 371
Morals, low state of, 32, 50, 52, 53, 98, 105, 106

N

Natural selection, 72, 89, 91, 117, 197, 325, 326, 336–338; counteracted, 72–74, 90, 91, 96, 117, 301, 325–332, 336–338

INDEX. 441

Nature, artificial control of, 378, 390, 391, 398, 408, 423
——, laws of, 195, 196, 390
Novels and poetry, the teaching of, 185, 186, 304, 305
Nursery, the, 198-202, 207, 208, 344, 365, 370, 371, 431
Nurture and nature, 365, 372

O

Oneida community, 208-210
Orbiston community, 237
Organization of labour, 239, 254, 289, 379, 380, 394, 395, 398, 407, 408, 433
Overcrowded professions, 113, 114
Overcrowding, 57, 58, 75, 83, 84, 98, 117, 257, 262, 263, 266, 301
Owen, Robert, 34, 35, 36, 107, 278, 285, 286

P

Parentage deferred, 302, 304, 307, 309, 310, 319-321, 429
Parental guardianship, 320
—— morals—the recalcitrant minority, 102, 336, 373, 405, 430
—— imprudence, 55, 56, 59, 66, 71, 75, 86, 93, 98, 117, 266, 301, 314, 428
—— prudence, 58, 64, 89, 94-96, 100, 102, 116, 118, 174-176, 288, 302, 307, 315, 320, 326, 339, 427
Pauperization, 75-79, 97, 99, 300, 301, 427, 434
Peasant proprietors, 251, 385
Pessimism, 5
"Peter Bell," 120, 121, 125
Philanthropy, popular, 9, 10, 36, 53, 63, 67-80, 97, 99, 101, 151, 259, 327, 400, 427
Philoprogenitiveness, 176, 177, 334, 340, 429
Physiology, teaching of, 170, 171, 177-179, 307, 321, 429
Podhalians, the, 123, 124
Political economy, 30, 82, 118, 383
—— reform, 101, 118, 246, 251, 252, 259, 388, 391, 399-401, 435

Population, increase of, 87, 93, 116, 175, 250, 277, 300, 302, 326, 384, 386, 427
——, the law of, 82, 87, 89, 93, 101, 173, 337, 386, 387
Positivism, 5, 209, 249, 389, 396, 408, 415
Poverty, the problem of, 37, 67, 71, 97, 101, 174, 276, 437
—— in America, 37, 263, 264
Power, the love of, 149, 153, 159, 184, 294, 300, 379, 387, 399, 433
Predatory instincts, 17, 28, 42, 46, 355, 431
Prevision, scientific, 396
Prisons, 366, 368, 369, 374-377
"Progress and Poverty" and Mr. H. George, 128, 129, 195, 249, 250, 251, 385, 387, 388
Property, the love of, 122-129, 267, 268, 305, 331, 332, 379, 388, 433
Prostitution, 58, 106, 144, 164, 303, 311, 316, 318, 425
Psychic or intellectual force, 134, 215, 247, 248, 381, 390-392, 395, 396, 398, 405, 408, 426
Public spirit, 204, 252, 253, 259, 372, 397, 398, 400, 407

R

Race deterioration, 266, 301, 325-332, 334, 373
—— regeneration, 332-342, 373, 391, 428, 429
Ragged schools, 76-78
Ralahine Agricultural Association, 239
Rational Dress Association, 217, 231
Recreation, capacity for, 53, 54, 66, 103, 104, 110, 115, 144, 220, 221, 279
Religion, definition of, 417, 418, 420, 421, 435, 436
—— of humanity, 249, 252, 275, 344, 345, 347, 392, 427
——, the modern controversy on, 413-417
Rent, 250, 380, 382
Revolution, the method of, 129, 401, 404

Rewards and punishments, 203, 204, 353-355, 358-360, 370, 371, 419, 431
Rich, a parasitic class, 257, 384, 402
—— necessary to civilization, 387, 402
Rights of animals, 195-197
—— of man, 195-201
Rousseau, the teaching of, 167, 168, 176, 178

S

School examinations, 353, 354, 359-361, 430, 431
—— terms and holidays, 361, 362
Schools, day, 268, 269, 294, 349-352, 430
——, public, 347, 349, 358, 359
Sex, differences of, 150, 151, 158, 159, 173, 179, 222, 233, 308, 310, 322
Sexes, disproportion of the, 163, 298-300, 322
Sexual instinct, 95, 96, 98, 117, 169, 172-181, 187, 193, 303, 304, 321, 334, 339, 341, 373, 428
Slavery, 379, 385, 397
Smoking, 329
Social equality, 112, 129, 138, 156, 198, 210, 250, 260, 289, 388, 395, 396, 400, 402, 403, 426; of sex, 112, 150, 155, 162, 165, 187, 188, 221, 233, 299, 300, 316, 435
Social evolution, the order of, 407, 408, 433, 434
—— intercourse, 219-243, 265, 269, 273, 291, 293, 307, 322, 428
—— life, qualifications for superior, 293, 294, 322, 351, 388, 398, 428-430
Socialism, 30, 36, 247-249, 283, 285, 292, 391-408, 430, 433; definition of, 396, 397, 430
Somerville, Mary, 271-275
State education, 405, 406
—— repression of the degraded class, 102, 336, 373, 405, 430
Statistics of births and marriages, 87, 88, 319, 386
—— of incomes, 257, 384, 401
Sterilization of the unfit, 373, 430

Stirpiculture. *See* Eugenics.
Story-books, evil effect of some, 21-23, 48, 431
Struggle for success, 112, 113, 115, 124, 277
Student marriages, 307-310
Subjection of women, 147, 150-152, 163, 179, 221, 224, 305, 345, 435
Survival of the unfit, 72-74, 90, 91, 117, 301, 325-332, 339, 373, 434
Sympathetic selection, 72, 90, 96, 117, 325, 326, 336-342, 429, 434
Sympathy with the young, 147, 148, 351, 355, 359, 360
Syphilis, mortality from, 328

T

Taine, H., 104
Taxation cumulative on incomes, 403, 435
Tea-drinking, 55
Teeth, degeneracy of the, 325, 327
Thrift, 60-62, 75, 123, 267, 270, 345
Trades unions, 43, 211, 214
Tribes, unwarlike, 13, 14, 40, 41, 122, 123
Truth, the love of, 134-137, 151, 162, 185, 211, 303, 393
Tytherly or Queenwood community, 239, 278, 286, 290

U

Unhappiness a cause of evil, 52, 55, 104, 106, 110
Unitary Homes. *See* Associated Homes.
Unwarlike tribes. *See* Tribes.

V

Vindictiveness, 345, 364, 365, 368, 369, 434
Vivisection, 197

W

Wages, 33, 43, 45, 123, 128, 380, 384
War, 35, 41, 49, 245, 379
Wealth, the pursuit of, 19, 31, 35, 46, 66, 115, 118, 124, 126-128, 267, 273, 277, 285, 322, 331, 332, 345, 346, 379-387, 400

Wealthy supported by the poor, 256, 257, 384, 385, 387, 401, 402
Wolstonecraft, Mary, 240, 241
Woman, uprise of, 43, 111, 112, 130, 135, 160, 163, 288, 299, 345, 435
Womanhood, the ideal of, 141, 143, 180, 186
Women, types of, 132, 141–151, 161, 162, 168, 221–224, 268, 273, 297, 307, 345, 356

Y

Young men, 110, 114, 151–159, 177, 178, 184–187, 208, 220, 225, 232, 283, 295, 307, 322, 347
—— women, 110, 112, 132–139, 141, 142, 145–165, 179–192, 221 224, 229, 241, 242, 262, 265, 273, 297–299, 305–310, 315, 316, 345, 356

THE END.

PRINTED BY WILLIAM CLOWES AND SONS, LIMITED, LONDON AND BECCLES.

www.ingramcontent.com/pod-product-compliance
Lightning Source LLC
Chambersburg PA
CBHW032008300426
44117CB00008B/950